T0299765

Cytogenetics and Molecular Cytogenetics

Genomic technologies provide the means of diagnosis and management of many human diseases. Without insights from cytogenetics, correct interpretation of modern high-throughput results is difficult, if not impossible. This book summarizes applications of cytogenetics and molecular cytogenetics for students, clinicians and researchers in genetics, genomics and diagnostics. The book combines the state-of-the-art knowledge and practical expertise from leading researchers and clinicians and provides a comprehensive overview of current medical and research applications of many of these technologies.

KEY FEATURES

- Provides clear summaries of fluorescence in situ hybridization technologies and others
- Comprehensively covers established and emerging methods
- Chapters from an international team of leading researchers
- Useful for students, researchers and clinicians

Medical Genomics and Proteomics

Series Editors
Scott O. Rogers
Science publisher
Charles R. Crumly—CRC Press/Taylor & Francis Group

Published Titles

Molecular Analyses
Edited by Scott O. Rogers

Cytogenetics and Molecular Cytogenetics
Edited by Thomas Liehr

For more information about this series, please visit: www.routledge.com/Medical-Genomics-and-Proteomics/book-series/CRCMOLGENPRO

Cytogenetics and Molecular Cytogenetics

Edited by
Thomas Liehr

CRC Press
Taylor & Francis Group
Boca Raton London New York

CRC Press is an imprint of the
Taylor & Francis Group, an **informa** business

Legend for figure in titlepage:
Central white Figure part consists (from left to right) of: Two schemes showing multicolor-banding and chromosomal translocations were provided by Leon Liehr (Jena, Germany; https://leonliehr. jimdofree.com/), the partial M-FISH karyogramm, the interphase nucleus and the chromosomal microarray figures are derived from chapters 14 (part of 1), 6 (part of 1D) and 21 (part of 2A), respectively.

First Edition published 2023
by CRC Press
6000 Broken Sound Parkway NW, Suite 300, Boca Raton, FL 33487–2742

and by CRC Press
4 Park Square, Milton Park, Abingdon, Oxon, OX14 4RN

CRC Press is an imprint of Taylor & Francis Group, LLC

© 2023 selection and editorial matter, Thomas Liehr; individual chapters, the contributors

Library of Congress Cataloging-in-Publication Data
Names: Liehr, Thomas, 1965-editor.
Title: Cytogenetics and molecular cytogenetics / edited by Thomas Liehr.
Description: First edition. | Boca Raton : CRC Press, 2023. | Series: Medical genomics and proteomics | Includes bibliographical references and index.
Identifiers: LCCN 2022026812 (print) | LCCN 2022026813 (ebook) | ISBN 9781032121628 (hardback) | ISBN 9781032122212 (paperback) | ISBN 9781003223658 (ebook)
Subjects: LCSH: Cytogenetics. | Molecular genetics.
Classification: LCC QH441.5 .C98 2023 (print) | LCC QH441.5 (ebook) | DDC 572.8—dc23/ eng/20220713
LC record available at https://lccn.loc.gov/2022026812
LC ebook record available at https://lccn.loc.gov/2022026813

ISBN: 978-1-032-12162-8 (hbk)
ISBN: 978-1-032-12221-2 (pbk)
ISBN: 978-1-003-22365-8 (ebk)

DOI: 10.1201/9781003223658

Typeset in Times
by Apex CoVantage, LLC

To my two teachers in Human Genetics and Chromosomics:
Prof. Erich Gebhart (Erlangen, Germany), who showed
to me the importance of publishing results, and
Prof. Uwe Claussen (Jena, Germany), who was
a role model for unconventional thinking.

Contents

Series Preface

In 2005, the *Encyclopedia of Medical Genomics and Proteomics* was published. It consisted of two large volumes of analytical methods and equipment that were applicable to medical and other studies of genomics and proteomics. While it was a comprehensive collection at the time, many variations, improvements, and new technological developments have occurred during the past two decades. Also, the alphabetical arrangements of the chapters (by title) made searching by topic somewhat cumbersome. Finally, the *Encyclopedia* was originally available in print version only. All of these factors necessitated an extensive revision in content, format, and organization. The new collection is entitled *Series in Medical Genomics and Proteomics*. The revised version will be offered as volumes, organized by topic, and will be published in print and electronic formats. Currently, at least one dozen volumes are planned.

This is the second volume in the Series, entitled *Cytogenetics and Molecular Cytogenetics*. Because of their importance in several fields, the methods described in this volume have been developed and used increasingly over the past several decades to study chromosomal loci, genetic alterations associated with diseases, chromosomal translocations, congenital abnormalities, tissue-specific changes, development, gene expression, molecular taxonomy, evolution, epigenetic changes associated with diseases (including cancer), and others. The chapters included in this volume provide a detailed comprehensive collection of descriptions of cytogenetic and molecular cytogenetic methods, with special emphasis on FISH (fluorescence in situ hybridization) procedures and applications. This includes an overview of FISH methods, as well as descriptions of techniques used to identify specific components of cells (e.g., centromeres, portions of synaptonemal complexes), disease diagnoses (e.g., leukemia, chromosome rearrangements), and protocols to characterize cellular components in various organisms (e.g., birds, fishes, plants). Additionally, chapters are included on basic cytogenetics, chromosome banding techniques, molecular karyotyping, and microdissection. This volume is poised to be a valuable resource for those in cell, molecular, and medical sciences.

Scott O. Rogers
Medical Genomics and Proteomics Book Series Editor

Preface

With this book, a contribution to the highly actual field of cytogenetics, cytogenomics and chromosomics is provided. The book presents applications and protocols in clinical and research applications. It is not restricted to human genetics, but also includes animal and plant cytogenetics and provides some insights in actual developments. This book is divided in following sections:

I. Basics on (molecular) cytogenetics, cytogenomics and chromosomics (Chapters 1 to 5);

II. (Molecular) cytogenetics in human genetic diagnostics (Chapters 6 to 13);

III. (Molecular) cytogenetics in clinical genetic research settings (Chapters 14 to 20);

IV. Molecular karyotyping (Chapter 21);

V. Molecular cytogenetics to detect mitochondrial DNA (Chapter 22);

VI. (Molecular) cytogenetics in animal cytogenomics and chromosomics (Chapters 23 to 27);

VII. (Molecular) cytogenetics in plant cytogenomics and chromosomics (Chapter 28); and

VIII. CRISPR/Cas9 and its role in cytogenomic and chromosomic research (Chapter 29).

Overall, the reader gets an up-to-date overview on possibilities of fluorescence in situ hybridization and related approaches, how they work and why they cannot be thought away from clinical genetic routine diagnostics and basic research.

Acknowledgements

The Editor thanks to all authors investing time and work to provide chapters to this book on cytogenetics and molecular cytogenetics. Only together could we provide this important contribution to the field of cytogenomics and chromosomics.

Editor

Thomas Liehr has been working in human genetics since 1991. He is a biologist by education and head of the molecular cytogenetic group in Jena, Germany. Research fields include clinical genetics, leukemia cytogenetics, interphase structure of human chromosomes, and research on chromosomal evolution. The results of his research are published in >10 books, ~100 book chapters, >800 referred papers, and ~1000 abstracts. His particular expertise includes small supernumerary marker chromosomes (sSMC), chromosomal heteromorphisms and uniparental disomy—see ChromosOmics databases (http://cs-tl.de/DB.html). He is active member of the European Board of Medical Genetics and received multiple prizes, two invited professorships and a Dr.h.c. as summarized at http://cs-tl.de/.

Contributors

Rouben Aroutiounian
Yerevan State University
Yerevan, Armenia

Sarah Breitenbach
Technische Universität Dresden
Dresden, Germany

Isabel Marques Carreira
Faculty of Medicine of the University
of Coimbra
Coimbra, Portugal

Hongyan Chai
Yale School of Medicine
New Haven, CT, USA

Chenghua Cui
Chinese Academy of Medical Sciences
& Peking Union Medical College
Tianjin, China

Marcelo de Bello Cioffi
Universidade Federal de São Carlos
São Carlos, SP, Brazil

Edivaldo Herculano Correa de Oliveira
Universidade Federal do Pará
Belém, Brazil

Ivanete de Oliveira Furo
Universidade Federal Rural da
Amazônia (UFRA)
Parauapebas, Brazil

Mariana de Souza
Instituto Nacional de Câncer José de
Alencar Gomes da Silva (INCA-RJ)
Rio de Janeiro, Brazil

Michelly da Silva dos Santos
Universidade Federal do Pará
Belém, Brazil

Gerson Ferreira
Instituto Nacional de Câncer José
de Alencar Gomes da Silva
(INCA-RJ)
Rio de Janeiro, Brazil

Susana Isabel Ferreira
Faculty of Medicine of the University
of Coimbra
Coimbra, Portugal

Amanda Figueiredo
Federal University of Rio de Janeiro
Rio de Janeiro, Brazil

Jelena Filipović Tričković
University of Belgrade
Belgrade, Serbia

Vladimir E. Gokhman
Moscow State University
Moscow, Russia

Tigran Harutyunyan
Yerevan State University
Yerevan, Armenia

Tony Heitkam
Technische Universität Dresden
Dresden, Germany

Galina Hovhannisyan
Yerevan State University
Yerevan, Armenia

Ivan Y. Iourov
Mental Health Research Center
Moscow, Russia

Ana Jardim
Faculty of Medicine of the University
of Coimbra
Coimbra, Portugal

Gordana Joksić
University of Belgrade
Belgrade, Serbia

Ivana Joksić
University of Belgrade
Belgrade, Serbia

Alla Krasikova
Saint Petersburg State University
Saint Petersburg, Russia

Rafael Kretschmer
1. Departamento de Genética e
   Evolução, Universidade Federal de
   São Carlos
   São Carlos, SP, Brazil
2. Departamento de Ecologia, Zoologia
   e Genética, Instituto de Biologia,
   Universidade Federal de Pelotas,
   Pelotas, RS, Brazil

Tatiana Kulikova
Saint Petersburg State University
Saint Petersburg, Russia

Valentina G. Kuznetsova
Russian Academy of Sciences
St. Petersburg, Russia

Marcelo Land
Federal University of Rio de Janeiro
Rio de Janeiro, Brazil

Peining Li
Yale School of Medicine
New Haven, CT, USA

Susan Liedtke
Technische Universität Dresden
Dresden, Germany

Roberto Matos
Instituto Nacional de Câncer José de
   Alencar Gomes da Silva (INCA-RJ)
Rio de Janeiro, Brazil

Eunice Matoso
Faculty of Medicine of the University
   of Coimbra
Coimbra, Portugal

Joana Barbosa Melo
Faculty of Medicine of the University
   of Coimbra
Coimbra, Portugal

Luís Miguel Pires
University of Coimbra
Coimbra, Portugal

Ilda Patrícia Ribeiro
Faculty of Medicine of the University
   of Coimbra
Coimbra, Portugal

Martina Rincic
School of Medicine, University
   of Zagreb
Zagreb, Croatia

Svetlana Romanenko
Russian Academy of Sciences
Novosibirsk, Russia

Anzhela Sargsyan
Yerevan State University
Yerevan, Armenia

Francisco de M. C. Sassi
Universidade Federal de São Carlos
São Carlos, SP, Brazil

Maria Luiza Silva
Instituto Nacional de Câncer José de
   Alencar Gomes da Silva
   (INCA-RJ)
Rio de Janeiro, Brazil

Victor Spangenberg
Vavilov Institute of General Genetics
   RAS
Moscow, Russia

Gustavo A. Toma
Universidade Federal de São Carlos
São Carlos, SP, Brazil

Vladimir Trifonov
Russian Academy of Sciences
Novosibirsk, Russia

Mariana Val
Faculty of Medicine of the University of
    Coimbra
Coimbra, Portugal

Svetlana G. Vorsanova
Mental Health Research Center
Moscow, Russia

Yuri B. Yurov
Mental Health Research Center
Moscow, Russia

1 Cytogenetics and Chromosomics

Thomas Liehr

CONTENTS

INTRODUCTION

This chapter is a general introduction to the book *Cytogenetics and Molecular Cytogenetics* published in this new "Medical Genomics and Proteomics" book series. The specific topic of this book on (molecular) cytogenetics is embedded in the field of chromosomics, which can be only realized based on cytogenomic approaches.[1, 2]

CHROMOSOMICS AND CYTOGENOMICS

The idea of chromosomics, as an overarching designation to integrate all research directions on the (human) genome, was introduced by Prof. Uwe Claussen (Jena, Germany) in 2005.[3] Chromosomic research is about all genetics-related presentations of life, including DNA-basepair to chromosome- and interphase-nucleus level, but also epigenetic aspects (including three-dimensional morphologically of nuclei, micro-RNAs, imprinting, etc.), breakpoint characteristics and interspecies genomic studies. Overall, chromosomics has the goal of introducing novel ideas and concepts in biology and medicine.[1–3]

To get closer to this noble goal, cytogenomic approaches are needed. In Table 1.1, a list of the most commonly used cytogenomic techniques is provided.[4] Cytogenetics and molecular cytogenetics were, together with classical approaches of

DOI: 10.1201/9781003223658-1

TABLE 1.1
Cytogenomic Approaches Adapted acc. to[4] Are Listed Here

Cytogenomic Field	Cytogenomic Technique
Cytogenetics	– Classical/solid staining
	– C-banding
	– NOR silver staining
	– Banding cytogenetics
Molecular Cytogenetics	– Fluorescence in situ hybridization
	– Primed in situ hybridization
	– Comparative genomic hybridization (CGH)
	– Molecular combing
Molecular Genetics	– Restriction fragment polymorphism analyses
(classical approaches)	– DNA-cloning in vectors
	– Blotting approaches like Southern blotting
	– DNA-fingerprinting
	– PCR approaches incl. MLPA
	– Sanger sequencing
Molecular Genetics	– Array-CGH/chromosomal microarray analyses (CMA)
(modern approaches)	– Second generation sequencing
	– Third generation sequencing
	– Gene editing (CRISP/CAS9)
Others	– Electron microscopy–based approaches
	– Laser scanning–based approaches
	– Optical mapping approaches
	– Uniparental disomy/Epigenetic changes oriented approaches—incl. studies on long non-coding RNAs, etc.
	– Bioinformatics

molecular genetics, the first possibilities of chromosomics. Groundbreaking insights into the secrets of inheritance were and are still provided by these basic cytogenomic tools.[1, 4] Without chromosome numbers being known, including information about sex-determination systems, modern (high-throughput) molecular genetic approaches are (if at all) only partly informative.[5, 6] It is a truism, which cannot be repeated enough especially in the human genetics field, that no classical cytogenomic technique is ever outdated.[7] Each approach has advantages and limitations, which can be substituted by each other in the optimal case. Thus, to answer questions in chromosomic research and/or diagnostics, a sensible combination of cytogenomic approaches to be applied is always necessary.[7]

APPLICATION FIELDS OF (MOLECULAR) CYTOGENETICS

Here the techniques listed in Table 1.1 for the cytogenomic fields (i) cytogenetics and (ii) molecular cytogenetics are treated, and it is shown in some examples where these approaches are necessary in routine diagnostics and chromosomic research.

CLASSICAL/SOLID STAINING

Classical solid Giemsa or Orcein staining—also called classical cytogenetics[8]—is the basic approach used to determine constitutional chromosome numbers in species previously not studied by cytogenetics.[9] Also, there are species in which chromosome banding is not applicable[10, 11]; here classical cytogenetics is necessary in research.

Solid chromosome stains are also applied in research settings of radiation or mutagenesis—and here also diagnostic applications are reported to determine number of chromosomal breaks per metaphases after irradiation and/or exposure to a mutagen.[12]

C-BANDING AND NOR SILVER STAINING

C-banding is applied to visualize the heterochromatic regions of genomes in cytogenetic preparations, including centromeres; the latter is eponymous for this approach. NOR silver staining highlights the location of active nucleolus organizing region(s) in a genome.[13]

Both techniques are done in initial, research-oriented cytogenetic studies to characterize the karyotype of new plant or animal species.[9] Also they are applied in many routine settings of cytogenetic pre- and postnatal diagnostics.[13] The latter helps defining if an aberrant chromosome banding pattern (see next) may be just due to a heteromorphism of heterochromatic DNA in pericentromeric, acrocentric-p-arm or Yq12 regions of the human genome.[14]

BANDING CYTOGENETICS

Banding cytogenetics was introduced by Lore Zech in the 1970s.[15] Nowadays most countries apply GTG-banding = G-bands by trypsin using Giemsa for routine banding cytogenetics in prenatal, postnatal and tumor cytogenetic diagnostics. Besides, banding cytogenetics is used in animal cytogenetics, in case the corresponding chromosomes allow for introduction of a protein-based banding pattern.[16]

FLUORESCENCE IN SITU HYBRIDIZATION (FISH)

Fluorescence in situ hybridization (FISH) is one of the major topics of this book; so here are just some general statements. FISH is an approach that enables the in situ localization and mapping of specifically defined DNA-sequences.[1, 2] It is indispensable in mapping of a genome, as sequencing alone is yet insufficient to reconstruct a karyotype.[5] The latter is due to the fact, that highly repetitive regions of genomes cannot be accessed by routine approaches, even though first tools are available also to solve that problem.[17] Still, in most cases based on NGS data, an end of chromosome cannot be distinguished from a high-repeat copy number region being typical for a centromere. Overall, FISH is a highly flexible approach, which can be adapted to many research and diagnostic questions, e.g. by multicolor-approaches that enable targeting many different DNA-sequences at a time.[1, 2, 18]

PRIMED IN SITU HYBRIDIZATION (PRINS)

Primed in situ hybridization (PRINS) is the second technique, besides FISH, originally included in the field of molecular cytogenetics. As this approach exclusively worked for repetitive sequences, it is nowadays practically no longer applied, even though it has been used both in research and diagnostics.[19]

COMPARATIVE GENOMIC HYBRIDIZATION (CGH)

Comparative genomic hybridization (CGH) is a variant of FISH.[20] However, originally here two differently labelled while human genomes are hybridized to a normal human metaphase plate. Thus, gains or losses in e.g. a tumor probe can be detected. In diagnostics, this approach is nowadays replaced by array-CGH/chromosomal microarray analyses (CMA), providing higher resolution and accuracy.[21]

In evolution research, CGH is still a great tool to, e.g. get insights into more stable compared to regions undergoing more rapid evolution in two closely related species.[6]

MOLECULAR COMBING

So-called fiber-FISH is a several decades–old variant of FISH, where DNA-probes are hybridized instead of on interphases or metaphases on to extended DNA-fibers, providing thus a higher in situ resolution. Recently, fiber-FISH on extremely stretched DNA-fibers became available as a standardized protocol, and is commercially available as "molecular combing". This enables applications in research and in diagnostics for studying repetitive DNA-stretches yet not reliably accessible by any other cytogenomic approach.[22]

CONCLUSIONS

Chromosomic research and diagnostics, in general are enabled and in parts motivated by new technical developments.[4] New cytogenomic approaches are always welcome; however, due to specific advantages of each technique, older ones should not be considered too hastily as outdated.[7] As shown exemplarily here and with no claim to completeness, cytogenetics and molecular cytogenetics are still, and also will be in future, the most relevant participants in the concert of actual cytogenomic approaches being necessary to get as comprehensive as possible chromosomic insights.[1, 2]

REFERENCES

1. Liehr, T. Molecular cytogenetics in the era of chromosomics and cytogenomic approaches. *Front. Genet.* **2021**, 12, 720507.
2. Liehr, T. From human cytogenetics to human chromosomics. *Int J Mol Sci.* **2019**, 20, E826.
3. Claussen, U. Chromosomics. Cytogenet. *Genome Res.* **2005**, 111, 101–106.
4. Liehr, T. A Definition for cytogenomics: Also may be called chromosomics. In: *Cytogenomics*; Liehr, T., Ed. Academic Press, London, **2021**, pp. 1–7.

5. Reichwald, K.; Petzold, A.; Koch, P.; Downie, B.R.; Hartmann, N.; Pietsch, S.; Baumgart, M.; Chalopin, D.; Felder, M.; Bens, M.; Sahm, A.; Szafranski, K.; Taudien, S.; Groth, M.; Arisi, I.; Weise, A.; Bhatt, S.S.; Sharma, V.; Kraus, J.M.; Schmid, F.; Priebe, S.; Liehr, T.; Görlach, M.; Than, M.E.; Hiller, M.; Kestler, H.A.; Volff, J.N.; Schartl, M.; Cellerino, A.; Englert, C.; Platzer, M. Insights into sex chromosome evolution and aging from the genome of a short-lived fish. *Cell*. **2015**, 163, 1527–1538.

6. Yano, C.F.; Sember, A.; Kretschmer, R.; Bertollo, L.A.C.; Ezaz, T.; Hatanaka, T.; Liehr, T.; Ráb, P.; Al-Rikabi, A.; Ferreira Viana, P.; Feldberg, E.; de Oliveira, E.A.; Toma, G.A.; Cioffi, M.d.B. Against the mainstream: Exceptional evolutionary stability of I ZW sex chromosomes across fish families Triportheidae and Gasteropelecidae (Teleostei: Characiformes). *Chromosome Res*. **2021**, 29, 391–416.

7. Liehr, T.; Mrasek, K.; Klein, E.; Weise, A. Modern high throughput approaches are not meant to replace 'old fashioned' but robust techniques. *J. Genet. Genomes*. **2017**, 1, e101.

8. Liehr, T. "Classical cytogenetics" is not equal to "banding cytogenetics". *Mol Cytogenet*. **2017**, 10, 3.

9. Chaiyasan, P.; Mingkwan, B.; Jantarat, S.; Suwannapoom, C.; Cioffi, M.d.B.; Liehr, T.; Talumphai, S.; Tanomtong, A.; Supiwong, W. Classical and molecular cytogenetics of Belontia hasselti (Perciformes: Osphronemidae): Insights into the ZZ/ZW sex chromosome system. *Biodiversitas*. **2021**, 22, 546–554.

10. D'Amato, G.; Bianchi, G.; Capineri, R.; Marchi, P. N-band staining in plant chromosomes with a HCl-Giemsa technique. *Caryologia*. **1979**, 32, 455–459.

11. Martínez-Lage, A.; González-Tizón, A.; Méndez, J. Characterization of different chromatin types in Mytilus galloprovincialis L. after C-banding, fluorochrome and restriction endonuclease treatments. *Heredity*. **1994**, 72, 242–249.

12. Natarajan, A.T. Chromosome aberrations: Past, present and future. *Mutat. Res*. **2002**, 504, 3–16.

13. Weise, A.; Liehr, T. Cytogenetics. In: *Cytogenomics*; Liehr, T., Ed. Academic Press, London, **2021**, pp. 25–34.

14. Liehr, T. Cases with heteromorphisms. http://cs-tl.de/DB/CA/HCM/0-Start.html [accessed on 01/12/2022].

15. Schlegelberger, B. In memoriam: Prof. Dr. rer. nat. Dr. med. h.c. Lore Zech; 24.9.1923–13.3.2013: Honorary member of the European Society of Human Genetics, Honorary member of the German Society of Human Genetics, Doctor laureate, the University of Kiel, Germany. *Mol. Cytogenet*. **2013**, 6, 20.

16. Claussen, U.; Michel, S.; Mühlig, P.; Westermann, M.; Grummt, U.W.; Kromeyer-Hauschild, K.; Liehr, T. Demystifying chromosome preparation and the implications for the concept of chromosome condensation during mitosis. *Cytogenet. Genome Res*. **2002**, 98, 136–146.

17. Nurk, S.; Koren, S.; Rhie, A.; Rautiainen, M.; Bzikadze, A.V.; Mikheenko, A.; Vollger, M.R.; Altemose, N.; Uralsky, L.; Gershman, A.; Aganezov, S.; Hoyt, S.J.; Diekhans, M.; Logsdon, G.A.; Alonge, M.; Antonarakis, S.E.; Borchers, M.; Bouffard, G.G.; Brooks, S.Y.; Caldas, G.V.; Cheng, H.; Chin, C.-S.; Chow, W.; de Lima, L.G.; Dishuck, P.C.; Durbin, R.; Dvorkina, T.; Fiddes, I.T.; Formenti, G.; Fulton, R.S.; Fungtammasan, A.; Garrison, E.; Grady, P.G.S.; Graves-Lindsay, T.-A.; Hall, I.M.; Hansen, N.F.; Hartley, G.A.; Haukness, M.; Howe, K.; Hunkapiller, M.W.; Jain, C.; Jain, M.; Jarvis, E.D.; Kerpedjiev, P.; Kirsche, M.; Kolmogorov, M.; Korlach, J.; Kremitzki, M.; Li, H.; Maduro, V.V.; Marschall, T.; McCartney, A.M.; McDaniel, J.; Miller, D.E.; Mullikin, J.C.; Myers, E.W.; Olson, N.D.; Paten, B.; Peluso, P.; Pevzner, P.A.; Porubsky, D.; Potapova, T.; Rogaev, E.I.; Rosenfeld, J.A.; Salzberg, S.L.; Schneider, V.A.; Sedlazeck, F.J.; Shafin, K.; Shew, C.J.; Shumate, A.; Sims, Y.; Smit, A.F.A.; Soto, D.C.; Sović, I.; Storer, J.M.; Streets, A.;

Sullivan, B.A.; Thibaud-Nissen, F.; Torrance, J.; Wagner, J.; Walenz, P.P.; Wenger, A.; Wood, J.M.D.; Xiao, C.; Yan, S.M.; Young, A.C.; Zarate, S.; Surti, U.; McCoy, R.C.; Dennis, M.Y.; Alexandrov, I.A.; Gerton, J.L.; O'Neill, R.J.; Timp, W.; Zook, J.M.; Schatz, M.C.; Eichler, E.E.; Miga, K.H.; Phillippy, A.M.The complete sequence of a human genome. *bioRxiv.* **2021**. preprint doi: https://doi.org/10.1101/2021.05.26.445798.

18. Liehr, T. Basics and literature on multicolor fluorescence in situ hybridization application. http://cs-tl.de/DB/TC/mFISH/0-Start.html [accessed on 01/12/2022].

19. Pellestor, F. Development and adaptation of the PRINS technology: An overview. *Methods Mol. Biol.* **2006**, 334, 211–220.

20. Kallioniemi, A.; Kallioniemi, O.P.; Sudar, D.; Rutovitz, D.; Gray, J.W.; Waldman, F.; Pinkel, D. Comparative genomic hybridization for molecular cytogenetic analysis of solid tumors. *Science.* **1992**, 258, 818–821.

21. Weise, A.; Liehr, T. Molecular karyotyping. In: *Cytogenomics*; Liehr, T., Ed. Academic Press, London, **2021**, pp. 72–85.

22. Bisht, P.; Avarello, M.D.M. Molecular combing solutions to characterize replication genetics and genome rearrangements. In: *Cytogenomics*; Liehr, T., Ed. Academic Press, London, **2021**, pp. 47–72.

2 Banding Cytogenetics

Hongyan Chai and Peining Li

CONTENTS

INTRODUCTION

Cytogenetics is an important field of genetics focusing on the study of chromosome structure and related functions in heredity and evolution. In 1902, W. Sutton and T. Boveri independently developed the chromosome theory of inheritance that identifies chromosomes as the carrier of genetic material.[1, 2] Earlier studies of human chromosomes struggled with the correct number of chromosomes.[3] In 1923, T.S. Painter analyzed chromosomes in human spermatogonial cells at mitotic prophase and made three important findings: i) the size and shape of the chromosomes were described, and the diploid chromosome number was determined to be in the range of 46–48 with an emphasis on 48; ii) human sex was found to be determined by an X/Y heteromorphic chromosomal pair mechanism; and iii) Caucasian and African-American males had indistinguishable chromosome sets.[4, 5] In 1952, T.C. Hsu introduced a hypotonic shock method to improve the quality of chromosome preparation from *in vitro* cultured cells.[6] In 1956, J.H. Tjio and A. Levan used this improved technique to recognize the correct number of 46 chromosomes in a human genome.[7] This certainty of chromosome number in normal individuals facilitated cytogenetic investigations

DOI: 10.1201/9781003223658-2

of numerical chromosomal abnormalities on different patients. In 1959, J. Lejeune and his coworkers discovered that trisomy 21 causes Down syndrome, C.E. Ford reported that monosomy X causes Turner syndrome, and P.A. Jacobs and J.A. Strong found out that presence of an extra sex chromosome XXY causes Klinefelter syndrome.[8–10] In 1960, P. Nowell and D. Hungerford identified a tiny chromosome termed 'Philadelphia chromosome' in patients with chronic myelogenous leukemia.[11] All these discoveries prompted great efforts from cytogenetic investigators to improve the cell culture techniques for various types of human tissues and to develop different staining methods for identifying individual chromosomes with recognizable banding patterns.[12] At the same time, the necessity of standardized documentation of human chromosomes was recognized and proposed. An International System for Human Cytogenetic Nomenclature (ISCN) was developed through a series of conferences and continuously updated to incorporate new techniques for extended clinical applications.[13] Banding cytogenetics has enabled the characterization of constitutional and somatic numerical and structural chromosomal abnormalities through different stages of human development and enriched the medical knowledge of these abnormalities on spontaneous abortion, developmental delay, intellectual disability, multiple congenital anomalies, infertility, tumorigeneses, and metastasis.

CHROMOSOME STRUCTURE AND VARIOUS BANDING TECHNIQUES

Human Chromosome Structure in Interphases and Metaphases

The double-stranded DNA of the human genome consisting of three billion nucleotides is packed into interphase chromatins and metaphase chromosomes through a multi-stage process during cell cycles. The first stage of compaction from naked double-stranded DNA with a width of 2 nm to a nucleosome fiber with a width of 11 nm is achieved by winding approximately 200-basepair (bp) of DNA around a core of eight histone molecules, two each for H2A, H2B, H3, and H4. In the next stage, the nucleosome fiber is twisted in a regular helix with about six nucleosomes per turn to form a 30 nm–width solenoid. The solenoid involves a loop arrangement radiating from a scaffold to form interphase chromatins. Chromatins form numerous compact dynamic domains such as topologically associating domain, contact domain, and loop domain to govern various functions. Further packing of chromatins to metaphase chromosomes with a width of 1,400 nm can be visualized by a light microscope.[14]

From earlier cytologic observation on interphase nuclei, condensed and darkly stained chromatins were termed heterochromatin, and decondensed and diffused chromatins were termed euchromatin.[15] In 1957, M.L. Barr discovered a dense mass by Giemsa staining of interphase nuclei in female cells from oral mucosal smear, skin biopsy, and blood specimen.[16] Later this dense stained mass termed 'Barr body' was found to be facultative heterochromatin from an inactivated X chromosome; this simple nuclear staining is a useful diagnostic tool for detecting X chromosome inactivation.[17, 18] Giemsa staining of a normal human metaphase showed solid stained 46 chromosomes of 23 pairs. Based on the size of chromosomes from large to small, the 22 pairs of autosomes were numbered chromosomes 1 to 22, and one pair of sex chromosomes was named XY for male and XX for female. Each chromosome has a primary constriction designated as centromere. Based on the position of the centromeres, these numbered autosomes were further divided into: group A of metacentric chromosomes 1, 2, and 3, group B of submetacentric chromosomes 4 and 5, group C of submetacentric chromosomes 6 to 12, group D of acrocentric chromosomes 13, 14, and 15, group E of submetacentric chromosomes 16, 17, and 18, group F of metacentric chromosomes 19 and 20, and group G of acrocentric chromosomes 21 and 22. Chromosomes were ordered according to their size, and finally it was determined that indeed chromosome 21 is the smallest human chromosome and not 22; still the designations were not changed. The constricted or narrow regions outside the centromere in some chromosomes are termed secondary constrictions. This original solid stained karyotype has been the reference for detecting numerical chromosomal abnormalities in patients, but the difficulty in distinguishing individual chromosomes within a group severely affects its accuracy.

Various Banding Techniques

Banding techniques for human chromosomes were pursued to identify individual chromosomes by recognizable and reproducible patterns. A band is defined as a specific locus or a part of a chromosome with clearly distinguishable darker or lighter staining from its adjacent segments. When metaphase chromosome preparations are treated successively with HCl, RNase, DNA denatured by NaOH, renatured by saline incubation, and then stained by Giemsa, a dark staining pattern in the centromere of each chromosome is shown. This specific C-band pattern for centromeres and the distal long arm of chromosome Y reveals constitutive heterochromatin in the human genome.[17] Hybridization of radioactive nucleic acids to mouse cytological preparations identified the location of satellite DNA in the centromeric heterochromatin regions.[19] Historically, satellite DNA was defined as genomic DNA fractions with different buoyant densities on cesium chloride or cesium sulfate gradients from the main fraction of genomic DNA.[20, 21] Inside nuclei, the nucleolus is the largest organelle with functions in producing and assembling cellular ribosomes. The nucleolus organizer regions (NORs) contain repeat units of ribosomal genes, which are located in the short arms of D and G group acrocentric chromosomes. Active NORs can be stained by silver nitrate solution to reveal the proteins in secondary constrictions on the short arms of chromosomes 13, 14, 15, 21, and 22.[22]

In 1968, L. Zech, coworker of T. Caspersson, discovered that a banding pattern on individual chromosomes by a fluorescent staining with quinacrine can be visualized under a fluorescent microscope.[23] This Q-banding technique, named after quinacrine staining, required a fluorescent microscope, and the fluorescent intensity quenched quickly; therefore, it was less optimal for routine diagnostic use. However, the Q-band pattern prompted the development of more banding techniques. By treating metaphase chromosome preparations with dilute trypsin and then staining with Giemsa, one can see G-band patterns along each individual chromosome under a light microscope.[24] G-banding is easy to perform, and the staining patterns on euchromatin of each chromosome are reliable and reproducible. Accordingly, G-banding was quickly accepted in clinical cytogenetics laboratories and has been the most widely used banding method in chromosome analysis. When incubating chromosome preparations in phosphate buffer at a high temperature followed by Giemsa staining, an R-band pattern with dark and light staining reverse of the G-band pattern was noted. Extending this treatment to a longer time or a higher temperature could erase most of the R bands but retain terminal bands, so called T-banding at the end of each chromosome. Further improvement in the banding technique involved the synchronization of cell cycle to capture late prophase and prometaphase and the integration of DNA intercalating agents such as ethidium bromide to elongate chromosomes for a high-resolution G-banding pattern.[25, 26] Standardized protocols for these traditional banding of chromosomes for cytogenetic analysis have been published.[27–29] A three-letter code with the first letter denoting the type of banding, the second letter for method of chromosome treatment, and the third letter for the staining dye is used to describe different banding techniques.[13] For example, GTG refers to G-Banding by Trypsin treatment and Giemsa staining, and CBG refers to C-banding by Barium hydroxide treatment and Giemsa staining. All these banding patterns reflect the organization of heterochromatin and euchromatin in human chromosomes and thus have different applications.

Molecular Basis of Banding in Human Chromosomes

Q-banding involves staining with quinacrine, which intercalates into chromosomal DNA and shows brighter fluorescence in regions of AT-rich DNA. G-banding in protease-treated chromosomes is resulted from interactions of both DNA and protein with the thiazine and eosin components of the Giemsa dye. The bright and dull Q-band patterns are consistent with the dark and light stained G-band patterns. R-banding involves denaturing of chromosomes in hot acidic saline followed by Giemsa staining. This treatment preferentially denatures AT-rich DNA and leaves under-denatured GC-rich regions for staining; therefore, the R-banding shows a reversed pattern of the Q/G-banding. T-banding is the staining of a subset of GC-richest R-banding resistant to more severe heat treatment and reflects the repeat sequences at the subtelomeric regions and telomeres.

Banding techniques reveal an inherent pattern of chromosome organization in the context of DNA sequences.[30, 31] Approximately one-third of the human genome is composed of short- and long-interspersed repeat elements abbreviated as SINEs and LINEs, respectively. The most abundant SINEs are Alu elements, and the predominant LINE is retrotransposon L1. Further molecular studies revealed that G-band positive regions are AT-rich, Alu poor, LINE rich, gene poor, and replicate late, while R-band positive regions are GC-rich, Alu rich, LINE poor, gene rich, and replicate early.[31] The structure of human metaphase chromosomes was also analyzed under atomic force microscopy; the G-band treated chromosomes showed a characteristic spiral pattern of chromatid fibers with ridges and grooves corresponding to the dark and light bands.[32] Using cryo-electron tomography and small-angle X-ray scattering analyses on metaphase chromosomes revealed an interdigitated multilayer structure where compact chromosomes are composed of many chromatin layers stacked along the chromosome axis.[33] Table 2.1 lists

TABLE 2.1
Various Banding Techniques and Related Staining Patterns and Genomic Sequences

Banding Technique	Staining Pattern	Genomic Sequences	Replication
Q-banding	Brighter Q band	AT-rich	
G-banding	G-band positive (dark stain)	AT-rich, Alu poor, LINE rich, gene poor	late
	G-band negative (light stain)	GC-rich, Alu rich, LINE poor, gene rich	early
R-banding	R-band positive (dark stain)	(similar to G-band negative)	
	R-band negative (light stain)	(similar to G-band positive)	
C-banding	C-band positive (dark stain)	centromeric satellite sequences and Yq12	
Silver staining	Positive on short arm of D/G group chromosomes	proteins on 5S/18S/28S ribosomal DNA at active NORs	
T-banding	Positive on both ends of each chromosome	subtelomeric and telomeric repeats	

the various banding techniques and their related staining patterns and genomic sequence organizations.

Studies from genome sequencing further defined the organization of heterochromatin and euchromatin in human chromosomes.[34, 35] The primary constriction or centromere in each chromosome contains heterochromatin consisting largely of high-order arrays of α-satellite DNA and other satellite repeats (beta, gamma, I, II, and III). The α-satellite DNA monomers, consisting of an approximately 171 bp repeat unit with 40% sequence divergence in different chromosomes, are tandemly organized into distinct high-order repeat units at the centromere. A 9 bp degenerate motif in the α-satellite serves as the binding site for centromere protein B. There are centromeric transition regions with euchromatic gene-containing segments duplicatively transposed toward the pericentric regions.[35] The secondary constrictions in the five acrocentric chromosome arms 13p, 14p, 15p, 21p, and 22p encode the 5S, 18S, and 28S ribosomal DNA genes, which lie on a 43-kb sequence present in approximately 50 tandem copies on each arm and are flanked by additional repeats arranged in a complex structure.[34] Silver staining reveals the protein components associated with these ribosomal gene transcriptions. Finally, there is a single large region on distal Yq composed primarily of thousands of copies of several repeat families. The heterochromatic regions all tend to be highly polymorphic in the human population. The euchromatin of the human genome are bounded proximally by heterochromatin and distally by a telomere consisting of several kilobases of the hexamer repeat TTAGGG.[34]

G-Banding in Normal Population

International System for Human Cytogenetic/Cytogenomic Nomenclature (ISCN)

In 1971, at the fourth International Congress of Human Genetics in Paris, a group of human cytogeneticists agreed upon a uniform system of human chromosome identification. The report from this Paris Conference proposed the basic ISCN for designating not only the short arm (p arm) and long arm (q arm) of individual chromosomes but also chromosome landmarks, regions, and bands. Landmarks include the ends of chromosome arms, the centromere, and certain bands in each chromosome. A region is defined as an area of a chromosome lying between two adjacent landmarks. Regions and bands are numbered consecutively from the centromere outward along each chromosome arm. When an existing band is subdivided in a high-resolution banding, sub-bands are numbered and placed after a decimal point.[13] For example, 1q31.1 stands for chromosome 1, long arm, region 3, band 1, and sub-band 1.

Current idiograms of G-banding for normal human chromosomes are presented in five different levels of resolution based on the number of bands. Given the size of three billion bp or 3,000 megabase (Mb) of human genome, karyotypes at approximately 300, 400, 550, 700, and 850 band levels have an average analytic resolution of 10-Mb, 7.5-Mb, 5.5-Mb, 4-Mb, and 3.5-Mb per band, respectively. A multicenter study on the subjectivity in chromosome band-level estimation recommended the use of representative bands from two to seven selected chromosomes to assess the banding level.[36] Symbols and formats for the description of numerical and structural chromosomal abnormalities have been standardized and continuously updated in the ISCN.[13] For examples, symbols +, -, t, del, dup, and der refer to gain of a chromosome, loss of a chromosome, translocation, deletion, duplication, and derivative chromosome, respectively. Therefore, 47,XY,+21 denotes a numerical chromosomal abnormality in a male with trisomy 21; 46,XX,t(9;22) (q34;q11.2) denotes a structural chromosomal abnormality in a female with a reciprocal translocation of the segment of chromosome 9 distal to band 9q34 and the segment of chromosome 22 distal to band 22q11.2; and 45,X,del(5)(p14) denotes compound numerical and structural abnormalities in a female with a loss of one copy of sex chromosome X and a deletion of the segment of chromosome 5 distal to band 5p14. With the integration of chromosome microarray analysis, region-specific assays, and genomic sequencing techniques, the ISCN was changed to an International System for Human Cytogenomic Nomenclature in the ISCN 2016 edition. The current ISCN 2020 provides not only a standardized nomenclature but also general principles applicable to multiple techniques for the documentation of numerical and structural chromosome abnormalities, genomic copy number variants, and gene rearrangements.[13] Furthermore, ISCN designation of chromosomes and bands has been a reference for the human genome sequence in the genome browser (https://genome.ucsc.edu/). Figure 2.1 shows a normal male karyotype, the designation of chromosomes, and common chromosome heteromorphisms.

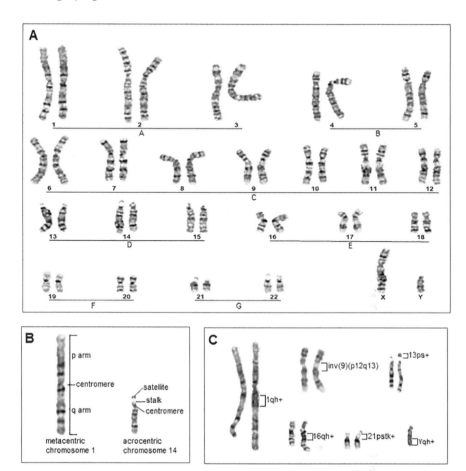

FIGURE 2.1 G-banded karyotype and chromosomal heteromorphisms.

A. A normal male karyotype.

B. Designation of short arm (p), long arm (q), and centromere in a metacentric chromosome 1, and short arm stalk (stk) and satellite (s) of an acrocentric chromosome 14.

C. Common chromosome heteromorphisms of 1qh+, inv(9)(p12q13), 13ps+, 16qh+, 21pstk+, and Yqh+.

Chromosome Heteromorphisms in Humans

Chromosome heteromorphisms refer to the observed variations of different size or staining on the pericentric regions of autosomes, the short arms of acrocentric chromosomes, and the distal long arm of Y chromosome.[37, 38] The biological and clinical implications of these variations in normal populations were first analyzed in a New Haven study of 4,482 consecutively born infants. This study by solid staining of metaphase chromosomes noted heterochromatin-associated polymorphisms in chromosomes 1, 9, 16, 18, and Y and estimated the relative frequency in 3,476 Caucasian and 807 African-American infants.[39] Further studies of chromosome variations in Q-band and C-band recognized increase (+) or decrease (-) in length of heterochromatin in the long arm (q) of chromosomes 1, 9, 16, and Y, and in the short arm (p) or centromere (cen) of chromosomes of 13, 14, 15, 21, and 22 and X.[40] The molecular basis of these variations related to the large arrays of tandemly repeated DNA sequences.

Another type of human chromosome heteromorphism is pericentric inversions likely resulting from the shuffling of satellite DNA in the heterochromatic regions. An early study on heterochromatin size and pericentric inversion by C-banding on 80 normal Caucasians noted that the frequency of size heteromorphisms for chromosomes 1, 9, and 16 were 11.3%, 47.5%, and 7.5%, respectively, and pericentric inversions of chromosomes 1 and 9 were 20% and 23%, respectively. These was no significant difference between males and females for size and position of these heteromorphisms.[41] A more recent large case series of 1,276 individuals from India detected pericentric inversion of chromosome 9, inv(9)(p11q13), and Robertsonian translocation of chromosomes 13 and 14, rob(13;14)(q10;q10), in 3.68% and 1.25% of individuals, respectively; while other chromosomal heteromorphisms of chromosomes 1, 9, 13, 14, 15, 21, 22, and Y were seen in 0.15%–1.95% of individuals.[42] A molecular cytogenetic study of 334 carriers of heterochromatic variants of chromosome 9 recognized the presence of different type of variants and absence of evidence linking these variants to infertility.[43] The pericentromeric region of chromosome 9 is the most obvious heteromorphism in human chromosomes; the proximal 5-Mb on 9p11 to distal 4-Mb on 9q12 comprise a mere 0.3% of the genome. These two regions have a high density of segmental duplications and a high degree of intrachromosomal sequence identity (98.7%). This high sequence similarity between the two regions is likely to be the reason for a polymorphic inv(9) present in approximately 1% of the human population.[34] A study of 2,970 prenatal cases in China showed an incidence of approximately 8.8% for chromosomal heteromorphisms. The most frequent was found to be in chromosome Yqh (5.28% males), followed by chromosome 1qh (1.65%), inv(9) (0.94%), 22p/cen (0.77%), 15p/cen (0.64 %), 9qh (0.58%), and 16qh (0.34%). The frequency of common chromosomal heteromorphisms and pericentric inversions from three large case series of different populations are reviewed and summarized in Table 2.2. The earlier New Haven study showed a higher frequency of heteromorphisms likely due to the solid staining and related size estimation for variant patterns.[39] Recent studies showed comparable frequencies of heteromorphisms between Indian and Chinese populations; furthermore, there was no evidence for an association between chromosome heteromorphisms and infertility or recurrent spontaneous abortions.[42, 44]

Chromosome heteromorphisms are inheritable traits within a family, so it is a useful biomarker to track parental origin of chromosomal abnormalities such as the maternal or paternal origin of trisomy 21, derivative chromosomes, and uniparental disomy of chromosome 15.[45–47] The reporting practice on chromosome heteromorphisms varies among clinical cytogenetics laboratories. Survey results from more than 200 cytogeneticists indicated that 61% of them would include selected heteromorphism data in a clinical report. More than 90% considered prominent short arms,

TABLE 2.2
The Estimated Frequency of Common Chromosomal Heteromorphisms from Different Populations

Heteromorphisms	Caucasian Infants[39] (n=3476, 1741 males) Frequency (%)	African-American[39] (n=807, 411 males) Frequency (%)	Indians[42] (n=1276, 638 males) Frequency (%)	Chinese[44] (n=2970, 1533 males) Frequency (%)
1qh	0.23	0.37	0.54	1.65
9qh			0.15	0.58
16qh	2.59	4.33		0.34
Yqh	15.27 (males)	15.31 (males)		5.28 (males)
13cenh/ps/pstk			0.31	0.17
14cenh/ps/pstk	16.47*	26.63*	0.15	0.24
15cenh/ps/pstk			1.95	0.64
21cenh/ps/pstk			0.54	0.49
22cenh/ps/pstk	5.16*	10.04*	1.17	0.77
inv(9)	0.06	1.24	3.68	0.94
inv(Y)	0.11 (males)		1.72 (males)	0.39 (males)

qh: long arm heterochromatin, cenh: centromeric heterochromatin, ps: short arm satellite, pstk: short arm stalk, inv: inversion

* frequency given by D or G groups

large or double satellites, or increased stalk length on acrocentric chromosomes to be heteromorphisms; 24% to 36% stated that they would include these in a clinical report. Heterochromatic regions on chromosomes 1, 9, 16, and Y were considered heteromorphisms by 97% of participants, and 24% indicated they would report these findings. Pericentric inversions of chromosomes 1, 2, 3, 5, 9, 10, 16, and Y were considered heteromorphisms, with more than 75% of respondents indicating they would report these findings.[48] More information on specific heteromorphisms can be found online at http://cs-tl.de/DB/CA/HCM/0-Start.html.

Inducible Fragile Sites in Human Chromosomes
Under microscopy, human chromosomes have apparent constrictions. The primary constriction at the centromere can be seen by C-banding, and the secondary constriction on a site of NORs can be seen by silver staining. Other inheritable non-random 'constrictions' of breaks or gaps induced by specific cell culture conditions are termed fragile sites.[49, 50] Fragile sites are classified as common or rare depending on their frequency within the population and are further subdivided into different groups based on their specific induction chemistry. The induction chemicals such as folate acid, aphidicolin, bromodeoxyuridine, or 5-azatidine function as

partial DNA replication inhibitors at constituent sites of chromatin and make them fail to compact during mitosis.[49] Examples of common fragile sites include aphidicolin-inducible FRA3B (3p14.2), FRA16D (16q23.2), and FRAXD (Xq27.2), while rare fragile sites include folate-sensitive FRA11B (11q23.3), FRA16A (16p13.11), FRAXA (Xq27.3), FRAXE (Xq28), and FRAXF (Xq28). Most fragile sites occur as normal variants. Some fragile sites such as FRA6E (6q26), FRA11B (11q23.3), FRA16D (16q23.2) have been suggested as 'hotspots' for constitutional and somatic terminal deletions in cancer.[49, 51] Sequencing analysis of several fragile sites revealed a CCG/CGG trinucleotide repeat sequence adjacent to a CpG island. FRAXA at Xq27.3 with an unstable CGG repeat within the fragile X mental retardation 1 gene (*FMR1*) causes fragile X syndrome (FXS, OMIM#300624). The CGG repeat at FRAXA is polymorphic with six to 52 repeats in normal people and is considered as a 'premutation' with 48 to 200 repeats in parental carriers. The affected males and females exhibit a massive expansion resulting in so-called 'full mutation' of 230 to more than 1,000 repeats. This dynamic expansion to full mutation is associated with aberrant methylation of the *FMR1* gene leading to transcriptional suppression. FXS is an X-linked dominant disorder with an estimated incidence of one in 1,500 males and one in 2,500 females. Patients with the full mutation show reduced penetrance of 80% in males and 30% in females. The understanding of unstable CGG repeats in the *FMR1* gene has led to the switch of its diagnosis from cytogenetic analysis of FRAXA to DNA testing of the trinucleotide expansion.[52]

CLINICAL APPLICATION OF BANDING CYTOGENETICS

Banding cytogenetics has been used to detect numerical and structural chromosomal abnormalities for spontaneous abortions, for prenatal and postnatal patients of various clinical indications, and for various types of cancers. Technical standards and guidelines for cytogenetic studies and reporting of constitutional (germline) and acquired (somatic) chromosomal abnormalities have been developed.[53, 54] Routine chromosome analysis is performed on 15–20 metaphase cells; increased cell count is required to detect constitutional mosaicism and somatic clones in a portion of cells. It is recommended that the analysis of 20, 50, and 100 metaphases could detect a mosaic pattern or a clonal abnormality in 14%, 6%, and 3% of cells with a 95% confidence, respectively.[55]

Detection of Constitutional Chromosome Abnormalities

Chromosomal abnormalities have been detected in different human development stages. Studies on products of conception from spontaneous abortions recognized that approximately 50% of the spontaneous abortions are resulted from aneuploidies, polyploidies, and structural abnormalities with a relative frequency of 77%, 18%, and 5%, respectively. The most frequently seen aneuploidies are monosomy X, trisomy 16, trisomy 21, trisomy 22, trisomy 15, and trisomy 18.[56] Loss of one copy of any autosome is rarely observed and incompatible to life. It is estimated that 94% of fetuses carrying chromosomal abnormalities will end in a spontaneous abortion, indicating a strong natural selection against major chromosomal abnormalities in human biology.

Clinical indications for prenatal cytogenetics testing include advanced maternal age, abnormal ultrasound finding, serum screening, and non-invasive prenatal screening (NIPS—also called non-invasive prenatal testing or NIPT) on cell-free fetal DNA from maternal serum. NIPS/NIPT has a high positive predictive value for aneuploidy, which have reduced invasive procedures by approximately 60% and almost doubled the diagnostic yield of chromosomal abnormalities in current prenatal diagnosis.[57, 58] The detection rate of chromosomal abnormalities is approximately 15%; the diagnostic efficacy measured by detected cases vs expected cases by newborn incidence reaches 92% for trisomy 21 (Down syndrome, OMIM#190685) and monosomy X (Turner syndrome), but only 0.1%–0.4% for other sex chromosome aneuploidies (XXY/XXX/XYY).[59] The low detection rate for sex chromosome aneuploidies is likely due to the variable positive predictive value from NIPS/NIPT, reduced follow-up rate for cytogenetic diagnosis, and less severe clinical implications. Since there is an estimated 3% to 20% chance of a mosaic pattern for sex chromosome aneuploidies and 3% chance for structural rearrangements of sex chromosomes, follow-up cytogenetic studies could provide accurate diagnosis for karyotype-phenotype correlation and for genetic counseling on parental studies for future pregnancies.[60] Prenatal diagnosis of common syndromic chromosomal abnormalities contributed significantly to reduce birth defects and improve clinical management.

For pediatric patients with intellectual disability, developmental delay, multiple congenital anomalies, and dysmorphic features, the detection rate for numerical and structural chromosomal abnormalities were approximately 5% and 3%, respectively. The diagnostic efficacy for Down syndrome, Turner syndrome, Klinefelter syndrome (47,XXY), 47,XXX, 47,XYY, and balanced Robertsonian translocations were 63%, 48%, 14%, 4%, 3%, and 2% respectively.[59] The low detection rate for the latter three abnormalities likely related to the mild to normal phenotypes at childhood and may be noted in adulthood due to reproductive problems. Aneuploidy has been observed in approximately 1–4% of sperms, 10–35% of eggs and polar bodies, 20–40% of preimplantation embryos, 50% of spontaneous abortions, and 0.3% of newborns.[61] Inter-chromosomal heterogeneity of centromeric features influences chromosome segregation fidelity and results in non-random aneuploidy.[62]

Various types of balanced and unbalanced structural chromosomal abnormalities have been detected in approximately 2% of prenatal and 3% of pediatric patients.[59] The detection of balanced translocations could bring attention to the risk of unbalanced derivative chromosomes for fetuses. A well-known example is the carriers of a Robertsonian translocation between the short arms of D/G group acrocentric chromosomes. With an incidence of 1 in 1,000 in the general population, Robertsonian translocations are likely resulted from fusions of the repeat sequences at NOR sites.[34] The risk of Down syndrome for carriers of a Robertsonian translocation involving a chromosome 21 and another D/G group chromosome was estimated to be 10%–15% for a maternal carrier and <1% for a paternal carrier. However, for a carrier of a Robertsonian translocation of 21q, rob(21;21), the risk for Down syndrome is 100%.

Unbalanced chromosomal structural abnormalities include chromosomally visible deletions, duplications, inversions, ring chromosomes, and complex rearrangements of several chromosomes. Chromosome microarray analysis provides further characterization of genomic imbalances and involved gene content in these structural abnormalities. For example, chromosomally observed deletions of 4q could be further defined as containing distal duplications and deletions.[63] Prenatal diagnosis and postnatal findings of a fetus with partial trisomy of 13q21.33-qter and partial monosomy of 10p15.3-pter led to the detection of a balanced translocation, t(10;13)(p15.3;q21.33) in the mother and explained a history of two neonatal deaths and one spontaneous abortion.[64] Ring chromosomes 21 showed unique genomic structure with a terminal deletion and interstitial duplication or deletion and distinct mitotic behaviors.[65] Prenatal diagnosis on recurrence of a duplication of Xq26.1-q26.3 in two male fetuses indicated that the normal mother could have gonadal mosaicism.[66] Further clinical, bioinformatic and functional analyses could provide precise genotype-phenotype correlations for dosage sensitive genes in the human genome.[51, 67] Online databases for interpreting gene dosage effects (ClinVar) and pathogenic copy number variants (ClinGen, DECIPHER) are available. Figure 2.2 shows various types of autosomal and sex chromosome abnormalities by G-banding.

Detection of Somatic Chromosome Abnormalities

Banding cytogenetics played an important role on detecting recurrent chromosomal rearrangements of diagnostic significance, prognostic risk stratification, and targeted therapy or treatments for various types of tumors.[68] The classical example is the so-called 'Philadelphia chromosome' observed initially by solid staining of metaphases from patients with chronic myelogenous leukemia (CML). Further banding analysis indicated that the Philadelphia chromosome is the derivative chromosome 22 from a reciprocal translocation t(9;22)(q34.1;q11.2). Molecular characterization of this translocation defined underlying *BCR-ABL1* gene fusion with increased expression of tyrosine kinase, which led to the development of tyrosine kinase inhibitor for the treatment of this leukemia.[69, 70]

Following World Health Organization (WHO) classification of tumors of hematopoietic and lymphoid tissues, many recurrent somatic numerical and structural chromosomal rearrangements are of diagnostic, prognostic, and therapeutic values.[71] For example, in pediatric patients with B-cell precursor acute lymphoblastic leukemia (B-ALL), a hyperdiploid clone with trisomies of chromosomes 4, 10, and 17, or a translocation (12;21)(p13;q22) with *ETV6-RUNX1* gene rearrangement predicts a good prognosis, while a translocation t(9;22)(q34.1;q11.2), rearrangements involving the *KMT2A* gene at 11q23, and intra-chromosomal amplification of chromosome 21 (iAMP21) are associated with an unfavorable prognosis.[72] Other specific rearrangements such as jumping translocations of 1q likely related to the heterochromatic region have been reported in myelodysplastic syndrome (MDS) and acute myeloid leukemia (AML).[73]

Specific recurrent chromosomal abnormalities have been seen in various type of solid tumors. For example, reciprocal and three-way translocations involving the *ETV6* gene at 12p13.2 and the *NTRK3* gene at 15q25.3 have been associated with congenital fibrosarcoma.[74] Complex rearrangements involving multiple

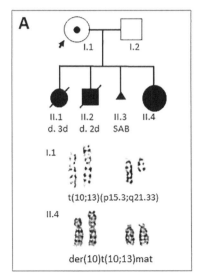

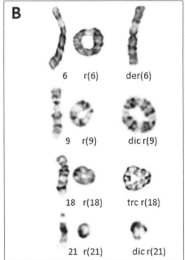

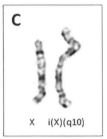

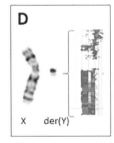

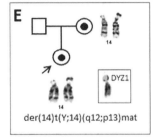

FIGURE 2.2   Various types of autosomal and sex chromosome abnormalities by G-banding.

A. A pedigree showing a mother (I.1) carrying a balanced translocation t(10;13) with a history of two neonatal deaths (II.1, II.2), a spontaneous abortion (SAB, II.3), and a daughter (II.4) carrying a derivative chromosome 10.

B. Cases of ring chromosomes 6, 9, 18 and 21 and derived variant ring broken chromosome, dicentric (dic) and tricentric (trc) ring chromosomes.

C. A female with an isochromosome of Xq.

D. A male with a derivative Y chromosome from a distal duplication and interstitial deletion in the short arm and deletion of the long arm as defined by a microarray analysis.

E. A pedigree showing mother to daughter transmission of a derivative chromosome 14 with enlarged short arm satellite region containing translocated Yq material as shown in the insert of a positive hybridization signal of DYZ1 probe to the derivative der(14).

chromosomes such as a giant ring or marker chromosome have been seen in liposarcoma.[75] A review of proficiency test results from over 200 cytogenetics laboratories for a 20-year period showed that more than 96% of laboratories provided correct interpretation of chromosomal abnormalities of hematological neoplasms.[76] Online databases such as the 'Atlas of Genetics and Cytogenetics in Oncology and

Haematology' http://atlasgeneticsoncology.org/ are available for interpreting somatic recurrent chromosomal abnormalities.

Report and Interpretation of Diagnostic Karyotype

Reporting and interpretation of constitutional or acquired chromosomal abnormalities follows the general principles of ISCN 2020 and standardized statements from diagnostic practice.[13, 77] For constitutional abnormalities, clinical impact on physical and intellectual development, congenital anomalies on organs and systems, reproductive fitness, cancer predisposition risk, and family history should be taken into consideration. Extended work up on additional metaphases should be pursued to differentiate true mosaicism from pseudomosaicism and to resolve mosaicism confined to the placenta in chorionic villus sampling.[55] Occasional metaphases with random gain or loss of one chromosome, inversion inv(14)(q11.2q32), inversion inv(7)(p14q35), or translocation t(7;14) in phytohemagglutinin-stimulated T cells could represent a culture artifact.[77] However, these inversions and translocations between chromosomes 7 and 14 in 10–45% of stimulated T cells is indicative for Nijmegen breakage syndrome (OMIM#251260), and further sequencing analysis on the *NBS1* gene should be recommended.

For acquired abnormalities, tumor classification and stratification, prognostic risk, and targeted therapy related to the clonal abnormalities should be presented in the report. Chromosomal abnormalities with impact on therapy such as translocation t(15;17)(q24;q21) with *PML-RARA* gene fusion in acute promyelocytic leukemia should be reported and communicated to referring physicians in a timely manner. Clonal abnormalities with a complex karyotype defined by the presence of three or more unrelated structural and numerical aberrations have been seen in a significant portion of cancers. The interpretation of complex karyotypes requires clinical evidence, expert consensus, and molecular and genomic findings. For example, a scoring practice counting the number of chromosome aberrations in complex karyotype has been proposed to correlate with the international prognosis scoring system for patients with myelodysplastic syndrome (MDS).[78] Furthermore, the poor prognosis associated with complex karyotypes in MDS could be driven by adverse *TP53* gene mutations.[79] Refinement of cytogenetic classification in acute myeloid leukemia revised the prognostic classification for rare recurring chromosomal abnormalities and complex karyotypes.[80] A retrospective analysis on a large case series of chronic lymphocytic leukemia defined cytogenetic complexity at levels of low or intermediate (3 to 4 aberrations) and high (≥5 aberrations) and presented their associated prognostic risk in combination with molecular findings.[81] For many patients, diagnostic karyotype provides the framework for further molecular analysis and clinical investigation.

CONCLUSIONS

Banding cytogenetics has provided structural basis for the understanding of sequence organization in human chromosomes and cytological landmarks for drafting the human genome. Technical standards and guidelines of G-band analysis have been developed and implemented for clinical cytogenetics. Current ISCN 2020 provides general principles applicable to multiple molecular and genomic techniques and standardized documentation of chromosomal abnormalities, pathogenic copy number variants, and gene rearrangements. Banding cytogenetics has been and continues to be an effective and highly proficient cell-based approach to detect constitutional numerical and structural chromosomal abnormalities in prenatal and pediatric patients and somatic clonal chromosomal abnormalities in various types of cancers.

WEB RESOURCES

Atlas of Genetics and Cytogenetics in Oncology and Haematology (http://atlasgeneticsoncology.org/)
Cases with heteromorphisms (http://cs-tl.de/DB/CA/HCM/0-Start.html)
ClinVar (www.ncbi.nlm.nih.gov/clinvar/)
ClinGen (https://clinicalgenome.org/)
DECIPHER (www.deciphergenomics.org/)
Human Genome Browser (https://genome.ucsc.edu/)

ACKNOWLEDGEMENT

Special thanks to Autumn DiAdamo and Jiadi Wen of Yale Clinical Cytogenetics Laboratory for helpful discussion and careful editing on the content of this chapter.

REFERENCES

1. Boveri, O.G.M. Über Mitosen bei einseitiger Chromosomenbindung. *Jenaische Zeitschrift für Naturwissenschaft*. **1903**, 37, 401–443.
2. Sutton, S.W. The chromosomes in heredity. *Biol. Bull.* **1903**, 4, 231–251.
3. Gartler, S.M. The chromosome number in humans: A brief history. *Nat. Rev. Genet.* **2006**, 7, 655–660.
4. Painter, S.T. Studies in mammalian spermatogenesis. II: The spermatogenesis of man. *J. Exp. Zool.* **1923**, 37, 291–338.
5. Ruddle, F.H. Theophilus painter: First steps toward an understanding of the human genome. *J. Exp. Zool. A. Comp. Exp. Biol.* **2004**, 301, 375–377.
6. Hsu, T.C. Mammalian chromosomes in vitro, I. Karyotype of man. *J. Heredity.* **1952**, 43, 167–172.
7. Tjio, J.H.; Levan, A. The chromosome number of man. *Hereditas.* **1956**, 42, U1–6.
8. Ford, C.E.; Jones, K.W.; Polani, P.E.; De Almeida, J.C.; Briggs, J.H. A sex-chromosome anomaly in a case of gonadal dysgenesis (Turner's syndrome). *Lancet.* **1959**, 1, 711–713.
9. Jacobs, P.A.; Strong, J.A. Case of human intersexuality having a possible XXY sex-determining mechanism. *Nature.* **1959**, 183, 302–303.
10. Lejeune, J.; Turpin, R.; Gautier, M. Chromosomic diagnosis of mongolism. *Arch. Fr. Pediatr.* **1959**, 16, 962–963.
11. Nowell, P.C.; Hungerford, D.A. A minute chromosome in human chronic granulocytic leukemia. *Science.* **1960**, 132, 1497.
12. Priest, J.H. *Medical Cytogenetics and Cell Culture*. Henry Kimpton Publisher, London; **1977**.

13. McGowan-Jordan, J.; Hastings, R.J.; Moore, S. *ISCN 2020: An International System for Human Cytogenomic Nomenclature.* Karger, Basel; **2020**.

14. Maeshima, K.; Iida, S.; Tamura, S. Physical nature of chromatin in the nucleus. Cold Spring Harb. *Perspect. Biol.* **2021**, 13, a040675.

15. Brown, S.W. Heterochromatin. *Science.* **1966**, 151, 417–425.

16. Barr, M.L. Dysgenesis of the seminiferous tubules. *Br. J. Urol.* **1957**, 29, 251–257.

17. Arrighi, F.E.; Hsu, T.C. Localization of heterochromatin in human chromosomes. *Cytogenetics.* **1971**, 10, 81–86.

18. Barr, M.L. Sex chromatin and phenotype in man. *Science.* **1959**, 130, 679–685.

19. Pardue, M.L.; Gall, J.G. Chromosomal localization of mouse satellite DNA. *Science.* **1970**, 168, 1356–1358.

20. Corneo, G.; Ginelli, E.; Polli, E. Repeated sequences in human DNA. *J. Mol. Biol.* **1970**, 48, 319–327.

21. Ginelli, E.; Corneo, G. The organization of repeated DNA sequences in the human genome. *Chromosoma.* **1976**, 56, 55–68.

22. Tantravahi, R.; Miller, D.A.; Dev, V.G.; Miller, O.J. Detection of nucleolus organizer regions in chromosomes of human, chimpanzee, gorilla, orangutan and gibbon. *Chromosoma.* **1976**, 56, 15–27.

23. Caspersson, T.; Farber, S.; Foley, G.E.; Kudynowski, J.; Modest, E.J.; Simonsson, E.; Wagh, U.; Zech, L. Chemical differentiation along metaphase chromosomes. *Exp. Cell. Res.* **1968**, 49, 219–222.

24. Seabright, M. A rapid banding technique for human chromosomes. *Lancet.* **1971**, 2, 971–972.

25. Hoo, J.J.; Jamro, H.; Schmutz, S.; Lin, C.C. Preparation of high resolution chromosomes from amniotic fluid cells. *Prenat. Diagn.* **1983**, 3, 265–267.

26. Yunis, J.J.; Sawyer, J.R.; Ball, D.W. The characterization of high-resolution G-banded chromosomes of man. *Chromosoma.* **1978**, 67, 293–307.

27. Arsham, M.S.; Barch, M.J.; Lawce, H.J. *The AGT Laboratory Manual.* 4th Edition. Wiley-Blackwell, Hoboken, NJ; **2017**.

28. Rooney, D.E.; Czepulkowski, B.H. *Human Cytogenetics: A Practical Approach Volume I: Constitutional Analysis.* 2nd Edition. Oxford University Press, Oxford; **1992**.

29. Rooney, D.E.; Czepulkowski, B.H. *Human Cytogenetics: A Practical Approach Volume II. Malignancy and Acquired Abnormalities.* 2nd Edition. Oxford University Press, Oxford; **1992**.

30. Bickmore, W.A. Patterns in the genome. *Heredity (Edinb).* **2019**, 123, 50–57.

31. Gardiner, K. Human genome organization. *Curr. Opin. Genet. Dev.* **1995**, 5, 315–322.

32. Ushiki, T.; Hoshi, O.; Iwai, K.; Kimura, E.; Shigeno, M. The structure of human metaphase chromosomes: Its histological perspective and new horizons by atomic force microscopy. *Arch. Histol. Cytol.* **2002**, 65, 377–390.

33. Chicano, A.; Crosas, E.; Oton, J.; Melero, R.; Engel, B.D.; Daban, J.R. Frozen-hydrated chromatin from metaphase chromosomes has an interdigitated multilayer structure. *EMBO J.* **2019**, 38, e99769.

34. Collins, F.S.; Lander, E.S.; Rogers, J.; Waterston, R.H.; Conso, I.H.G.S. Finishing the euchromatic sequence of the human genome. *Nature.* **2004**, 431, 931–945.

35. She, X.W.; Horvath, J.E.; Jiang, Z.S.; Liu, G.; Furey, T.S.; Christ, L.; Clark, R.; Graves, T.; Gulden, C.L.; Alkan, C.; Bailey, J.A.; Sahinalp, C.; Rocchi, M.; Haussler, D.; Wilson, R.K.; Miller, W.; Schwartz, S.; Eichler, E.E. The structure and evolution of centromeric transition regions within the human genome. *Nature.* **2004**, 430, 857–864.

36. Geiersbach, K.B.; Gardiner, A.E.; Wilson, A.; Shetty, S.; Bruyère, H.; Zabawski, J.; Saxe, D.F.; Gaulin, R.; Williamson, C.; Van Dyke, D.L. Subjectivity in chromosome band-level estimation: A multicenter study. *Genet. Med.* **2014**, 16, 170–175.

37. Bishop, A.; Blank, C.E.; Hunter, H. Heritable variation in the length of the Y chromosome. *The Lancet.* **1962**, 280, 18–20.

38. Starkman, M.N.; Shaw, M.W. Atypical acrocentric chromosomes in Negro and Caucasian Mongols. *Am. J. Hum. Genet.* **1967**, 19, 162–173.

39. Lubs, H.A.; Ruddle, F.H. Chromosome polymorphism in American Negro and White populations. *Nature.* **1971**, 233, 134–136.

40. McKenzie, W.H.; Lubs, H.A. Human Q and C chromosomal variations: Distribution and incidence. *Cytogenet. Cell Genet.* **1975**, 14, 97–115.

41. Verma, R.S.; Dosik, H.; Lubs, H.A. Size and pericentric inversion heteromorphisms of secondary constriction regions (h) of chromosomes 1, 9, and 16 as detected by CBG technique in Caucasians: Classification, frequencies, and incidence. *Am. J. Med. Genet.* **1978**, 2, 331–339.

42. Rawal, L.; Kumar, S.; Mishra, S.R.; Lal, V.; Bhattacharya, S.K. Clinical manifestations of chromosomal anomalies and polymorphic variations in patients suffering from reproductive failure. *J. Hum. Reprod. Sci.* **2020**, 13, 209–215.

43. Kosyakova, N.; Grigorian, A.; Liehr, T.; Manvelyan, M.; Simonyan, I.; Mkrtchyan, H.; Aroutiounian, R.; Polityko, A.D.; Kulpanovich, A.I.; Egorova, T.; Jaroshevich, E.; Frolova, A.; Shorokh, N.; Naumchik, I.V.; Volleth, M.; Schreyer, I.; Nelle, H.; Stumm, M.; Wegner, R.D.; Reising-Ackermann, G.; Merkas, M.; Brecevic, L.; Martin, T.; Rodríguez, L.; Bhatt, S.; Ziegler, M.; Kreskowski, K.; Weise, A.; Sazci, A.; Vorsanova, S.; Cioffi, M.deB.; Ergul, E. Heteromorphic variants of chromosome 9. *Mol. Cytogenet.* **2013**, 6, 14.

44. Zhu, J.J.; Qi, H.; Cai, L.; Wen, X.H.; Zeng, W.; Tang, G.D.; Luo, Y.; Meng, R.; Mao, X.Q.; Zhang, S.Q. C-banding and AgNOR-staining were still effective complementary methods to indentify chromosomal heteromorphisms and some structural abnormalities in prenatal diagnosis. *Mol. Cytogenet.* **2019**, 12, 41.

45. Ceylaner, G.; Ceylaner, S.; Danişman, N.; Ergün, A.; Ekici, E.; Schinzel, A.; Baumer, A. Chromosomal heteromorphisms may help for the diagnosis of uniparental disomy (UPD): A case report. *Prenat. Diag.* **2007**, 27, 1072–1074.

46. Magenis, R.E.; Overton, K.M.; Chamberlin, J.; Brady, T.; Lovrien, E. Parental origin of the extra chromosome in Down's syndrome. *Hum. Genet.* **1977**, 37, 7–16.

47. Magenis, R.E.; Sheehy, R.R.; Brown, M.G.; McDermid, H.E.; White, B.N.; Zonana, J.; Weleber, R. Parental origin of the extra chromosome in the cat eye syndrome: Evidence from heteromorphism and in situ hybridization analysis. *Am. J. Med. Genet.* **1988**, 29, 9–19.

48. Brothman, A.R.; Schneider, N.R.; Saikevych, I.; Cooley, L.D.; Butler, M.G.; Patil, S.; Mascarello, J.T.; Rao, K.W.; Dewald, G.W.; Park, J.P.; Persons, D.L.; Wolff, D.J.; Vance, G.H.; Cytogenetics Resource Committee.; College of American Pathologists/American College of Medical Genetics. Cytogenetic heteromorphisms: Survey results and reporting practices of Giemsa-band regions that we have pondered for years. *Arch. Pathol. Lab. Med.* **2006**, 130, 947–949.

49. Lukusa, T.; Fryns, J.P. Human chromosome fragility. *Biochim. Biophys. Acta.* **2008**, 1779, 3–16.

50. Sutherland, G.R.; Richards, R.I. The molecular basis of fragile sites in human chromosomes. *Curr. Opin. Genet. Dev.* **1995**, 5, 323–327.

51. Xie, X.; Chai, H.; DiAdamo, A.; Grommisch, B.; Wen, J.; Zhang, H.; Li, P. Genotype-phenotype correlations for putative haploinsufficient genes in deletions of 6q26-q27: Report of eight patients and review of literature. *Glob. Med. Genet.* **2022**, 9, 166–174.

52. Warren, S.T.; Nelson, D.L. Advances in molecular analysis of fragile X syndrome. *JAMA.* **1994**, 271, 536–542.

53. Mikhail, F.M.; Heerema, N.A.; Rao, K.W.; Burnside, R.D.; Cherry, A.M.; Cooley, L.D. Section E6.1–6.4 of the ACMG technical standards and guidelines: Chromosome studies of neoplastic blood and bone marrow-acquired chromosomal abnormalities. *Genet. Med.* **2016**, 18. 635–642.

54. Silva, M.; de Leeuw, N.; Mann, K.; Schuring-Blom, H.; Morgan, S.; Giardino, D.; Rack, K.; Hastings, R. European guidelines for constitutional cytogenomic analysis. *Eur. J. Hum. Genet.* **2019**, 27, 1–16.

55. Hook, E.B. Exclusion of chromosomal mosaicism: Tables of 90%.; 95% and 99% confidence limits and comments on use. *Am. J. Hum. Genet.* **1977**, 29, 94–97.

56. Zhou, Q.H.; Wu, S.Y.; Amato, K.; DiAdamo, A.; Li, P.N. Spectrum of cytogenomic abnormalities revealed by array comparative genomic hybridization on products of conception culture failure and normal karyotype samples. *J. Genet Genomics.* **2016**, 43, 121–131.

57. Chai, H.; DiAdamo, A.; Grommisch, B.; Boyle, J.; Amato, K.; Wang, D.; Wen, J.; Li, P. Integrated FISH, karyotyping and aCGH analyses for effective prenatal diagnosis of common aneuploidies and other cytogenomic abnormalities. *Med. Sci. (Basel).* **2019**, 7, 16.

58. Meng, J.; Matarese, C.; Crivello, J.; Wilcox, K.; Wang, D.; DiAdamo, A.; Xu, F.; Li, P. Changes in and efficacies of indications for invasive prenatal diagnosis of cytogenomic abnormalities: 13 years of experience in a single center. *Med. Sci. Monit.* **2015**, 21, 1942–1948.

59. Chai, H.; DiAdamo, A.; Grommisch, B.; Xu, F.; Zhou, Q.; Wen, J.; Mahoney, M.; Bale, A.; McGrath, J.; Spencer-Manzon, M.; Li, P.; Zhang, H. A retrospective analysis of 10-year data assessed the diagnostic accuracy and efficacy of cytogenomic abnormalities in current prenatal and pediatric settings. *Front. Genet.* **2019**, 10, 1162.

60. Xie, X.; Tan, W.; Li, F.; Ramirez, P.; DiAdamo, A.; Grommisch, B.; Amato, K.; Chai, H.; Wen, J.; Li, P. Diagnostic cytogenetic testing following positive noninvasive prenatal screening results of sex chromosome abnormalities: Report of five cases and systematic review of evidence. Mol. Genet. *Genomic Med.* **2020**, 8, e1297.

61. Nagaoka, S.I.; Hassold, T.J.; Hunt, P.A. Human aneuploidy: Mechanisms and new insights into an age-old problem. *Nat. Rev. Genet.* **2012**, 13, 493–504.

62. Dumont, M.; Gamba, R.; Gestraud, P.; Klaasen, S.; Worrall, J.T.; De Vries, S.G.; Boudreau, V.; Salinas-Luypaert, C.; Maddox, P.S.; Lens, S.M.; Kops, G.J.; McClelland, S.E.; Miga, K.H.; Fachinetti, D. Human chromosome-specific aneuploidy is influenced by DNA-dependent centromeric features. *EMBO J.* **2020**, 39, e102924.

63. Rossi, M.R.; DiMaio, M.S.; Xiang, B.; Lu, K.; Kaymakcalan, H.; Seashore, M.; Mahoney, M.J.; Li, P. Clinical and genomic characterization of distal duplications and deletions of chromosome 4q: Study of two cases and review of the literature. *Am. J. Med. Genet. A.* **2009**, 149A, 2788–2794.

64. Wei, Y.; Gao, X.; Yan, L.; Xu, F.; Li, P.; Zhao, Y. Prenatal diagnosis and postnatal followup of partial trisomy 13q and partial monosomy 10p: A case report and review of the literature. *Case Rep. Genet.* **2012**, 2012, 821347.

65. Zhang, H.Z.; Xu, F.; Seashore, M.; Li, P. Unique genomic structure and distinct mitotic behavior of ring chromosome 21 in two unrelated cases. *Cytogenet. Genome Res.* **2012**, 136, 180–187.

66. Cook, S.; Wilcox, K.; Grommisch, B.; Li, P.; Xu, F. Prenatal diagnosis of Xq26.1-q26.3 duplication in two fetuses of a woman with gonadal mosaicism. *North Am. J. Med. Sci.* **2014**, 7, 176–179.

67. Xu, F.; Li, L.; Schulz, V.P.; Xiang, B.; Zhao, H.; Li, P. Cytogenomic mapping and bioinformatic mining reveal interacting brain expressed genes for intellectual disability. *Mol. Cytogenet.* **2014**, 7, 4.

68. Qumsiyeh, M.B.; Li, P. *Chapter 5: Molecular Biology of Cancer: Cytogenetics.* Philadelphia: Lippincott Williams & Wilkins Publishers; **2001**.

69. Nowell, P.C.; Hungerford, D.A. Chromosome studies on normal and leukemic human leukocytes. *J. Natl. Cancer. Inst.* **1960**, 25, 85–109.

70. Rowley, J.D. Genetics: A story of swapped ends. *Science.* **2013**, 340, 1412–1413.

71. Swerdlow, S.H.; Campo, E.; Pileri, S.A.; Harris, N.L.; Stein, H.; Siebert, R.; Advani, R.; Ghielmini, M.; Salles, G.A.; Zelenetz, A.D.; Jaffe, E.S. The 2016 revision of the World Health Organization classification of lymphoid neoplasms. *Blood.* **2016**, 127, 2375–2390.

72. Moorman, A.V. New and emerging prognostic and predictive genetic biomarkers in B-cell precursor acute lymphoblastic leukemia. *Haematologica.* **2016**, 101, 407–416.

73. Couture, T.; Amato, K.; DiAdamo, A.; Li, P. Jumping translocations of 1q in myelodysplastic syndrome and acute myeloid leukemia: Report of three cases and review of literature. *Case Rep. Genet.* **2018**, 2018, 8296478.

74. Marino-Enriquez, A.; Li, P.; Samuelson, J.; Rossi, M.R.; Reyes-Mugica, M. Congenital fibrosarcoma with a novel complex 3-way translocation t(12;15;19) and unusual histologic features. *Hum. Pathol.* **2008**, 39, 1844–1848.

75. Chai, H.; Xu, F.; DiAdamo, A.; Grommisch, B.; Mao, H.; Li, P. Cytogenomic characterization of giant ring or rod marker chromosomes on four cases of well-differentiated and dedifferentiated liposarcomas. *Case Rep. Genet.* **2022**, 2022, 6341207 .

76. Larson, D.P.; Akkari, Y.M.; Van Dyke, D.L.; Raca, G.; Gardner, J.A.; Rehder, C.W.; Kaiser-Rogers, K.A.; Eagle, P.; Yuhas, J.A.; Gu, J.; Toydemir, R.M.; Kearney, H.; Conlin, L.K.; Tang, G.; Dolan, M.M.; Ketterling, R.P.; Peterson, J.F. Conventional cytogenetic analysis of hematologic neoplasms: A 20-year review of proficiency test results from the college of American pathologists/American college of medical genetics and genomics cytogenetics committee. *Arch. Pathol. Lab. Med.* **2021**, 145, 176–190.

77. Giersch, A.B.S.; Bieber, F.R.; Dubuc, A.M.; Fletcher, J.A.; Ligon, A.H.; Mason-Suares, H.; Morton, C.C.; Weremowicz, S.; Xiao, S.; Cin, P.D. Reporting of diagnostic cytogenetic results. Curr. Protoc. *Hum. Genet.* **2016**, 89, A 1D 1-A 1D 23.

78. Chun, K.; Hagemeijer, A.; Iqbal, A.; Slovak, M.L. Implementation of standardized international karyotype scoring practices is needed to provide uniform and systematic evaluation for patients with myelodysplastic syndrome using IPSS criteria: An International Working Group on MDS cytogenetics study. *Leuk. Res.* **2010**, 34, 160–165.

79. Grimwade, D.; Hills, R.K.; Moorman, A.V.; Walker, H.; Chatters, S.; Goldstone, A.H.; Wheatley, K.; Harrison, C.J.; Burnett, A.K; National Cancer Research Institute Adult Leukaemia Working Group. Refinement of cytogenetic classification in acute myeloid leukemia: Determination of prognostic significance of rare recurring chromosomal abnormalities among 5876 younger adult patients treated in the United Kingdom Medical Research Council trials. *Blood.* **2010**, 116, 354–365.

80. Baliakas, P.; Jeromin, S.; Iskas, M.; Puiggros, A.; Plevova, K.; Nguyen-Khac, F.; Davis, Z.; Rigolin, G.M.; Visentin, A.; Xochelli, A.; Delgado, J.; Baran-Marszak, F.; Stalika, E.; Abrisqueta, P.; Durechova, K.; Papaioannou, G.; Eclache, V.; Dimou, M.; Iliakis, T.; Collado, R.; Doubek, M.; Calasanz, M.J.; Ruiz-Xiville, N.; Moreno, C.; Jarosova, M.; Leeksma, A.C.; Panayiotidis, P.; Podgornik, H.; Cymbalista, F.; Anagnostopoulos, A.; Trentin, L.; Stavroyianni, N.; Davi, F.; Ghia, P.; Kater, A.P.; Cuneo, A.; Pospisilova,

S.; Espinet, B.; Athanasiadou, A.; Oscier, D.; Haferlach, C.; Stamatopoulos, K.; ERIC.; the European Research Initiative on CLL. Cytogenetic complexity in chronic lymphocytic leukemia: Definitions, associations, and clinical impact. *Blood*. **2019**, 133(11), 1205–1216.

81. Haase, D.; Stevenson, K.E.; Neuberg, D.; Maciejewski, J.P.; Nazha, A.; Sekeres, M.A.; Ebert, B.L.; Garcia-Manero, G.; Haferlach, C.; Haferlach, T.; Kern, W.; Ogawa, S.; Nagata, Y.; Yoshida, K.; Graubert, T.A.; Walter, M.J.; List, A.F.; Komrokji, R.S.; Padron, E.; Sallman, D.; Papaemmanuil, E.; Campbell, P.J.; Savona, M.R.; Seegmiller, A.; Adès, L.; Fenaux, P.; Shih, L.Y.; Bowen, D.; Groves, M.J.; Tauro, S.; Fontenay, M.; Kosmider, O.; Bar-Natan, M.; Steensma, D.; Stone, R.; Heuser, M.; Thol, F.; Cazzola, M.; Malcovati, L.; Karsan, A.; Ganster, C.; Hellström-Lindberg, E.; Boultwood, J.; Pellagatti, A.; Santini, V.; Quek, L.; Vyas, P.; Tüchler, H.; Greenberg, P.L.; Bejar, R.; International Working Group for MDS Molecular Prognostic Committee. TP53 mutation status divides myelodysplastic syndromes with complex karyotypes into distinct prognostic subgroups. *Leukemia*. **2019**, 33, 1747–1758.

3 Generation of Microdissection-Derived Painting Probes from Single Copy Chromosomes

Svetlana Romanenko and Vladimir Trifonov

CONTENTS

INTRODUCTION

The method of chromosome microdissection with a subsequent DNA isolation was developed over 40 years ago. It allows physically isolating either a whole or a large fragment of a chromosome. After early works on *Drosophila* polytene,[1] murine,[2] and human[3] metaphase chromosomes, the method was considerably improved by Senger and colleagues,[4] who first employed a combination of an inverted microscope equipped with a rotating plate, including a deproteinization step, followed by DNA amplification. From 1992 onwards, most paints for fluorescence in situ hybridization (FISH) have been made by direct polymerase chain reaction (PCR) amplification, using a degenerate universal primer (DOP-PCR).[5] Subsequent major

DOI: 10.1201/9781003223658-3

technological improvements included the addition of a second left-handed micro-manipulator equipped with a pipette, containing a collection drop, which enabled a more convenient handling of dissected chromosomal parts.[6] Chromosome-specific libraries for multicolor FISH and multicolor banding constructed by microdissection have found wide and successful application in clinical genetics, comparative inter-species studies, and interphase cytogenetics.[7, 8] Semi-archived material can also be used for chromosome microdissection.[9]

However, one of the significant limitations of microdissection method was the need to obtain multiple copies of the chromosome or its fragment of interest in order to create high-quality paints. The development of multiple displacement amplifi-cation (MDA) whole-genome amplification methods[10] makes it possible yet to reduce the amount of starting material and work with a single chromosome copy. The approach showed high efficiency in obtaining samples for FISH in compara-tive genomic studies[11, 12] and in medical genomics.[13] Moreover, it was shown that sequencing of DNA from single isolated chromosomes (ChromSeq) can be done and is suitable to determine the chromosome content and assign genomic scaffolds to chromosomes, thus bridging the gap between cytogenetics and genomics.[14]

Here we provide an updated protocol for generating paint probes by an MDA-based system from microdissected chromosomes or chromosomal segments.

MATERIALS

INSTRUMENTS

- Inverted microscope the IX Series (Olympus Shinjuku, Tokyo, Japan) or Axiovert 10 or 135 (Zeiss, Jena, Germany)
- Right-handed and left-handed micromanipulators (Narishige, Tokyo, Japan or Eppendorf, Germany)
- Pipette puller (Narishige, Japan)
- Glass rods, 2mm diameter (Schott Glas, Mainz, Germany)
- Pasteur pipettes, 230mm (Deltalab, Barcelona, Spain, or Assistent, Sondheim, Germany)
- Aspirator with trap flask ("Grant", UK).

REAGENTS

- Glycerol (Sigma-Aldrich)
- Proteinase K from Tritirachium album (Sigma-Aldrich, USA)
- Trypsin (Sigma-Aldrich, USA)
- 2×SSC
- 1×PBS
- Giemsa (Merck)
- Carbon tetrachloride (Sigma-Aldrich, USA)
- Dimethyldichlorosilane (Sigma-Aldrich, USA)
- GenomePlex® Whole Genome Amplification (WGA) Kit, WGA1 (Sigma-Aldrich, USA)

- GenomePlex® WGA Reamplification Kit, WGA3 (Sigma-Aldrich, USA)
- QIAquick PCR Purification Kit (Quiagen, USA)

SOLUTIONS

- Collection drop solution: 30% Glycerol, 10mM Tris-HCl pH 7.5, 10mM NaCl, 0.1% SDS, 1mM EDTA, 0.1% Trition X100, 1.44mg/ml Proteinase K.

DESCRIPTION OF METHODS

MICRO-NEEDLE AND MICROPIPETTE PREPARATION

1. Prepare micro-needles from 2mm glass rods on a vertical two-step puller, to get very sharp but not very long edges.
2. Expose the needles to UV light for at least 30min.
3. Prepare micropipettes on the same puller from Pasteur pipettes.
4. Break the micropipette tips carefully to obtain a small round opening (~30–70µm).
5. Siliconize micropipettes by immersing into 1% dimethyldichlorosilane in carbon tetrachloride using an aspirator with a trap flask ("Grant").
6. Air dry and wash twice in 1mM EDTA (pH 7.5).
7. Incubate micropipettes for 30min at 100°C.
8. Keep micropipettes and micro-needles in closed boxes prior to use.
   Caution: Dimethyldichlorosilane and carbon tetrachloride are highly toxic—all manipulations should be done under the draft hood. Two ml of siliconization solution is enough to siliconize more than 30 micropipettes.

SLIDE PREPARATION

Traditionally, preparations for microdissection were dropped on coverslips.[15] However, in our experience, the use of microscope slides is more convenient. To achieve a good spreading of chromosomes, the conditions for slides preparation should be optimized. In some cases, for better spreading, it is recommended to keep the slide over a water bath, or add more fixative over the freshly dropped preparation.

1. Put clear 76×26 mm microscope slides (Menzel-Glaeser, Braunschweig, Germany) in cold DNA-free distilled water (i.e. PCR water).
2. Take a wet slide and drop 15µl of metaphase preparation in 3:1 methanol/acetic acid fixative onto the slide.
3. Let slide air dry at room temperature.

CHROMOSOME STAINING

Traditionally, chromosome staining was made using trypsin and phosphate buffer incubation.[15] However, conventional GTG-banding[16] allows for better staining of chromosomes and does not affect the quality of probes. In cases where the desired chromosome is easily identified (the smallest or the biggest chromosome in karyotype, the only acrocentric, and etc.), the trypsin treatment step can be skipped. Obtaining a high-quality GTG-banding is required for an accurate identification of chromosomes or their regions.

1. Incubate the slide in trypsin solution (0.12% or 0.25% trypsin in 1×PBS) for 30 to 90s (the required time will vary depending on chromosome preparation).
2. Rinse in 2×SSC.
3. Incubate in Giemsa solution (35ml phosphate buffer or sterile water and 3.5ml Giemsa) for 0.5–2min (the time will vary depending on chromosome preparation).
4. Rinse in sterile water.
5. Air-dry the slide.
6. Assess the quality of the metaphase spreading and banding achieved under the microscope.
7. Keep the slides refrigerated until the microdissection step.

MICRODISSECTION

1. Prepare 10–20µl of collection drop solution.
2. Use a 10× or 20× objective.
3. Move the objective down to adjust the needle.
4. Load a needle into the holder, and then load onto the micromanipulator, so that the tip of the needle is close to the center of visual field of the objective.
5. Find the needle in the visual field and center it; then switch to 40× objective and ensure that the tip of the needle can be seen with the 40× objective.
6. Put the slide on the specimen stage and find a suitable metaphase spread. By rotating the table, place the chromosome to be cut at a right angle to the needle.
7. Bring the needle down carefully until it is just above the chromosome. Forward movement of the tip will lead to the excision of the chromosome material.
8. Touch the excised fragment with the tip of the needle several times until it is attached to the needle. Carefully elevate the needle.
9. Switch to a 10× or 20× objective.
10. Remove the microscope slides.
11. Take a siliconized pipette; gently touch the surface of the collection drop solution using the siliconized pipette, so that a tiny amount of collection drop solution will be sucked up by capillary force.
12. Suspend the pipette above the objective.
13. Place the needle to the same plane as the pipette tip using micromanipulator. Transfer the chromosome fragment by touching the solution inside of the pipette with the needle, and then withdraw the needle. Large fragments will remain visible in the collection drop for several seconds.
14. Put the pipette in a humidified tray at 60°C for 1–2h.

GENERATION OF PAINT PROBE BY WHOLE GENOME AMPLIFICATION (WGA)

Contamination is one of the major problems when performing a sequence-independent DNA amplification starting from low amounts of DNA. We recommend using separate micropipettes for pre- and post-primary DOP-PCR procedures. UV treatment of the microdissection room and instruments as well as DNA-EX (Genaxis) cleaning of working surfaces is highly recommended.

1. Aliquot 5μl DNA-free water into 0.2ml tubes.
2. Transfer the collection drop with the microdissected material from the Pasteur pipette into the tube containing the water by breaking off the pipette tip into the tube.
3. Briefly spin tube. The solution can be kept on ice for several hours before DNA amplification (for longer periods store at −20°C).
4. Use Whole Genome Amplification Kit (WGA1) for DNA amplification. Follow manufacturer's protocol.
5. Purify the amplification product using any available PCR purification kit (e.g. QIAquick PCR Purification Kit). Determine the concentration of the purified amplification product using NanoDrop spectrophotometry.
6. Run 2μl of PCR products on a 1.5% agarose gel. The products should look as a smear with an average size of 0.2 to 0.8kb.
7. Use WGA Reamplification Kit (WGA3) with 0.1μM of added modified nucleotides (bio-dUTP or dig-dUTP) for DNA labeling. Follow manufacturer's protocol.
8. Purify the amplification product using a PCR purification kit (e.g. QIAquick PCR Purification Kit).
9. Run 2μl of PCR products on a 1% agarose gel. The products should appear as a smear with an average size of 0.2 to 0.8kb.
10. For FISH use, precipitate the PCR product in ethanol and resuspend in 30–40μl hybridization buffer.

According to the manufacturer's protocol, starting material for WGA1 is 10ng of DNA in 10μl volume of the reaction mixture. In our protocol, the starting material is well below 1pg; therefore, we recommend transferring the collection drop content into 5μl of sterile water and proportionally halving all other reagents when working with WGA1. The number of amplification cycles should be increased up to 30.

EXPECTED RESULTS

The resulted libraries may demonstrate different qualities depending on the repetitive DNA content in studied genomes and chromosomes. Therefore, we recommend using COT DNA for repetitive DNA suppression during FISH.[17] Figure 3.1 demonstrates expected results of FISH with painting probes obtained from single-copies of chromosomes 15 and 16 of the Nile crocodile (*Crocodylus niloticus*). Although both chromosomes are specifically painted by the respective libraries, heterochromatic

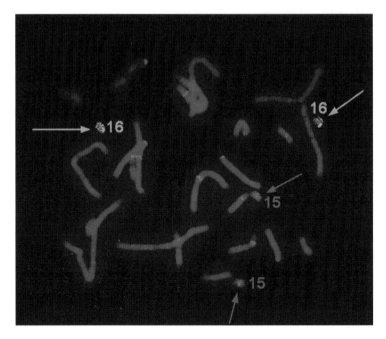

FIGURE 3.1 FISH-result on a metaphase spread from Nile crocodile CNO (*Crocodylus niloticus*) hybridized with painting probes obtained from single-copies of chromosomes 15 (red) and 16 (green) of this species. Painting probe CNO15 additionally paints heterochromatic regions of most CNO chromosomes, despite suppression with sonicated total genomic CNO-DNA.

regions on other chromosomes are also highlighted with the CNI15 probe, despite suppression with sonicated total genomic DNA of *Crocodylus niloticus*.

ACKNOWLEDGEMENTS

While this protocol was written, SR and VT were supported by RFBR grant # 19-54-26017.

REFERENCES

1. Scalenghe, F.; Turco, E.; Edstrom, J.E.; Pirrotta, V.; Melli, M. Microdissection and cloning of DNA from a specific region of *Drosophila melanogaster* polytene chromosomes. *Chromosoma*. **1981**, 82, 205–216.
2. Röhme, D.; Fox, H.; Herrmann, B.; Frischauf, A.M.; Edström, J.E.; Mains, P.; Silver, L.M.; Lehrach, H. Molecular clones of the mouse t complex derived from microdissected metaphase chromosomes. *Cell*. **1984**, 36, 783–788.
3. Bates, G.P.; Wainwright, B.J.; Williamson, R.; Brown, S.D. Microdissection and microcloning from the short arm of human chromosome 2. *Mol. Cell. Biol*. **1986**, 6, 3826–3830.

4. Senger, G.; Lüdecke, H.J.; Horsthemke, B.; Claussen, U. Microdissection of banded human chromosomes. *Hum. Genet.* **1990**, 84, 507–511.

5. Telenius, H.; Carter, N.P.; Bebb, C.E.; Nordenskjold, M.; Ponder, B.A.; Tunnacliffe, A. Degenerate oligonucleotide-primed PCR: General amplification of target DNA by a single degenerate primer. *Genomics.* **1992**, 13, 718–725.

6. Weimer, J.; Kiechle, M.; Senger, G.; Wiedemann, U.; Ovens-Raeder, A.; Schuierer, S.; Kautza, M.; Siebert, R.; Arnold, N. An easy and reliable procedure of microdissection technique for the analysis of chromosomal breakpoints and marker chromosomes. *Chromosome. Res.* **1999**, 7, 355–362.

7. Liehr, T.; Heller, A.; Starke, H.; Rubtsov, N.; Trifonov, V.; Mrasek, K.; Weise, A.; Kuechler, A.; Claussen, U. Microdissection based high resolution multicolor banding for all 24 human chromosomes. *Int. J. Mol. Med.* **2002**, 9, 335–339.

8. Liehr, T.; Starke, H.; Heller, A.; Kosyakova, N.; Mrasek, K.; Gross, M.; Karst, C.; Steinhaeuser, U.; Hunstig, F.; Fickelscher, I.; Kuechler, A.; Trifonov, V.; Romanenko, S.A.; Weise, A. Multicolor fluorescence in situ hybridization (FISH) applied to FISH-banding. *Cytogenet. Genome Res.* **2006**, 114, 240–244.

9. Al-Rikabi, A.B.H.; Cioffi, M.B.; Liehr, T. Chromosome microdissection on semi-archived material. *Cytometry A.* **2019**, 95, 1285–1288.

10. Höckner, M.; Erdel, M.; Spreiz, A.; Utermann, G.; Kotzot, D. Whole genome amplification from microdissected chromosomes. *Cytogenet. Genome Res.* **2009**, 125, 98–102.

11. Andreyushkova, D.A.; Makunin, A.I.; Beklemisheva, V.R.; Romanenko, S.A.; Druzhkova, A.S.; Biltueva, L.B.; Serdyukova, N.A.; Graphodatsky, A.S.; Trifonov V.A. Next generation sequencing of chromosome-specific libraries sheds light on genome evolution in paleotetraploid sterlet (*Acipenser ruthenus*). *Genes.* **2017**, 8, 318.

12. Rajičić, M.; Romanenko, S.A.; Karamysheva, T.V.; Blagojević, J.; Adnadević, T.; Budinski, I.; Bogdanov, A.S.; Trifonov, V.A.; Rubtsov, N.B.; Vujošević, M. The origin of B chromosomes in yellow-necked mice (*Apodemus flavicollis*): Break rules but keep playing the game. *PLoS ONE.* **2017**, 12, e0172704.

13. Lemskaya, N.A.; Romanenko, S.A.; Rezakova, M.A.; Filimonova, E.A.; Prokopov, D.Y.; Dolskiy, A.A.; Perelman, P.L.; Maksimova, Y.V.; Shorina, A.R.; Yudkin, D.V. A rare familial rearrangement of chromosomes 9 and 15 associated with intellectual disability: A clinical and molecular study. *Mol. Cytogenet.* **2021**, 14, 47.

14. Iannucci, A.; Makunin, A.I.; Lisachov, A.P.; Ciofi, C.; Stanyon, R.; Svartman, M.; Trifonov, V.A. Bridging the gap between vertebrate cytogenetics and genomics with single-chromosome sequencing (ChromSeq). *Genes.* **2021**, 12, 124.

15. Yang, F.; Trifonov, V.; Ng, B.L.; Kosyakova, N.; Carter, N.P. Generation of paint probes from flow-sorted and microdissected chromosomes. In: *Fluorescence In Situ Hybridization (FISH)*; Liehr, T., Ed. Springer Protocols Handbooks, Berlin, **2016**, pp. 63–79.

16. Seabright, M. A rapid banding technique for human chromosomes. *The Lancet.* **1971**, 298(7731), 971–972.

17. Trifonov, V.A.; Vorobieva, N.V.; Serdyukova, N.A.; Rens, W. FISH with and without COT1 DNA. In: *Fluorescence In Situ Hybridization (FISH)*; Liehr, T., Ed. Springer Protocols Handbooks, Berlin, **2016**, pp. 123–132.

4 FISH—An Overview

Thomas Liehr

CONTENTS

INTRODUCTION

FISH = fluorescence in situ hybridization is one of two basic approaches originally being considered to constitute the field of molecular cytogenetics.[1] FISH is the approach that at present is widely applied in chromosomic research, while its sister approach PRINS = primed in situ hybridization (also sometimes called primed in situ amplification, or primed in situ labeling) is practically no more in use.[1, 2] FISH can be used to visualize DNA or RNA in tissues, interphase cells and metaphases.[1, 3] Here only DNA-oriented FISH is considered.

FISH is in general single cell–oriented; single to multiple loci in a genome can be accessed by single to multicolor-FISH approaches. The possibilities of FISH are only limited by ideas whose questions can be solved by this approach.[1] Thus, since the first radioactive ISH experiments in the 1970s[4] and FISH-experiments in the 1980s,[5] practically each year new applications in human genetic, animal- and plant-genetics and even in single-cell eukaryotic and prokaryotic research are reported.[1, 6] The FISH-principle is depicted in Figure 4.1. Necessary for FISH are only a target-DNA and a probe-DNA to hybridize.[3]

In the following, basics on DNA-probes being used in FISH and some examples for application of FISH—mainly in human genetics—are provided.

DOI: 10.1201/9781003223658-4

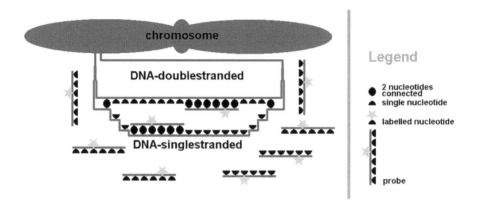

FIGURE 4.1 FISH-principle in a scheme: A chromosome is depicted from which schematically and at greater magnitude the DNA-double strand is emanating. After denaturation, it is single-stranded and brought together with single-stranded, labeled probe-DNA, which hybridizes at homologous regions in the target region.

FISH PROBES

FISH probes = probe-DNA in human genetics are nowadays mainly commercially available ones, being ready to use and labeled already with fluorophores or other haptens like biotin or digoxigenin.[1, 3] Besides, there also can be used homemade/homebrewed probes being labeled either by PCR-methods[7] or Nick-translation.[8] Here five basic probe DNA types suitable and used for FISH are listed:

REPETITIVE PROBES

Repetitive and/or heterochromatic DNA stretches can be easily visualized by FISH.[9] Basic units of the repeats are 1 to 200 base pairs; however, they become detectable by FISH as soon as they span 10s to 1000s of kilo base pairs.[9] In FISH, randomly selected 2n to 6n repeats, telomeric repeats (6n), centromeric repeats (~170n) and repeats being present in nucleolus organizing regions (NOR) are applied as FISH-probes in general.[10] As in human chromosomes most centromeres have a private satellite-DNA sequence, these centromeric/alphoid besides telomeric probes are popular, commercially available probes.[3]

LOCUS-SPECIFIC PROBES

Locus-specific probes (LSPs) are normally euchromatic DNA-stretches cloned into genetic vectors; nowadays in most cases bacterial artificial chromosome (BAC) clones are used as basis for LSPs.[3] To produce good and reproducible FISH-signals, a vector needs to carry inserts of at least 12 kb,[11] while for mapping purposes exceptionally 0.44 kb–sized cDNA probes can be useful.[12]

PARTIAL CHROMOSOME PAINTS

A "painting probe" stains at least one or two euchromatic chromosomal subbands and not only a single locus or a heterochromatic block.[13] To establish a partial chromosome paint = pcp, only glass needle–based chromosome microdissection (midi) is suited.[14] The maximum size of a pcp is spanning the arm of a meta- or submetacentric chromosome,[15] and they can be as small as the diameter of the extended glass needle, i.e. a part of a chromosomal subband,[16] or, in case of doing midi lampbruch chromosome, even in LSP size.[17] Just for completeness, it must be mentioned that also radiation hybrid cells[18] or ZOO-FISH probes like applied for RX-FISH[19] have been used to create kinds of pcps.

WHOLE CHROMOSOME PAINTS

A whole chromosome paint = wcp can be established in two ways: either, as pcps, by midi,[14] or by chromosome flow sorting.[20] Also, interspecies hybrids (e.g. mouse/human somatic cell hybrid) have been used to set up wcp probes.[21]

WHOLE GENOME

FISH is normally used to access single cells. However, when doing comparative genomic hybridization (CGH),[22] the single cell–derived metaphase chromosomes of a normal individual is just used as a kind of matrix, or low-resolution DNA-chip. Towards this matrix, DNA extracted from millions of cells derived from two different body tissues or even two closely related species is hybridized. Certainly tested DNAs are labeled in two different colors. Thus, it is possible to detect copy number gains and losses between the two probes. This CGH approach has been used in human genetic research for solid tumors and has been further evolved as an important tool in diagnostics nowadays as chromosome microarray (CMA) technique.[23, 24] In addition, CGH is currently applied in evolution-research.[10]

FISH PROBE SETS

All different kinds of FISH-probes can be used for one- to multicolor-FISH (mFISH) experiments.[3, 6] Important mFISH probe sets in diagnostics are all 24 human wcp probes taken into one probe set,[25, 26] LSP/centromeric probe sets being tumor-type specific[27] or for prenatal testing.[28] LSPs and/or repetitive probes are the probes normally being applied in molecular combing, where FISH is possible in almost base pair resolution on maximally stretched DNA-fibers.[29]

BASIC FISH-PROTOCOL

As each laboratory uses a specific protocol to do FISH, here just basic steps to be included in each metaphase FISH-protocol are listed—see also ref[3]:

- Slides with metaphase spreads (e.g. obtained from a cytogenetic laboratory) have to air-dry at room temperature (RT) overnight.
- To obtain optimal results later, a pretreatment of the slides is recommended; here a three step procedure led to best plasma-free metaphases with strong FISH-signals and optimal inverted 4′,6-diamidino-2-phenylindole (DAPI) banding pattern:
  - Step 1: incubate slides in a 3% paraformaldehyde solution for ~5 min.
  - Step 2: incubate slides in a 0.5% pepsin/1% 1N HCl solution for ~2–3 min.
  - Step 3: repeat step 1.
  - Dry slides in a 70%/90%/100% ethanol series at RT.
- Do washing in between steps 1–3 with 2×SSC or 1×PBS; also consider environment protection when discarding formaldehyde-contaminated solutions.
- Denature slides in a 70% formamide solution in 2×SSC—formamide reduces the melting point of DNA; SSC-salts stabilize DNA structure—at 75°C for 3–5 min.
- Dry slides in a 70%/90%/100% ethanol series at RT—only 70% ethanol should be at −20°C.
- Denature FISH-probe (here a commercial probe is suggested, which has a ready to use) at 95°C for 5 min. Then either do prehybridization at 37°C for 10–30 min or place directly on ice.
- Pipette denatured FISH-probe on slide with denatured target-DNA, cover with a coverslip, and seal with rubber cement.
- Incubate in a humid chamber over night at 37°C.
- Remove coverslip and do post-hybridization washings acc. to probe used.
- If necessary, a hapten detection must be done in case of a probe labeled e.g. by biotin.
- Also, consider environment protection when discarding formamide-contaminated solutions.
- Add antifade with DAPI and a coverslip.
- Evaluate slide under a fluorescence microscope with suited filter sets.

YIELD = FISH APPLICATIONS

Here a subjective list of possible FISH-application without claim of completeness is provided:

  (i) mosaic characterization[30, 31]
  (ii) diagnostics in clinical genetics and acquired diseases[1, 3]
  (iii) microdissection, reverse FISH and sequencing can be combined[32]
  (iv) detection of radio- or mutagen-sensitivity[33]

    (v) chromothripsis detection[34–36]
    (vi) CGH in ZOO-FISH experiments[10, 37]
    (vii) gene mapping in metaphase[12] or by fiber-FISH Nguyen[29]
    (viii) nucleomics = interphase architecture studies[38–40]
    (ix) in situ visualization of uniparental disomy[41]

CONCLUSIONS

As a conclusion, still best suited is a statement of Serakinci and Koelvraa from 2009:

> FISH techniques were originally developed as extra tools in attempts to map genes and a number of advances were achieved with this new technique. However, it soon became apparent that the FISH concept offered promising possibilities also in a number of other areas in biology and its use spread into new areas of research and also into the area of clinical diagnosis. In very general terms the virtues of FISH are in two areas of biology, namely genome characterization and cellular organization, function and diversity. . . . To what extend FISH technology will be further developed and applied in new areas of research in the future remains to be seen.[42]

REFERENCES

1. Liehr, T. Molecular cytogenetics in the era of chromosomics and cytogenomic approaches. *Front. Genet.* **2021**, 12, 720507.
2. Pellestor, F. Development and adaptation of the PRINS technology: An overview. *Methods Mol. Biol.* **2006**, 334, 211–220.
3. Liehr, T. (Ed.). *Fluorescence in situ Hybridization (FISH): Application Guide.* 2nd Edition. Springer, Berlin; **2017**.
4. Gall, J.G.; Pardue, M.L. Formation and detection of RNA-DNA hybrid molecules in cytological preparations. Proc. Natl. Acad. Sci. U. S. A. **1969**, 63, 378–383.
5. Langer, P.; Waldrop, A.; Ward, D. Enzymatic synthesis of biotin-labeled olynucleotides: Novel nucleic acid affinity probes. Proc. Natl. Acad. Sci. U. S. A. **1981**, 78, 6633–6637.
6. Liehr, T. Basics and literature on multicolor fluorescence in situ hybridization application. http://cs-tl.de/DB/TC/mFISH/0-Start.html [accessed on 01/12/2022].
7. Telenius, H.; Carter, N.P.; Bebb, C.E.; Nordenskjöld, M.; Ponder, B.A.J.; Tunnacliffe, A. Degenerate oligonucleotide-primed PCR: General amplification of target DNA by a single degenerate primer. *Genomics.* **1992**, 13, 718–725.
8. Rigby, P.W.J.; Dieckmann, M.; Rhodes, C.; Berg, P. Labeling deoxyribonucleic acid to high specific activity in vitro by nick translation with DNA polymerase I. *J. Mol. Biol.* **1977**, 113, 237–251.
9. Liehr, T. Repetitive elements in humans. *Int. J. Mol. Sci.* **2021**, 22, 2072.
10. Xu, D.; Sember, A.; Zhu, Q.; de Oliveira, E.A.; Liehr, T.; Al-Rikabi, A.B.H.; Xiao, Z.; Song, H.; de Bello Cioffi, M. Deciphering the origin and evolution of the X1X2Y system in two closely-related oplegnathus species (Oplegnathidae and Centrarchiformes). *Int. J. Mol. Sci.* 2019, 20, 3571.
11. Liehr, T.; Thoma, K.; Kammler, K.; Gehring, C.; Ekici, A.; Bathke, K.; Grehl, H.; Rautenstrauss, B. Direct preparation of uncultured EDTA-treated or heparinized blood for interphase FISH analysis. *Appl. Cytogenet.* **1995**, 21, 185–188.

12. Pröls, F.; Liehr, T.; Rinke, R.; Rautenstrauss, B. Assignment of the microvascular endothelial differentiation gene 1 (MDG1) to human chromosome band 14q24.2→q24.3 by fluorescence in situ hybridization. Cytogenet. *Genome Res.* **1997**, 79, 149–150.

13. Liehr, T. *Human Genetics: Edition 2020: A Basic Training Package.* Epubli, Berlin; **2020**.

14. Al-Rikabi, A.B.H.; Cioffi, M.d.B.; Liehr, T. Chromosome microdissection on semi-archived material. *Cytometry Part A.* **2019**, 95, 1285–1288.

15. Kytölä, S.; Rummukainen, J.; Nordgren, A.; Karhu, R.; Farnebo, F.; Isola, J.; Larsson, C. Chromosomal alterations in 15 breast cancer cell lines by comparative genomic hybridization and spectral karyotyping. *Genes Chromosomes Cancer.* **2000**, 28, 308–317.

16. Senger, G.; Lüdecke, H.J.; Horsthemke, B.; Claussen, U. Microdissection of banded human chromosomes. *Hum. Genet.* **1990**, 84, 507–511.

17. Zlotina, A.; Kulikova, T.; Kosyakova, N.; Liehr, T.; Krasikova, A. Microdissection of lampbrush chromosomes as an approach for generation of locus-specific FISH-probes and samples for high-throughput sequencing. *BMC Genomics.* **2016**, 17, 126.

18. Guyon, R.; Rakotomanga, M.; Azzouzi, N.; Coutanceau, J.; Bonillo, C.; D'Cotta, H.; Pepey, E.; Soler, L.; Rodier-Goud, M.; D'Hont, A.; Conte, M.; van Bers, N.; Penman, D.; Hitte, C.; Crooijmans, R.; Kocher, T.; Ozouf-Costaz, C.; Baroiller, J.; Galibert, F. A high-resolution map of the Nile tilapia genome: A resource for studying cichlids and other percomorphs. *BMC Genomics.* **2012**, 13, 222.

19. Müller, S.; Wienberg, J. "Bar-coding" primate chromosomes: Molecular cytogenetic screening for the ancestral hominoid karyotype. *Hum. Genet.* **2001**, 109, 85–94.

20. Ferguson-Smith, M.; Yang, F.; Rens, W.; O'Brien, P. The impact of chromosome sorting and painting on the comparative analysis of primate genomes. *Cytogenet. Genome Res.* **2005**, 108, 112–121.

21. Sabile, A.; Poras, I.; Cherif, D.; Goodfellow, P.; Avner, P. Isolation of monochromosomal hybrids for mouse chromosomes 3, 6, 10, 12, 14, and 18. *Mamm. Genome.* **1997**, 8, 81–85.

22. Kallioniemi, A.; Kallioniemi, O.P.; Sudar, D.; Rutovitz, D.; Gray, J.W.; Waldman, F.; Pinkel, D. Comparative genomic hybridization for molecular cytogenetic analysis of solid tumors. *Science.* **1992**, 258, 818–821.

23. Pinkel, D.; Segraves, R.; Sudar, D.; Clark, S.; Poole, I.; Kowbel, D.; Collins, C.; Kuo, W.L.; Chen, C.; Zhai, Y.; Dairkee, S.H.; Ljung, B.M.; Gray, J.W.; Albertson, D.G. High resolution analysis of DNA copy number variation using comparative genomic hybridization to microarrays. *Nat. Genet.* **1998**, 20, 207–211.

24. Brady, P.D.; Vermeesch, J.R. Genomic microarrays: A technology overview. *Prenat. Diagn.* **2012**, 32, 336–343.

25. Schröck, E.; du Manoir, S.; Veldman, T.; Schoell, B.; Wienberg, J.; Ferguson-Smith, M.; Ning, Y.; Ledbetter, D.; Bar-Am, I.; Soenksen, D.; Garini, Y.; Ried, T. Multicolor spectral karyotyping of human chromosomes. *Science.* **1996**, 273, 494–497.

26. Speicher, M.; Gwyn Ballard, S.; Ward, D. Karyotyping human chromosomes by combinatorial multi-fluor FISH. *Nat. Genet.* **1996**, 12, 368–375.

27. Nagai, T.; Okamura, T.; Yanase, T.; Chaya, R.; Moritoki, Y.; Kobayashi, D.; Akita, H.; Yasui, T. Examination of diagnostic accuracy of UroVysion fluorescence in situ hybridization for bladder cancer in a single community of Japanese hospital patients. *Asian Pac. J. Cancer. Prev.* **2019**, 20, 1271–1273.

28. Eiben, B.; Trawicki, W.; Hammans, W.; Goebel, R.; Pruggmayer, M.; Epplen, J. Rapid prenatal diagnosis of aneuploidies in uncultured amniocytes by fluorescence in situ hybridization: Evaluation of >3,000 cases. *Fetal. Diagn. Ther.* **1999**, 14, 193–197.

29. Nguyen, K.; Broucqsault, N.; Chaix, C.; Roche, S.; Robin, J.; Vovan, C.; Gerard, L.; Mégarbané, A.; Urtizberea, J.; Bellance, R.; Barnérias, C.; David, A.; Eymard, B.; Fradin, M.; Manel, V.; Sacconi, S.; Tiffreau, V.; Zagnoli, F.; Cuisset, J.M.; Salort-Campana, E.; Attarian, S.; Bernard, R.; Lévy, N.; Magdinier, F. Deciphering the complexity of the 4q and 10q subtelomeres by molecular combing in healthy individuals and patients with facioscapulohumeral dystrophy. *J. Med. Genet.* **2019**, 56, 590–601.

30. Vorsanova, S.G.; Zelenova, M.A.; Yurov, Y.B.; Iourov, I.Y. Behavioral variability and somatic mosaicism: A cytogenomic hypothesis. *Curr. Genomics.* **2018**, 19, 158–162.

31. Mkrtchyan, H.; Gross, M.; Hinreiner, S.; Polytiko, A.; Manvelyan, M.; Mrasek, K.; Kosyakova, N.; Ewers, E.; Nelle, H.; Liehr, T.; Bhatt, S.; Thoma, K.; Gebhart, E.; Wilhelm, S.; Fahsold, R.; Volleth, M.; Weise, A. The human genome puzzle: The role of copy number variation in somatic mosaicism. *Curr. Genomics.* **2010**, 11, 426–431.

32. Jancuskova, T.; Plachy, R.; Stika, J.; Zemankova, L.; Hardekopf, D.W.; Liehr, T.; Kosyakova, N.; Cmejla, R.; Zejskova, L.; Kozak, T.; Zak, P.; Zavrelova, A.; Havlikova, P.; Karas, M.; Junge, A.; Ramel, C.; Pekova, S. A method to identify new molecular markers for assessing minimal residual disease in acute leukemia patients. *Leuk. Res.* **2013**, 37, 1363–1373.

33. Hovhannisyan, G.G. Fluorescence in situ hybridization in combination with the comet assay and micronucleus test in genetic toxicology. *Mol. Cytogenet.* **2010**, 3, 17.

34. Shimizu, N. Extrachromosomal double minutes and chromosomal homogeneously staining regions as probes for chromosome research. Cytogenet. *Genome Res.* **2009**, 124, 312–326.

35. Koltsova, A.; Pendina, A.; Efimova, O.; Chiryaeva, O.; Kuznetzova, T.; Baranov, V.S. On the complexity of mechanisms and consequences of chromothripsis: An update. Front. *Genet.* **2019**, 10, 393.

36. Nazaryan-Petersen, L.; Bjerregaard, V.; Nielsen, F.; Tommerup, N.; Tümer, Z. Chromothripsis and DNA repair disorders. *J. Clin. Med.* **2020**, 9, 613.

37. Spangenberg, V.; Arakelyan, M.; Cioffi, M.; Liehr, T.; Al-Rikabi, A.; Martynova, E.; Danielyan, F.; Stepanyan, I.; Galoyan, E.; Kolomiets, O. Cytogenetic mechanisms of unisexuality in rock lizards. *Sci. Rep.* **2020**, 10, 8697.

38. Lemke, J.; Claussen, J.; Michel, S.; Chudoba, I.; Mühlig, P.; Westermann, M.; Sperling, K.; Rubtsov, N.; Grummt, U.; Ullmann, P.; Kromeyer-Hauschild, K.; Liehr, T.; Claussen, U. The DNA-based structure of human chromosome 5 in interphase. *Am. J. Hum. Genet.* **2002**, 71, 1051–1059.

39. Cremer, T.; Cremer, M.; Hübner, B.; Silahtaroglu, A.; Hendzel, M.; Lanctôt, C.; Strickfaden, H.; Cremer, C. The interchromatin compartment participates in the structural and functional organization of the cell nucleus. *Bioessays.* **2020**, 42, e1900132.

40. Daban, J. Supramolecular multilayer organization of chromosomes: Possible functional roles of planar chromatin in gene expression and DNA. *FEBS Lett.* **2020**, 594, 395–411.

41. Weise, A.; Othman, M.A.K.; Bhatt, S.; Löhmer, S.; Liehr, T. Application of BAC-probes to visualize copy number variants (CNVs). *Meth. Mol. Biol. (Clifton, N.J.).* **2015**, 1227, 299–307.

42. Serakinci, N.; Koelvraa, S. Molecular cytogenetic applications in diagnostics and research: An overview. In *In Fluorescence In Situ Hybridization (FISH): Application Guide*; Liehr, T., Ed. 1st Edition. Springer, Berlin, **2009**.

5 FISH—Microscopy and Evaluation

Ivan Y. Iourov

CONTENTS

INTRODUCTION

In the postgenomic era, microscopic (visual) analysis of human chromosomes is often regarded as an optional procedure in genomic research. However, the availability of fluorescence-based microscopic (genomic) techniques (like fluorescence in situ hybridization = FISH) gradually expands the area of genetic and genomic research.[1–3] Moreover, practical issues in medical genetics cannot be properly addressed without FISH, the essence of which lies in the fluorescence microscopy, visualization approaches, and DNA-based chromosome research.[1, 3, 4] Accordingly, to understand the role of FISH in current bio- and medical science,[5–7] fluorescence microscopy is an area to start with. Currently, FISH-based assays are required for analyzing structure and numbers of human chromosomes in interphase (the longest part of the cell cycle or the most usual state of a cell of an eukaryotic organism),[7–9] genome organization at the chromosomal level,[10, 11] developing single-cell genomic approaches,[12] and cytogenetic genotoxicological studies.[13] In total, there is a huge and increasing amount of FISH-based assays applicable for different biomedical tasks.[5, 7, 8, 14] Fluorescence microscopy is the basis for evaluation of FISH results, which determines the resolution and efficiency of the molecular cytogenetic analysis. Therefore, fluorescence microscopy and evaluation are of prime importance for successful application of FISH-based molecular cytogenetic technologies. Here, a brief overview of fluorescence microscopy and FISH results evaluation is provided.

DOI: 10.1201/9781003223658-5

FLUORESCENCE MICROSCOPY

The first key concept in fluorescence microscopy is the resolution. The smallest visible molecular target in the field of the vision determines the microscopic resolution of FISH. Usually, FISH analysis is referred to as the evaluation of small fluorescing objects of roundish, globular, or spherical shape (i.e. FISH signals or fluorescence signals). The diffraction may cause problems in distinguishing the closest fluorescence signals. Accordingly, fluorescence microscopy resolution is related to the determination of the closest distinguishable visible objects. The resolution depends on the wavelength (λ), the numerical aperture (NA), the magnification, and detection devices. The smallest diameter of a visible FISH signal is determined by the following ratio: $d = 1.22\ \lambda/NA$. Visible FISH signals are still distinguished when the distance between fluorescing objects is more than half a diameter, determined by the aforementioned ratio.[5]

The second key concept in fluorescence microscopy is the background autofluorescence. The related phenomena affect actually all the visualization-based florescence assays. The decrease of the effective contrast of the adjacent FISH signals impeding distinguishing between the signals may help to autofluorescence-related problems in visualization of FISH results. However, reducing the image contrast would result in resolution reduction. This dependence determines interplay between autofluorescence and resolution (i.e. the first and the second key concept of fluorescence microscopy). Actually, the background may be removed by imaging software (imaging techniques). A number of procedures for pre-treatment of microscopic slides before or after FISH is able to increase the signal removing background autofluorescence are available, as well.[5]

The third key concept in fluorescence microscopy is the detection. The use of specific detection systems or algorithms is the way to increase the microscopic resolution of FISH-based techniques. Generally, a digital camera is used as a device for FISH detection. The resolution may be modified by an increase in the quantum efficiency of the detection system. Automatically or manually, the increase of the quantum efficiency reduces the noise and optical NA of the microscope and camera. The dependence between NA and d is as follows: increasing NA decreases d. In the detection context, FISH resolution may also be affected by the maximum number of distinguishable intensity levels. This number is defined by the parameters of the detection system; exposure time may be adjusted manually or automatically. Alternatively, detection using digitalization for quantitative analysis allows differentiation between FISH signals and background autofluorescence. The digitalization and quantitative analysis underlie quantitative FISH (QFISH), which is discussed hereafter. For details about fluorescence microscopy concepts considered in the FISH context, readers are recommended to address following sources.[5–7]

The essential device for FISH-based microscopic analysis is the fluorescence microscope. The device is required to include a fluorescence light source and a set of fluorescence filters, which are composed of dichroic mirror and excitation and emission filters. For FISH-based molecular cytogenetic analyses, a high-pressure Hg (mercury) lamp is generally the most preferable to be used as a light source. The latter provides suitable high-energy excitations at wavelengths specific for the majority of FISH-based techniques. The lamp produces heat. Accordingly, lamp cooling is systematically required. The hot lamp should not be turned on. Usually, the overall working time of these lamps is less than 200 burning hours. Since recently, light-emitting diodes (LED) are an Hg-free alternative as light source, which also have longer 'burning times' in the range of 2000 hours or more.[15] FISH specimens are not to be exposed to the light. To avoid fading, light shutter is used to block the exposure. Multicolor nature of FISH is realized through the application of fluorescence filters. The combination of filters has to provide proper contrast between fluorochromes used for probe labeling.[5]

IMAGING AND IMAGE ANALYSIS

Imaging and image analysis are aimed at enhancing the efficiency of FISH. Technologically, FISH imaging is available when three steps are performed: (i) acquisition of a microscopic (FISH) image, (ii) image pre-processing, and (iii) digital analysis. To perform the analysis, a number of devices are to be used. These are a fluorescence microscope, image acquisition devise (the image sensor, e.g., a charge coupled device (CCD) camera, a device of choice for FISH), and hardware/software for digitizing and processing acquired microscopic images.[4–7]

The key element of FISH image acquisition is focusing. Numerous opportunities of manual and automatic focusing are provided by modern imaging systems. Specific FISH approaches require autofocusing (e.g. FISH analysis merging colors obtained by processing intensities of signals; volumetric FISH analyses for 3D studies of chromosomal organization in interphase nuclei[7, 8, 10]), which may cause a problem referred to differing between FISH signals and autofluorescence particles. An adequate solution for different FISH image acquisition is to apply both manual focus and autofocusing.

FISH image analysis is not generally associated with extreme difficulties either due to availability of specific digital imaging systems or to DNA probe specificity and high hybridization efficiency. However, raw FISH images may cause some interpretational problems.[8, 14] To solve these, interactive or manual removal of fluorescence background by thresholding and contrast normalization is performed.

Canonical FISH protocols (i.e. FISH using site-specific and chromosome-enumerating DNA probes) are usually designed for visual/express analysis. More sophisticated FISH techniques are impossible to perform without digital analysis. The latter is specifically designed for each corresponding FISH assay.[5, 14] Regardless of being secondary to basic FISH procedures (sample preparation, hybridization, and detection), imaging digital analysis may underlie sophisticated molecular cytogenetic analyses, which are focused on detection of multicolor FISH results.[5] An important example of digital imaging, which opens new opportunities for FISH-based chromosomal analysis, is QFISH (quantitative FISH).

QFISH has been demonstrated to be useful for a variety of molecular cytogenetics studies of chromosomes. The approach is applicable for analyses of somatic chromosomal mosaicism and chromosome instability in interphase nuclei. It allows detecting chromosomal DNA variations (e.g. chromosomal variants or pericentromeric

DNA variations) between homologous human chromosomes at the single cell level.[16] Through the last two decades, QFISH has been demonstrated to be valuable for different basic research and diagnostic purposes in medical genetic and cancer research.[16–18] Thus, it allows specific chromosomal organization in interphase nuclei to be depicted.[19] More importantly, QFISH is applicable for the definition of the parental origin of homologous chromosomes.[20, 21] The ratio of relative intensities obtained by integrating intensity profiles of FISH signals are compared to differ between chromosomal associations and chromosome loss in interphase. These measurements of FISH-painted hypervariable chromosomal loci are used for defining homologous chromosome parental origins. In total, FISH result evaluations using QFISH may be highly helpful as an additional tool for a molecular cytogenetic study.[22]

HUMAN EYE VERSUS ELECTRIC EYE

The aforementioned advantages offered by FISH result evaluations using digital imaging may be used for promoting ideas suggesting that microscopic analysis performed by researcher's eyes is outdated. Accordingly, one can propose to diminish the significance of visual analysis as a contributor to successful molecular cytogenetic research. This might be further supported by a number of multicolor FISH techniques, which are useless without digitalization of FISH signals. Probably, the most important FISH-based approach to be mentioned in this context is multicolor banding (MCB). This is an important technique for basic and applied chromosomal analysis at molecular resolution.[23] MCB is based almost exclusively on microscopic evaluation using digital imaging and is systematically proven effective in molecular cytogenetic studies.[24] Moreover, the approach may be used for developing methods of interphase chromosomal analysis, providing an unprecedented resolution for visualization of an interphase chromosome at any cell cycle stage and at molecular resolutions using digital image analysis.[25, 26] However, one can argue that microscopic analyses performed by researcher's eyes are outdated. The confrontation between digital and visual (researcher's eye) analyses is to be recognized as insignificant. Human eye microscopic evolution ensures efficiency through researcher's experience, ability of immediate decision considering the significance of FISH results, and the selection of FISH results for further digital analysis. Digital microscopic evolution (= 'electric eye' microscopic evolution) provides for specific FISH image analysis, which undoubtedly increases the resolution and/or gives an opportunity of DNA-based chromosomal analysis unavailable to researcher eye chromosomal evaluations. Thus, the combination of both is the most effective way to succeed (as usual).

CONCLUSIONS

Microscopy is the basis of FISH analysis. Actually, a number of biomedical areas require FISH to have an important place in the methodology. Somatic human genetics/genomics, cytogenomics, and cancer research are certainly among these areas.[27–29] Therefore, knowledge of basic principles underlying FISH may be useful for researchers undertaking related biomedical studies. Since FISH and microscopy are intimately linked to each other, it is to acknowledge the importance of knowing opportunities and limitations of fluorescence microscopy and evaluations in the context of this molecular cytogenetic platform.

ACKNOWLEDGMENTS

The chapter is dedicated to professors Svetlana G. Vorsanova and Yuri B. Yurov and to Dr. Ilia V. Soloviev. The author's work is supported by the Government Assignment of the Russian Ministry of Science and Higher Education, Assignment no. AAAA-A19-119040490101-6 and by the Government Assignment of the Russian Ministry of Health, Assignment no. 121031000238-1.

REFERENCES

1. Liehr, T. From human cytogenetics to human chromosomics. *Int J Mol Sci.* **2019**, 20, 826.
2. Ye, C.J.; Stilgenbauer, L.; Moy, A.; Liu, G.; Heng, H.H. What is karyotype coding and why is genomic topology important for cancer and evolution? *Front Genet.* **2019**, 10, 1082.
3. Iourov, I.Y.; Yurov, Y.B.; Vorsanova, S.G. Chromosome-centric look at the genome. In: *Human Interphase Chromosomes: Biomedical Aspects*; Iourov, I.Y.; Vorsanova, S.G.; Yurov, Y.B., Eds. Springer, Berlin, **2020**, pp. 157–170.
4. Gersen, S.L.; Keagle, M.B. *The Principles of Clinical Cytogenetics.* 2nd Edition. Humana, Totowa, NJ; **2005**.
5. Liehr, T. (Ed.). *Fluorescence In Situ Hybridization (FISH): Application Guide.* Springer, Berlin, Heidelberg, Germany; **2017**.
6. Garini, Y.; Young, I.T.; McNamara, G. Spectral imaging: Principles and applications. *Cytometry A.* **2006**, 69, 735–747.
7. Yurov, Y.B.; Vorsanova, S.G.; Iourov, I.Y. *Human Interphase Chromosomes: Biomedical Aspects.* Springer, Berlin; **2013**.
8. Vorsanova, S.G.; Yurov, Y.B.; Iourov, I.Y. Human interphase chromosomes: A review of available molecular cytogenetic technologies. *Mol Cytogenet.* **2010**, 3, 1.
9. Bakker, B.; van den Bos, H.; Lansdorp, P.M.; Foijer, F. How to count chromosomes in a cell: An overview of current and novel technologies. *Bioessays.* **2015**, 37, 570–577.
10. Manvelyan, M.; Hunstig, F.; Bhatt, S.; Mrasek, K.; Pellestor, F.; Weise, A.; Simonyan, I.; Aroutiounian, R.; Liehr, T. Chromosome distribution in human sperm: A 3D multicolor banding-study. *Mol Cytogenet.* **2008**, 1, 25.
11. McClelland, S.E. Single-cell approaches to understand genome organisation throughout the cell cycle. *Essays Biochem.* **2019**, 63, 209–216.
12. Iourov, I.Y.; Vorsanova, S.G.; Yurov, Y.B. Single cell genomics of the brain: Focus on neuronal diversity and neuropsychiatric diseases. *Curr Genomics.* **2012**, 13, 477–488.
13. Hovhannisyan, G.G. Fluorescence in situ hybridization in combination with the comet assay and micronucleus test in genetic toxicology. *Mol Cytogenet.* **2010**, 3, 17.
14. Iourov, I.Y.; Vorsanova, S.G.; Yurov, Y.B. Recent patents on molecular cytogenetics. *Recent Pat DNA Gene Seq.* **2008**, 2, 6–15.
15. Lang, D.S.; Zeiser, T.; Schultz, H.; Stellmacher, F.; Vollmer, E.; Zabel, P.; Goldmann, T. LED-FISH: Fluorescence microscopy based on light emitting diodes for the molecular analysis of Her-2/neu oncogene amplification. *Diagn Pathol.* **2008**, 3, 49.

16. Iourov, I.Y.; Soloviev, I.V.; Vorsanova, S.G.; Monakhov, V.V.; Yurov, Y.B. An approach for quantitative assessment of fluorescence in situ hybridization (FISH) signals for applied human molecular cytogenetics. *J Histochem Cytochem.* **2005**, 53, 401–408.

17. Konsti, J.; Lundin, J.; Jumppanen, M.; Lundin, M.; Viitanen, A.; Isola, J. A public-domain image processing tool for automated quantification of fluorescence in situ hybridisation signals. *J Clin Pathol.* **2008**, 61, 278–282.

18. Amakawa, G.; Ikemoto, K.; Ito, H.; Furuya, T.; Sasaki, K. Quantitative analysis of centromeric FISH spots during the cell cycle by image cytometry. *J Histochem Cytochem.* **2013**, 61, 699–705.

19. Iourov, I.Y.; Vorsanova, S.G.; Yurov, Y.B. Fluorescence intensity profiles of in situ hybridization signals depict genome architecture within human interphase nuclei. *Tsitol Genet.* **2008**, 42, 3–8.

20. Vorsanova, S.G.; Iourov, I.Y.; Beresheva, A.K.; Demidova, I.A.; Monakhov, V.V.; Kravets, V.S.; Bartseva, O.B.; Goyko, E.A.; Soloviev, I.V.; Yurov, Y.B. Non-disjunction of chromosome 21, alphoid DNA variation, and sociogenetic features of Down syndrome. *Tsitol Genet.* **2005**, 39, 30–36.

21. Weise, A.; Gross, M.; Mrasek, K.; Mkrtchyan, H.; Horsthemke, B.; Jonsrud, C.; Von Eggeling, F.; Hinreiner, S.; Witthuhn, V.; Claussen, U.; Liehr, T. Parental-origin-determination fluorescence in situ hybridization distinguishes homologous human chromosomes on a single-cell level. *Int J Mol Med.* **2008**, 21, 189–200.

22. Iourov, I.Y. Quantitative fluorescence in situ hybridization (QFISH). *Methods Mol Biol.* **2017**, 1541, 143–149.

23. Liehr, T.; Heller, A.; Starke, H.; Rubtsov, N.; Trifonov, V.; Mrasek, K.; Weise, A.; Kuechler, A.; Claussen, U. Microdissection based high resolution multicolor banding for all 24 human chromosomes. *Int J Mol Med.* **2002**, 9, 335–339.

24. Alhourani, E.; Aroutiounian, R.; Harutyunyan, T.; Glaser, A.; Schlie, C.; Pohle, B.; Liehr, T. Interphase molecular cytogenetic detection rates of chronic lymphocytic leukemia-specific aberrations are higher in cultivated cells than in blood or bone marrow smears. *J Histochem Cytochem.* **2016**, 64, 495–501.

25. Iourov, I.Y.; Liehr, T.; Vorsanova, S.G.; Kolotii, A.D.; Yurov, Y.B. Visualization of interphase chromosomes in postmitotic cells of the human brain by multicolour banding (MCB). *Chromosome Res.* **2006**, 14, 223–229.

26. Iourov, I.Y.; Liehr, T.; Vorsanova, S.G.; Yurov, Y.B. Interphase chromosome-specific multicolor banding (ICS-MCB): A new tool for analysis of interphase chromosomes in their integrity. *Biomol Eng.* **2007**, 24, 415–417.

27. Iourov, I.Y.; Vorsanova, S.G.; Yurov, Y.B.; Kutsev, S.I. Ontogenetic and pathogenetic views on somatic chromosomal mosaicism. *Genes.* **2019**, 10, 379.

28. Ye, C.J.; Sharpe, Z.; Heng, H.H. Origins and consequences of chromosomal instability: From cellular adaptation to genome chaos-mediated system survival. *Genes.* **2020**, 11, 1162.

29. Iourov, I.Y.; Vorsanova, S.G.; Yurov, Y.B. Systems cytogenomics: Are we ready yet? *Curr Genomics.* **2021**, 22, 75–78.

6 FISH—in Routine Diagnostic Settings

Chenghua Cui and Peining Li

CONTENTS

INTRODUCTION

Fluorescence in situ hybridization (FISH) is a cell-based macromolecule recognition technique based on the complementary nature of DNA double strands. Fluorophore-coupled nucleotides can be enzymatically incorporated into cloned or amplified DNA fragments. These fluorescence-labeled DNA fragments can be used as probes to hybridize onto the complementary sequences in interphase nuclei and metaphase chromosomes, and then visualized through a fluorescence microscope. This technique was initially developed to replace radioactive probes for physical mapping of genes to chromosomes.[1–4] Human genomic DNA fragments of 100 to 300 kilobases (kb) cloned in bacterial artificial chromosomes (BAC) were compiled in a database and mapped to chromosomes by FISH.[5] These mapped BAC clones provided firstly cytogenetic landmarks for the Human Genome Project and later a resource for developing locus-specific and gene-targeted FISH probes. FISH assays have higher analytical resolution than karyotyping and high sensitivity and

DOI: 10.1201/9781003223658-6

specificity on detecting targeted DNA. The design of multicolor probes enabled simultaneous detection of chromosomes 13, 18, 21, X, and Y in uncultured amniotic fluid cells; this FISH assay was immediately accepted as the first clinical application for rapid prenatal screening of common aneuploidies.[6, 7] The introduction of FISH into clinical cytogenetics excited the black-white banding pattern with fluorescent colors and evolved the field into molecular cytogenetics with expanded diagnostic applications.[8, 9]

ANALYTIC AND CLINICAL VALIDITY OF FISH ASSAYS

DESIGN OF LOCUS-SPECIFIC, GENE-TARGETED, REGIONAL OR WHOLE CHROMOSOME PAINTING PROBES

BAC clones mapped across the human genome in the genome browser (https://genome.ucsc.edu/) are a reliable resource for developing FISH probes targeting to specific loci and genes.[10] A wide variety of fluorophores are available, and several enzymatic reactions such as nick translation, random priming, and degenerative oligonucleotide primed polymerase chain reaction have been used to incorporate fluorophore-coupled nucleotides.[11] Locus-specific, gene-targeted, regional or whole chromosome painting probes have been developed for diagnostic uses. Enumerative locus-specific probes containing highly repetitive α-satellite sequences at centromere and unique DNA sequences at pericentric or specific chromosome loci are used to detect numerical chromosome abnormalities.[6, 10] Probes containing repetitive sequences at telomeres and subtelomeric sequences are used to detect telomeric integrity and subtelomeric rearrangements.[12] Gene-targeted design of dual-color double fusion probes for two genes are used to detect conjugated signals for a reciprocal translocation involving the two genes, while dual-color break apart probes juxtaposing the 5′ and 3′ regions of a gene are used to detect separated 5′ and 3′ signals for rearrangements of the gene.[13, 14] Contig probes for a chromosome band or collection of probes for a whole chromosome can be labeled in a multicolor pattern to generate a painting effect for an individual chromosome or a whole metaphase.[15–17] Due to the standardized probe design and widespread diagnostic usage, all these probes are available commercially or can be made custom. Figure 6.1 shows various designs of FISH probes and typical signal patterns on interphase nuclei and metaphase cells.

VALIDATION AND VERIFICATION OF FISH ASSAYS FOR CLINICAL TESTING

The analytical resolution of FISH is defined as the ability to distinguish two points along the length of interphase chromatins or metaphase chromosomes. The limit of this resolution is determined as 200–500 nm by the light microscope. The width of naked DNA double strands, packed chromatins, and metaphase chromosomes is 2 nm, 30 nm, and 1,400 nm, respectively, which explains the resolvable visibility only for chromosomes but not for chromatins or naked DNA under a light microscope. The length of 1 kb DNA double helix is 340 nm. Considering the labeling efficiency of fluorophore-couple nucleotide incorporation and hybridization efficiency

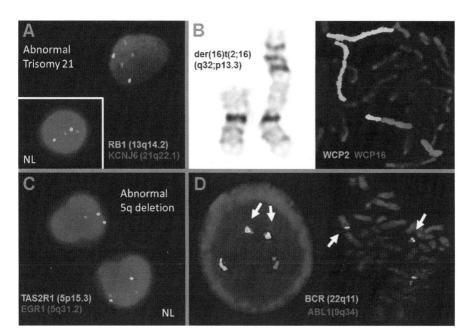

FIGURE 6.1 FISH probes designs and typical signal patterns on interphase nuclei and metaphase chromosomes. A. FISH using locus-specific probes targeted to chromosomes 13 and 21 shows an abnormal pattern of three signals for the *KCNJ6* probe indicating trisomy 21 in a patient and a normal two signals for both *RB1* and *KCNJ6* probes in a normal control (NL). B. FISH using whole chromosome painting (WCP) probes for chromosomes 2 and 16 shows translocated segment of 2q onto a 16p. C. FISH using locus-specific probes for 5p and 5q detects a deletion of 5q in an abnormal cell and a disomic pattern in a normal cell. D. FISH using gene-specific dual-color double fusion probes for the *ABL1* and *BCR* genes shows one independent signal and two conjugated signals for *BCR-ABL1* rearrangement (pointed by arrow).

of FISH probes onto nuclei or chromosomes, the targeted region of unique sequence FISH probes for clinical use is in the range of 100–200 kb. This has been generally accepted as the analytic resolution for FISH assays using a fluorescence microscope. The size of G-band in the 550-band high resolution level is in an average of 5–6 megabases (Mb) per band. The analytical resolution of FISH is approximately 50 times higher than karyotyping.

To comply with clinical laboratory standards and guidelines, FISH assays need to be verified and validated before their usage on clinical specimens.[18, 19] Verification refers to analytical accuracy by ensuring the location of hybridized probes on the expected chromosome loci and analytical precision by measuring quantitative agreement between repeated assessments of the same sample.[18] Accuracy of FISH probes can be evaluated by a hybridization to normal control metaphases for a correct signal pattern onto the expected loci in chromosomes and to abnormal metaphases for the intended abnormal signal pattern. The latter approach has the advantage of confirming both chromosomal location and numerical or structural rearrangements. Accuracy can be evaluated by scoring a minimum of five metaphase cells to verify the presence of hybridized probes to the appropriate chromosome loci and absence of cross-hybridization to other chromosomes. Precision of FISH assays measures the reproducibility by performing consecutive three to ten repeat FISH assays over three to ten days on a selected specimen with a known proportional of normal and abnormal cells. The precision is calculated as the mean, standard deviation, and range of the results; intra-assay and inter-assay concordance for signal pattern scoring should be ≥ 95%.[20, 21]

Validation refers to the assessment of analytical sensitivity, specificity, reportable range, and other performance characteristics required to ensure assay performance. For the examination of a FISH signal at the targeted location, probe sensitivity and specificity are equivalent to analytical sensitivity and specificity.[18] Probe sensitivity is the percentage of scorable metaphase chromosomes or interphase nuclei with the expected probe signal pattern. Probe specificity is the percentage of all scored signals that occur at the expected location at metaphase chromosomes or at the expected signal pattern at interphase nuclei. Probe sensitivity and specificity should be established by analyzing signal patterns of probe hybridization to at least 40 chromosomes in 20 metaphases or 100 interphase nuclei. Reportable range is defined by a cut-off value for normal signal patterns.[20, 21] This cut-off value for the upper limit at the 95% and 99% confidence intervals can be easily ascertained using the β inversion function (BETAINV) in a Microsoft excel spreadsheet.[22] In addition to the pre-clinical verification and validation, quality control, quality assurance, and quality management procedures should be implemented in the pre-analytic, analytic, and post-analytic phases for each FISH assay. Interpretation and reporting of FISH results should follow the technical standards and guidelines.[18, 19] The results of FISH assays should be described following the current International System for Human Cytogenomic Nomenclature (ISCN 2020).[23] Proficiency testing for laboratories performing FISH assays showed a consensus of 91.7% of participants with the same diagnostic conclusion, indicating the reliability of current FISH assays in detecting constitutional and somatic chromosomal abnormalities.[24]

DETECTION OF CONSTITUTIONAL CYTOGENOMIC ABNORMALITIES

The newborn incidence for numerical and structural chromosomal abnormalities is 1/154 and for recurrent microdeletions and microduplications and other pathogenic copy number variants is estimated as 1/291; together these cytogenomic abnormalities occur in approximately 1% of newborns.[25] Karyotyping, FISH, and chromosome microarray analysis (CMA) have been routinely used in clinical cytogenetics laboratories to detect these abnormalities. FISH has the advantages of being directly performed on uncultured interphase nuclei for a short turn-around time, accurately detecting cryptic chromosomal imbalances and rearrangements, and reliably tracking mosaic patterns at a single cell level. It plays an integral role in the detection of constitutional cytogenomic abnormalities.

RAPID PRENATAL SCREENING OF COMMON ANEUPLOIDY

Numerical chromosomal abnormalities involving autosomes 13, 18, and 21 and sex chromosomes X and Y are considered common aneuploidies with syndromic phenotypes. For example, trisomy 21 causes Down syndrome (OMIM #190685), trisomy 18 causes Edwards syndrome, and monosomy X causes Turner syndrome. In current prenatal clinics, advanced maternal age and results from ultrasound examination, maternal serum screening, and non-invasive prenatal testing (NIPT) of cell-free fetal DNA in maternal serum plasma are used to recognize fetuses highly suspected for common aneuploidies and other cytogenomic abnormalities. Invasive procedures will be performed on high-risk pregnancies to collect amniotic fluid or chorionic villus for diagnostic testing. AneuVysion assay, using tri-color centromeric probes for chromosomes X, Y, and 18 and dual-color locus-specific probes for chromosomes 13 and 21, was the first FISH assay approved by the US Food and Drug Administration (FDA) for rapid prenatal aneuploidy screening in uncultured amniocytes.[6, 7] A cut-off value from consecutive normal cases should be calculated to differentiate a normal disomic pattern from abnormal trisomic or monosomic patterns.[26] A multi-center retrospective study showed an extremely high concordance rate of 99.8% between FISH and karyotyping results.[27] This highly effective FISH assay was further extended to the analysis of uncultured cells from chorionic villi. AneuVysion FISH assay on uncultured cells can provide results within 24 hours, while chromosome analyses on cultured amniocytes or villi cells take seven to twelve days. This rapid screening can release the psychological distress from pregnant women. However, the risk of misdiagnosis for as low as 0.2–0.4% by AneuVysion could have significant impact in a prenatal setting.[26–28] False negative or positive results could be caused by technical errors in processing and scoring, sampling-related issues of twin or triplet pregnancies and maternal cell contamination, as well as occasionally inherent factors of low-level mosaicism, mosaicism confined to the placenta, structural chromosome abnormalities, and cryptic rearrangements. It is recommended that the results from AneuVysion should be used exclusively as a preamble to full chromosome analysis, and irreversible clinical actions should not be performed solely based on FISH result.[26, 28] Approximately 10% of cases with common aneuploidy showed a mosaic pattern, and mosaicism for sex chromosome aneuploidy varied from 3%–20%; structural rearrangements affecting the FISH interpretation, such as isochromosome of Xq, isodicentric Y, and derivative chromosome from a familial translocation t(Y;14)(q12;p13), were reported.[29, 30] An integrated FISH, karyotyping, and CMA approach should be applied for safe and accurate diagnosis of common aneuploidies and other cytogenomic abnormalities in current prenatal cytogenetic testing.[29]

DETECTION OF MICRODELETION AND MICRODUPLICATION SYNDROMES

High-resolution prometaphase chromosome banding has allowed the detection of subtle interstitial or terminal deletions of 8q23.3/24.1 for Langer-Giedion syndrome, 11p13 for WAGR (Wilms tumor aniridia gonadoblastoma retardation) syndrome (OMIM #612469), 15q11/12 for Prader-Willi syndrome (PWS, OMIM #176270) or Angelman syndrome (AS, OMIM #105830), 17p13.3 for Miller-Dieker lissencephaly syndrome (MDLS, OMIM #247200), and 22q11.2 for DiGeorge syndrome (DGS,

OMIM #188400).[31] These microdeletions could be missed by routine karyotyping but are reliably detected by targeted FISH probes.[32, 33] FISH analysis enabled the detection of a microduplication at 22q11.2 as an emerging syndrome, and further confirmed that low copy repeats (LCRs) on 22q mediated nonallelic homologous recombination (NAHR) for recurrent microdeletions and microduplications.[34, 35] Microdeletion and microduplication syndromes are a group of genomic disorders caused by LCR-mediated NAHR in the human genome. The estimated newborn incidence for the most frequently seen DiGeorge syndrome and all microdeletions and microduplications were approximately 1/4000 and 1/400, respectively. The diagnostic efficacy in current prenatal and pediatric settings was estimated as 8% and 92% for DiGeorge syndrome and 3% and 46% for all microdeletions and microduplications, respectively.[25] To improve the diagnostic efficacy in current prenatal setting, enhanced NIPT to screen for microdeletions and microduplications followed by diagnostic FISH testing and CMA on amniocytes or villi cells should be implemented. Table 6.1 lists commonly used FISH probes for detecting aneuploidy, microdeletions, and microduplications in current prenatal and pediatric settings.

TABLE 6.1
FISH Probes for Common Aneuploidies and Microdeletion/Duplication Syndromes

Diseases	Chromosomal Abnormalities	FISH Probes
Aneuploidy		
Sex chromosome aneuploidies	45,X, 47,XXX, 47,XXY, 47,XYY	DXZ1, DYZ3
Pätau syndrome	trisomy 13	*RB1, KCNJ6*
Down syndrome	trisomy 21	
Edwards syndrome	trisomy 18	D18Z1
Microdeletion/microduplication syndromes		
Kallman syndrome	del(X)(p22.31p22.31)	*KAL1*
X-linked ichthyosis	del(X)(p22.31p22.31)	*STS*
Cri-du-Chat syndrome	del(5)(p15.2p15.2)	*CTNND2*
Sotos syndrome	del(5)(q35q35)	*NSD1*
Saethre-Chotzen syndrome	del(7)(p21.1p21.1)	*TWIST1*
Williams-Beuren syndrome	del(7)(q11.23q11.23)	*ELN*
Williams-Beuren region duplication syndrome	dup(7)(q11.23q11.23)	*ELN*
Langer-Giedion Syndrome	del(8)(q23.3q23.3)	*TRPS1*
Multiple hereditary exotoses	del(8)(q24.1q24.1)	*EXT1*
Prader-Willi/Angelman syndromes	del(15)(q11q13)	*SNRPN, UBE3A*
15q11-q13 duplication syndrome	dup(15)(q11q13)	*SNRPN, UBE3A*
Smith-Magenis syndrome	del(17)(p11.2p11.2)	*RAI1*
Miller-Dieker syndrome	del(17)(p13.3p13.3)	*LIS1*
DiGeorge syndrome	del(22)(q11.21q11.21)	*TUPLE1*
22q11.2 duplication syndrome	dup(22)(q11.21q11.21)	*TUPLE1*
Phelan-McDermid syndrome	del(22)(q13q13)	*SHANK3*

Abbreviations: del: deletion, dup: duplication

DETECTION OF NUMERICAL AND STRUCTURAL CHROMOSOMAL ABNORMALITIES

Unique numerical chromosomal abnormalities such as small supernumerary marker chromosomes (sSMC) posed an interpretation challenge because the nature of these markers cannot be unambiguously identified by their morphology and banding pattern. The newborn incidence of sSMC was estimated as 0.043%, and approximately 34% of the sSMC cases are correlated with known clinical syndromes. The isochromosome of 12p, i(12p), detected preferentially in skin fibroblasts and less likely in peripheral blood lymphocytes, is diagnostic for Pallister-Killian syndrome (PKS; OMIM #601803). The isochromosome of 18p, i(18p), causes i(18p) syndrome (OMIM #614290). The derivative chromosome 22 from malsegregation of parental carriers with constitutional balanced translocation t(8;22)(q24.13;q11.2) or t(11;22)(q23;q11.2) caused supernumerary der(22)t(8;22) syndrome (OMIM #613700) or Emanuel syndrome (OMIM #609029) with distinct phenotype of severe intellectual and physical disabilities. An inverted duplication of 22q, inv dup(22), is diagnostic for cat eye syndrome (CES; OMIM #115470). The i(12p), der(22)t(11;22), inv dup(22), and i(18p) accounted for approximately 11%, 10%, 7%, and 6% of sSMC, respectively.[36] Multiplex centromeric and pericentromeric probes and multicolor probes have been used to characterize sSMC.[10, 37, 38] A study of 28,000 prenatal cases from four diagnostic laboratories detected 23 cases with sSMC (0.082%); 10 pregnancies were terminated due to ultrasound abnormalities and syndromic associated sSMC by FISH assays, and 13 pregnancies resulted in normal healthy neonates.[39] Approximately 50% of sSMC cases showed different levels of somatic mosaicism in different tissues; the vast majority of these somatic mosaicism had no direct clinical effect, but a de novo sSMC in a somatic state may be a hint of uniparental disomy (UPD).[40] UPD of chromosomes 6, 7, 14, 15, 16, and 20 have been reported in cases with sSMC in those chromosomes.[41] Collected data from literature also demonstrated that the presence of sSMC raises clinical concern of infertility, including spermatogenesis impairment, amenorrhea, premature ovarian failure, implantation difficulties, and recurrent pregnancy loss.[42]

Unbalanced and balanced structural chromosomal abnormalities could be further characterized using FISH assays. Subtelomeric rearrangements, including familial cases with carriers of balanced cryptic subtelomeric translocations, may be missed by routine banding analysis but can be detected by multiplex FISH using a panel of subtelomeric probes.[12] Of 11,688 individuals with developmental disabilities analyzed by a panel of subtelomeric FISH probes, the detection rate for clinically significant subtelomeric abnormalities was approximately 2.5% with an additional 0.5% for presumed familial variants.[43]

Constitutional ring chromosomes are a rare type of structural abnormalities with an estimated newborn incidence of 1 in 50,000. Ring chromosomes present possible genomic imbalances from the ring formation, variable levels of dynamic mosaicism through mitosis, and selective karyotype evolution in different tissues. This cytogenomic heterogeneity is likely correlated with clinical heterogeneity resulting from compound effects of generalized features of 'ring syndrome' and chromosome-specific and dosage-sensitive gene-related phenotypes, risks of infertility, and cancer predisposition.[44] FISH assays are a valuable tool to visualize the deletions and duplications in the ring chromosomes and to track the dynamic mosaic patterns in interphase cells. For example, FISH mapping identified the presence of terminal deletions in a ring chromosome 6 and the association of a deletion of the *FOXC1* gene with eye anomalies in the patient.[45] Cases of ring chromosomes 13 and 21 with terminal deletions and interstitial duplication detected by CMA were further confirmed by FISH.[46, 47] Figure 6.2A shows the FISH assays performed on a case with a ring chromosome 13 and the tracking of a variant ring and ring chromosome loss in interphase cells.

CMA has been the method of choice to define genomic imbalances, but karyotyping and FISH are needed to construct these imbalances into the rearranged chromosomes.[48, 49] For example, in a case with a reciprocal translocation t(5;8)(p15.3;p23.1) and derivative chromosomes 11 and 16 from likely an 11q/16q rearrangement by banding analysis and deletions in an 11q and a 16q defined by CMA, FISH assays using whole chromosome painting probes, subtelomeric probes, and locus-specific probes on metaphase chromosomes revealed multiple-breakpoint inter-/intra-chromosome rearrangements during the formation of these derivative chromosomes and identified *FZD4* haploinsufficiency for exudative vitreoretinopathy in this patient.[50] In a case of spontaneous abortion with a 1.5 Mb deletion at 5p15.33 detected by CMA, chromosome analysis and FISH assay revealed inverted duplication, triplication, and quintuplication resulted from sequential breakage–fusion–bridge events (Figure 5.2B).[51] For balanced translocations with clinical symptoms, physical mapping of breakpoints by FISH probes could lead to the identification of candidate genes. For example, in a unique case with a balanced translocation t(9;13), the delineation of the breakpoint and

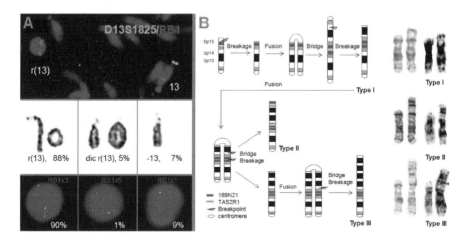

FIGURE 6.2 FISH analysis for chromosomal structural abnormalities. A. FISH using probes for the *RB1* gene at 13q14.2 and the D13S1825 locus at 13q34 on a metaphase shows a ring chromosome 13 with an interstitial duplication and a terminal deletion. Further examination of signal pattern on interphase nuclei noted this ring 13, a dicentric ring 13, and loss of this ring chromosome 13 for monosomy 13 in 90%, 1%, and 9% of cells. B. FISH using dual-color probes for distal 5p detects three types of rearrangements. A diagram shows a sequential breakage–fusion–bridge process through mitosis in the formation of this mosaic pattern.

further expression analysis on the gene at the junction identified that increased expression of the *α-Kloth* gene results in hypophosphatemic rickets and hyperparathyroidism.[52] These cases demonstrated that FISH is an integral part of cytogenomic analysis to facilitate accurate reporting and interpretation.

DETECTION OF SOMATIC NUMERICAL AND STRUCTURAL CHROMOSOMAL ABNORMALITIES

The identification of Philadelphia chromosome in patients with chronic myelogenous leukemia (CML) and further characterization of this reciprocal translocation t(9;22) (q34.1;q11.2) with underlying *BCR-ABL1* gene fusion marked the start of cancer cytogenetics.[53, 54] A FISH assay using dual-color probes detected one conjugated signal for *BCR-ABL1* gene fusion in metaphase derivative chromosome 22 and interphase nuclei.[55] Follow up FISH assays detected trisomy 12 in chronic lymphocytic leukemia (CLL) and trisomy 8 in myeloid leukemia and thus initiated diagnostic application of FISH on various types of tumors.[56–58] A multicenter investigation validated the dual-color FISH assay using *BCR/ABL1* probes with a clinical sensitivity of 94% and high concordance for well-trained FISH readers.[59] With the development of professional guidance for FISH testing and specific panels of FISH probes for recurrent clonal abnormalities of various types of cancers, FISH assays have been an indispensable tool with expanding applications in cancer diagnostics.[60–61]

DETECTION OF RECURRENT REARRANGEMENTS IN HEMATOPOIETIC AND LYMPHOID TISSUES

The World Health Organization (WHO) continuously updated the classification of tumors of hematopoietic and lymphoid tissues and correlated recurrent numerical and structural chromosomal rearrangements.[62] Most of these chromosome rearrangements have distinctive diagnostic and prognostic significance, and some of them can guide targeted therapy.[63] Conventional karyotyping and FISH assays are used to establish the cytogenetic profile at the initial diagnosis and to monitor the clonal evolution, disease progression, and therapeutic effect in follow-up studies.

CML is classified under myeloproliferative neoplasms. The translocation t(9;22) (q34.1;q11.2) with *BCR-ABL1* gene fusion is seen in 90–95% of CML, and the remaining 5%–10% have variant translocations involving three or four chromosomes and/or cryptic translocation involving *BCR-ABL1* gene fusion. The cryptic translocations are hard to identify in the banding analysis but can be detected by FISH. Approximately 43% of variant translocations were observed with deletions in the formation of three-way translocations and worse response to treatment.[64] Even FISH may miss some rare cases carrying insertional translocation in *BCR-ABL1* positive CML. The chimeric *BCR-ABL1* protein with more potent tyrosine kinase activity are subjected to FDA-approved targeted therapy by small molecule tyrosine kinase inhibitors such as imatinib, dasatinib, nilotib, and ponatinib.[63]

Myeloid and lymphoid neoplasms with eosinophilia are often a result of rearrangements of the *PDGFRA*, *PDGFRB*, and *FGFR1* genes fusing with multiple partner genes; these abnormal gene fusions encode constitutively activated tyrosine kinases. Routine karyotype can detect most of these rearrangements but escape cryptic rearrangements such as a submicroscopic deletion del(4)(q12q12) for *FIP1L1-PDGFRA* gene fusion. FISH assays using dual-color break apart probes for these genes can detect cryptic rearrangements and infer the partner genes in chromosomally observed or unobserved rearrangements. Imatinib is the first line therapy for patients with rearrangements of the *PDGFRA* and *PDGFRB* genes, where patients with *FGFR1* gene rearrangements are resistant to this therapy and carry a poor prognosis.[65]

A panel of 20 probes for 10 FISH assays was utilized to detect recurrent numerical and structural rearrangements for myelodysplastic syndromes (MDS) and acute myeloid leukemia (AML); the concordance rate between karyotyping and FISH was more than 98%.[66] Dual-color two loci enumeration probes for 5q, 7q, and 8q/20q detect −5/5q−, −7/7q−, +8, and −20/20q−, respectively. Dual-color double fusion probes for the *RUNX1-RUNX1T1*, *BCR-ABL1*, *PML-RARA*, and *CBFB-MYH11* genes can detect translocations t(8;21)(q22;q22), t(9;22)(q34.1;q11.2), t(15;17)(q24;q21), and inversion inv(16)(p13q22), respectively. Dual-color break apart probes for the *MECOM* gene at 3q26.2, the *KMT2A* gene at 11q23, and the *NUP98* gene at 11p15.4 can detect translocations t(3q;var), t(11q;var), and t(11p;var) for rearrangements of these genes with multiple partner genes. Patients with translocations t(8;21), t(15;17), and inversion inv(16) have a favorable prognosis, trisomy 8 for an intermediate prognosis, and 5q-, 7q-, and translocation t(11;var), and translocation t(9;22) for unfavorable prognosis. Rapid FISH can give a result within a few hours, which could be vital for specific leukemias, such as acute promyelocytic leukemia (APL). APL has characteristic reciprocal translocation t(15;17) and variant rearrangements involving the *PML-RARA* gene fusion. APL is prone to intra-vascular coagulation and causes a high risk of mortality due to cerebral hemorrhage. All-trans retinoic acid therapy is essential for complete remission in these patients. Rapid FISH test can provide important evidence for the correct diagnosis and treatment of APL, which leads to more possibilities for patients to survive.

B-cell precursor acute lymphoblastic leukemia (pre B-ALL) is a heterogeneous disease with associated recurrent chromosomal abnormalities for diagnostic subtyping, prognostic stratification, and treatment strategy.[67] Current FISH testing for B-ALL uses two panels of FISH probes and additional FISH probes for specific rearrangements. The primary panel has four sets of probes including the *ETV6-RUNX1* genes for translocation t(12;21)(p13;q22) and an intrachromosomal amplification of 21q (iAMP21), the CEP4/CEP10/CEP17 centromeric probes for hyperdiploidy and low hypodiploidy, the *KMT2A* gene for an 11q23 translocation t(11q;var) with various partners, and the *BCR-ABL1* genes for translocation t(9;22). The detection of iAMP21 by FISH is defined by five or more signals for the *RUNX1* probe in interphases or three or more signals in a single chromosome 21. Additionally, FISH using probes for the *TCF3-PBX1* genes for translocation t(1;19)(q23;p13.3) and the *TCF3-HLF* genes for translocation t(17;19)(q22;p13.3) can be performed. The secondary panel uses dual-color break apart probes for the *ABL2* gene at 1q25, the *PDGFRB* gene at 5q32, the *JAK2* gene at 9p24.1, the *ABL1* gene at 9q34, and the *CRLF2* gene at Xp22.3 or Yp11.2 to detect rearrangements for *BCR-ABL1*-like B-ALL (Ph-like

B-ALL). Patients detected with iAMP21, hypodiploidy, translocations t(17;19), t(9;22), and t(11q,var) have a high risk stratification. Patients with t(1;19) have an intermediate risk. Patients with hyperdiploidy and translocation t(12;21) have a good prognosis. T-cell acute lymphoblastic leukemia (T-ALL) showed clonal rearrangements of the T-cell receptor genes. FISH assays using dual-color break apart probes for the *TCRA/D* gene at 14q11.2, the *TCRB* gene at 7q34, the *TCRG* gene at 7p14, and the *TCL* gene at 14q32 were introduced as a sensitive and accurate approach for rapid detection of rearrangements involving the T-cell receptor genes.[68]

Mature B-cell neoplasms include CLL/small lymphocytic lymphoma (SLL), plasma cell neoplasms, and various types of B-cell lymphomas. FISH assays on interphase nuclei are an essential part of the clinical evaluation for CLL patients. A CLL research consortium tested a panel of FISH probes for CLL and noted a 2% false negative rate and 3% false positive rate.[69] Current FISH panel include probes of the *ATM* and *TP53* genes for deletions of 11q and 17p, the D12Z3/*DLEU1/CDC16* loci for trisomy 12 and deletions of 13q or monosomy 13, and the *IGH* gene for translocation t(14q;var). An additional FISH application using probes of the *MYB/* D6Z3 loci for a deletion of 6q can be performed. CLL patients with deletions of 13q have a favorable prognosis, with trisomy 12 for an intermediate prognosis, and with deletions of 11q and 17p for a poor prognosis.[70]

Plasma cell neoplasms include monoclonal gammopathy of undetermined significance (MGUS) and plasma cell myeloma (PCM). Because of the low percentage of plasma cells in bone marrow specimen and low mitotic index of abnormal cells under in vitro cell culture, chromosomal abnormalities could escape routine karyotyping or present in a very low percentage of cells by interphase FISH. Enrichment of plasma cells by CD138 magnetic micro bead sorting has been introduced to improve the diagnostic yield.[71] Current FISH arrays for MGUS/PCM is ideally performed on enriched plasma cells (CD138+) using a panel of probes of the *CDKN2C/CKS1B* genes for deletion of 1p and duplication of 1q, the D9Z3/*PML/TP53* loci for hyperdiploidy and a deletion of 17p, the DLEU1/*CDC16* loci for deletions of 13q or monosomy 13, the IGH gene for translocation t(14q;var) with various partner genes, and the *MYB* gene for a deletion of 6q. Further characterization of partner genes for positive *IGH* break apart can be performed by using dual-color double fusion probes of the *IGH-FGFR3, IGH-CCND3, IGH-CCND1, IGH-MAF,* and *IGH-MAFB* genes for translocations t(4;14)(p16;q32), t(6;14)(p21;q32), t(11;14)(q13;q32), t(14;16)(q32;q23), and t(14;20)(q32;q12), respectively. Patients with deletions of 13q and 17p, hypodiploidy, translocations t(4;14), t(14;16), or t(14;20) have unfavorable risk. Patients with absence of unfavorable abnormalities and presence of hyperdiploidy and translocations t(11;14) or t(6;14) have favorable risk.[71]

A panel of FISH probes for various types of lymphomas include dual-color break-apart probes for the *BCL6* gene at 3q27, the *MYC* gene at 8q24, the *IGH* gene at 14q32, and the *BCL2* and *MALT1* genes at 18q21. More specifically, extranodal marginal zone lymphoma of mucosa-associated lymphoid tissue (MALT lymphoma) associated with translocations t(11;18)(q21;q21), t(1;14)(p22;q32), t(14;18)(q32;q21), and t(3;14)(p14.1;q32) can be indicated by FISH assays using the *IGH* and *MALT1* probes. Follicular lymphoma (FL) caused by translocation t(14;18)(q32;q21) can be detected using dual-color double fusion *IGH-BCL2* probes. Diffuse large B-cell lymphoma (DLBCL) with *BCL6* gene rearrangement and Burkitt lymphoma (BL) with *MYC* gene rearrangement can also be detected by this FISH panel for lymphoma. For cutaneous T-cell lymphoma (CTCL), a panel of 11 FISH probes was developed and validated to capture gene copy number alterations. This panel includes enumeration probes for genes *ARID1A* at 1p35.3, *DNMT3A* at 2p23, *CARD11* at 7p22, *MYC* at 8q24.2, *CDKN2A* at 9p21.3, *ZEB1* at 10p11.2, *FAS* at 10q24.1, *ATM* at 11q22.3, *RB1* at 13q14.2, *TP53* at 17p13.1, and *STAT3/5B* at 17q21.3.[72]

In general, FISH is an effective way to detect recurrent rearrangements with significant clinical value on diagnosis, prognosis, and treatment. FISH is mandatory when the karyotype is normal or analyzable metaphases are not present. Different kinds of specimens such as bone marrow, blood, and lymph node are available for FISH testing in hematology and lymphoma. Selecting a suitable specimen is critical for a precise result. It is better to use bone marrow for myeloid diseases, bone marrow or blood for CLL, CD138+ sorted plasma cells for MGUS/PCM, and lymph nodes for lymphoma. FISH assays offer a rapid detection of recurrent clonal rearrangements, many of which are cryptic and escape the detection by banding analysis, such as translocation t(4;14)(p16.3;q32) with *IGH-FGFR3* gene fusions, translocation t(12;21)(p13;q22) with *ETV6-RUNX1* gene fusions, gene deletions of 9p21(*CDKN2A*), 12p13(*ETV6*), 13q14.2(*RB1*), and 17p13.1(*TP53*), and rearrangements leading to Ph-like or *ETV6-RUNX1*-like ALL. Some chromosome rearrangements involve genes with many partner genes, such as *KMT2A*, *MYC*, *IGH*, and *ETV6*. FISH is recommended to detect rearrangements for these genes. For example, *KMT2A* rearrangement is one of the major drivers and generally regarded as a poor prognostic factor in acute leukemias. There are more than 120 partner genes of *KMT2A* rearrangement identified currently, many of which are cryptic and unidentified by chromosome analysis. FISH is a good approach to detect *KMT2A* rearrangement using the break-apart probe. Additionally, FISH assay using probes for sex chromosomes X and Y has been used to check engraftment of donor cells from sex-mismatched bone marrow and blood stem cell transplantation. Online resources such as the 'Atlas of Genetics and Cytogenetics in Oncology and Haematology' (http://atlasgeneticsoncology.org/) are very helpful in interpreting chromosome and FISH results. Table 6.2 lists the routinely used FISH panels for tumors of hematopoietic and lymphoid tissues in clinical cytogenetics.

DETECTION OF RECURRENT REARRANGEMENTS IN SOLID TUMORS

Conventional cytogenetics has detected recurrent chromosomal abnormalities from various types of solid tumors. However, this cell culture–dependent karyotyping has experienced reduced success rate in many solid tumor specimens. Interphase FISH assays performed directly on tumor cells have been a supplement to overcome this difficulty and applied to different types of tumors such as brain tumors, malignant melanoma, and head and neck squamous cell carcinoma.[73–75] The feasibility of FISH assays on sections of paraffin-embedded tissues was evaluated and then quickly applied to detect numerical aberrations from renal cortical neoplasms and gonadal yolk sac tumors.[76–78] This development extended FISH assays onto fixed tumor tissues.

TABLE 6.2

FISH Probes for Various Tumors of Hematopoietic and Lymphoid Tissues

Disease	Recurrent Rearrangements	Gene Rearrangement	FISH Probes	Prognosis	Targeted Therapy
Myeloproliferative neoplasms (chronic myelogenous leukemia, CML)					
	t(9;22)(q34.1;q11.2)	BCR-ABL1	BCR, ABL1	good following therapy	BCR-ABL1 tyrosine kinase inhibitor
Myeloid/lymphoid neoplasms with eosinophilia					
	del(4q12) or t(4q12;var)	FIP1L1-PDGFRA	FIL1L1, CHIC2, PDGFRA	favorable prognosis	tyrosine kinase inhibitor
	t(5q32;var)	PDGFRB	PDGFRB	good following therapy	tyrosine kinase inhibitor
	t(8p11;var)	FGFR1	FGFR1	poor	
	t(8;9)(p22;p24.1)	PCM1-JAK2	JAK2	survival highly variable	
Myelodysplastic syndrome (MDS)					
	del(5q)/-5		TAS2R1, EGR1	good	
	del(7q)/-7		RELN, TES	intermediate or poor	
	trisomy 8, del(20q)		MYC, PTPRT	intermediate, good	
	del(17p)		TP53	poor	
Acute myeloid leukemia (AML)					
	t(8;21)(q22;q22)	RUNX1-RUNX1T1	RUNX1, RUNX1T1	good	
	t(9;22)(q34.1;q11.2)	BCR-ABL1	BCR, ABL1	poor	
	t(11p15.4;var)	NUP98	NUP98	intermediate	
	t(11q23;var)	KMT2A	KMT2A		
	t(15;17)(q24;q21)	PML-RARA	PML, RARA	good	tretinoin and arsenic trioxide
	inv(16)(p13q22), t(16;16)	CBFB-MYH11	CBFB, MYH11	good	
	inv(3)(q21.3q26.2), t(3;3)	MECOM	MECOM	poor	

(Continued)

TABLE 6.2
(Continued)

Disease	Recurrent Rearrangements	Gene Rearrangement	FISH Probes	Prognosis	Targeted Therapy
B-cell precursor acute lymphoblastic leukemia (B-ALL)					
	t(9;22)(q34.1;q11.2)	BCR-ABL1	BCR, ABL1	poor	tyrosine kinase inhibitors
	t(11q23;var)	KMT2A	KMT2A	poor	
	t(12;21)(p13.2;q22.1), iAMP21	ETV6-RUNX1	ETV6, RUNX1	very favorable	
	Hyperdiploid 4/10/17		D4Z1, D10Z1, D17Z1	very favorable	
	t(1;19)(q23;p13.3)	TCF3-PBX1	TCF3, PBX1	intermediate risk	intensive therapy
	t(17;19)(q22;p13.3)	TCF3-HLB	TCF3, HLB		
Ph-like B-cell ALL					
	t(1q25;var)	ABL2	ABL2	high risk	ABL-class tyrosine kinase inhibitors
	t(5q32;var)	PDGFRB	PDGFRB	high risk	tyrosine kinase inhibitors
	t(9p24.1;var)	JAK2	JAK2	high risk	tyrosine kinase inhibitors
	t(9q34;var)	ABL1	ABL1		tyrosine kinase inhibitors
	t(Xp22.3/Yp11.2;var)	CRLF2-P2RY8	CRLF2, P2RY8	high risk	tyrosine kinase inhibitors
T-lymphoblastic leukemia/lymphoma					
	t(14q11.2;var)	TCRA/D	TCRA/D		
	t(7q35;var)	TCRB	TCRB		
	t(7p14;var)	TCRG	TCRG		
	t(14q32;var)	TCL	TCL		
Mature B-cell neoplasms (chronic lymphocytic leukemia, CLL)					
	del(11q), del(17p)		ATM, TP53	poor	
	trisomy 12, del(13q)		D12Z6, DELU1, CDC16	intermediate, favorable	
	t(14q32;var)	IGH	IGH		
	del(6q)		MYB, D6Z3		

Plasma cell neoplasms/multiple myeloma

del(1p), dup(1q)		CDKN2C, CKS1B	
hyperdiploid, 9/15/17		D9Z3, PML, TP53	
del(13q)		DLEU1, CDC16	
t(14q32;var)	IGH	IGH	
t(4;14)(p16.3;q32)	IGH-FGFR3	IGH, FGFR3	intermediate risk
t(6;14)(p21;q32)	IGH-CCND3	IGH, CCND3	standard risk
t(11;14)(q13;q32)	IGH-CCND1	IGH, CCND1	standard risk
t(14;16)(q32;q23)	IGH-MAF	IGH, MAF	high risk
t(14;18)(q32;q21)	IGH-BCL2	IGH, BCL2	BCL2 inhibitors
t(14;19)(q32;q13)	IGH-BCL3	IGH, BCL3	
t(14;20)(q32;q11.2)	IGH-MAFB	IGH, MAFB	high risk
t(8;14)(q24.2;q32)	IGH-MYC	IGH, MYC	

Lymphomas

t(3q27;var)	BCL6	BCL6	
t(8q24;var)	MYC	MYC	
t(14q32;var)	IGH	IGH	
t(18q21;var)	BCL2	BCL2	
t(18q21;var)	MALT1	MALT1	

Follicular lymphoma

t(14;18)(q32;q21)	IGH-BCL2	IGH, BCL2	

Mantle cell lymphoma

t(11;14)(q13;q32)	IGH-CCND1	IGH, CCND1	median survival 3–5 years

ALK-positive large B-cell lymphoma

t(2;17)(p23;q23)	CLTC-ALK	ALK	
t(2;5)(p23;q35)	ALK-NPM1	ALK	inhibitors of ALK kinase

Abbreviations: var: variant partner genes

Recurrent chromosomal rearrangements involving tumorigenesis gene fusions have been detected in various types of solid tumors.[79] For example, rearrangements involving the *EWSR* gene at 22q12 with various partner genes, translocation t(22q;var), have been diagnostic for Ewing sarcoma, angiomatoid fibrous histocytoma, clear cell sarcoma, desmoplastic round cell tumor, myxoid round cell liposarcoma, and extraskeletal myxoid chondrosarcoma. Rearrangements involving the *ALK* gene at 2p23, translocation t(2p;var), are diagnostic for inflammatory myofibroblastic tumor. Rearrangements involving the *FOXO1A, PAX3, NCOA1* and *PAX7* genes are diagnostic for alveolar rhabdomyosarcoma. Translocations of t(7;16)(q33;p11.2) for *FUS-CREB3L2* gene fusion and t(11;16)(p11;p11.2) for *FUS-CREB3L1* gene fusion are diagnostic for low-grade fibromyxoid sarcoma. Translocations of t(X;18)(p11.2;q11.2) for *SS18-SSX1/2/4* gene fusions and t(X;20)(p11.2;q13) for S*S18L1-SSX1* gene fusion are diagnostic for synovial sarcoma. Specific translocation t(X;17)(p11.2;q25) with *ASPSCR1-TFE3* gene fusion is diagnostic for alveolar soft-part sarcoma and rearrangements of *TFE3* with other genes are diagnostic for renal cell carcinoma. The rearrangements involving the *TMPRSS2* gene at 22q22.3 are indicative of early stage of prostate cancer. The translocation t(17;22) with *COL1A1-PDGFRB* gene fusion and its variants is diagnostic for dermatofibrosarcoma protuberans.[80] FISH probes are available for the detection of these rearrangements on metaphase chromosomes and interphases nuclei from tumor specimens.

Recurrent numerical chromosome abnormalities with histologic and clinical association to various solid tumors have been characterized.[79] Genomic approaches like CMA can detect genome-wide somatic copy number alterations, but FISH is still the gold standard to determine losses and gains of specific genes and chromosome regions at a single-cell level.[81] For renal cell carcinoma, deletions of 3p, gains of chromosomes 7 and 17, losses of chromosomes 1 and 17, and losses of chromosomes 1 and 14 are for differential diagnosis of clear cell carcinoma, papillary carcinoma, chromophobe carcinoma, and oncocytoma, respectively. Amplification of the *EGFR* gene distinguishes glioblastoma with small cell phenotype. For bladder cancer, an UroVysion FISH panel, including centromeric probes for chromosomes 3, 7, and 17 and a locus-specific probe for the *CDKN2A* gene at 9p21, was approved by the FDA for surveillance and determining therapy effectiveness.[82] Additional deletions of the *RB1* gene at 13q14.2, the *PTEN* gene at 10q23, and the *TP53* gene at 17p13.1, and amplification of the *ERBB2* gene at 17q21 predict poor outcome for bladder cancer. Amplification of *ERBB2* gene (*HER2*) and *EGFR* gene at 17p12 have been seen in breast carcinoma. Guidelines for *HER2* FISH testing in breast cancer have been updated for accurate scoring of *HER2*-postive breast cancer for anti-*HER2* targeted therapy.[83]

CONCLUSIONS

In summary, FISH assays are a clinically validated and cell-based diagnostic tool capable of detecting numerical and structural chromosomal abnormalities and genomic copy number alterations. FISH assays have been routinely used in rapid prenatal screening of common aneuploidies, detection of microdeletion and microduplication syndromes, and further characterization of sSMC, subtelomeric rearrangements, ring chromosome structure, and complex chromosome rearrangements. FISH assays play an essential role in cancer cytogenetics by detecting recurrent numerical aberrations and gene rearrangements of clinical significance in diagnosis, prognosis, and targeted therapy for various types of tumors.

ACKNOWLEDGEMENTS

Special thanks to Autumn DiAdamo and Jiadi Wen of Yale Clinical Cytogenetics Laboratory (USA) for helpful discussion and careful editing on the content of this chapter.

REFERENCES

1. Gall, J.G.; Pardue, M.L. Formation and detection of RNA-DNA hybrid molecules in cytological preparations. *Proc. Natl. Acad. Sci. U. S. A.* **1969**, 63, 378–383.
2. Rudkin, G.T.; Stollar, B.D. High resolution detection of DNA-RNA hybrids in situ by indirect immunofluorescence. *Nature.* **1977**, 265, 472–473.
3. Lichter, P.; Cremer, T.; Borden, J.; Manuelidis, L.; Ward, D.C. Delineation of individual human chromosomes in metaphase and interphase cells by in situ suppression hybridization using recombinant DNA libraries. *Hum. Genet.* **1988**, 80, 224–234.
4. Lichter, P.; Tang, C.J.; Call, K.; Hermanson, G.; Evans, G.A.; Housman, D.; Ward, D.C. High-resolution mapping of human chromosome 11 by in situ hybridization with cosmid clones. *Science.* **1990**, 247, 64–69.
5. Cheung, V.G.; Nowak, N.; Jang, W.; Kirsch, I.R.; Zhao, S.; Chen, X.N.; Furey, T.S.; Kim, U.J.; Kuo, W.L.; Olivier, M.; Conroy, J.; Kasprzyk, A.; Massa, H.; Yonescu, R.; Sait, S.; Thoreen, C.; Snijders, A.; Lemyre, E.; Bailey, J.A.; Bruzel, A.; Burrill, W.D.; Clegg, S.M.; Collins, S.; Dhami, P.; Friedman, C.; Han, C.S.; Herrick, S.; Lee, J.; Ligon, A.H.; Lowry, S.; Morley, M.; Narasimhan, S.; Osoegawa, K.; Peng, Z.; Plajzer-Frick, I.; Quade, B.J.; Scott, D.; Sirotkin, K.; Thorpe, A.A.; Gray, J.W.; Hudson, J.; Pinkel, D.; Ried, T.; Rowen, L.; Shen-Ong, G.L.; Strausberg, R.L.; Birney, E.; Callen, D.F.; Cheng, J.F.; Cox, D.R.; Doggett, N.A.; Carter, N.P.; Eichler, E.E.; Haussler, D.; Korenberg, J.R.; Morto, C.C.; Albertson, D.; Schuler, G.; de Jong, P.J.; Trask, B.J.; BAC Resource Consortium. Integration of cytogenetic landmarks into the draft sequence of the human genome. *Nature.* **2001**, 409, 953–958.
6. Ried, T.; Landes, G.; Dackowski, W.; Klinger, K.; Ward, D.C. Multicolor fluorescence in situ hybridization for the simultaneous detection of probe sets for chromosomes 13, 18, 21, X and Y in uncultured amniotic fluid cells. *Hum. Mol. Genet.* **1992**, 1, 307–313.
7. Weise, A.; Liehr, T. Rapid prenatal aneuploidy screening by fluorescence in situ hybridization (FISH). *Methods Mol. Biol.* **2019**, 1885, 129–137.
8. Speicher, M.R.; Ward, D.C. The coloring of cytogenetics. *Nat. Med.* **1996**, 2, 1046–1048.
9. Cui, C.; Shu, W.; Li, P. Fluorescence in situ hybridization: Cell-based genetic diagnostic and research applications. *Front. Cell. Dev. Biol.* **2016**, 4, 89.
10. Castronovo, C.; Valtorta, E.; Crippa, M.; Tedoldi, S.; Romitti, L.; Amione, M.C.; Guerneri, S.; Rusconi, D.; Ballarati, L.; Milani, D.; Grosso, E.; Cavalli, P.; Giardino, D.; Bonati, M.T.; Larizza, L.; Finelli, P. Design and validation of a pericentromeric BAC clone set aimed at improving diagnosis and phenotype prediction of supernumerary marker chromosomes. *Mol. Cytogenet.* **2013**, 6, 45.
11. Morrison, L.E.; Ramakrishnan, R.; Ruffalo, T.M.; Wilber, K.A. Labeling fluorescence in situ hybridization probes for genomic targets. *Methods Mol. Biol.* **2002**, 204, 21–40.

12. Ning, Y.A.R.; Smith, A.C.M.; Macha, M.; Precht, K.; Riethman, H.; Ledbetter, D.H.; Flint, J.; Horsley, S.; Regan, R.; Kearney, L.; Knight, S.; Kvaloy, K.; Brown, W.R.A. A complete set of human telomeric probes and their clinical application: National Institutes of Health and Institute of Molecular Medicine collaboration. *Nat. Genet.* **1996**, 14, 86–89.

13. Bentz, M.; Cabot, G.; Moos, M.; Speicher, M.R.; Ganser, A.; Lichter, P.; Döhner, H. Detection of chimeric BCR-ABL genes on bone marrow samples and blood smears in chronic myeloid and acute lymphoblastic leukemia by in situ hybridization. *Blood.* **1994**, 83, 1922–1928.

14. van der Burg, M.; Beverloo, H.B.; Langerak, A.W.; Wijsman, J.; van Drunen, E.; Slater, R.; van Dongen, J.J. Rapid and sensitive detection of all types of MLL gene translocations with a single FISH probe set. *Leukemia.* **1999**, 13, 2107–2113.

15. Pinkel, D.; Landegent, J.; Collins, C.; Langerak, A.W.; Wijsman, J.; van Drunen, E.; Slater, R.; van Dongen, J.J. Fluorescence in situ hybridization with human chromosome-specific libraries: Detection of trisomy 21 and translocations of chromosome 4. *Proc. Natl. Acad. Sci. U. S A.* **1988**, 85, 9138–9142.

16. Meltzer, P.S.; Guan, X.Y.; Burgess, A.; Trent, J.M. Rapid generation of region specific probes by chromosome microdissection and their application. *Nat. Genet.* **1992**, 1, 24–28.

17. Speicher, M.R.; Gwyn Ballard, S.; Ward, D.C. Karyotyping human chromosomes by combinatorial multi-fluor FISH. *Nat. Genet.* **1996**, 12, 368–375.

18. Mascarello, J.T.; Hirsch, B.; Kearney, H.M.; Ketterling, R.P.; Olson, S.B.; Quigley, D.I.; Rao, K.W.; Tepperberg, J.H.; Tsuchiya, K.D.; Wiktor, A.E; Working Group of the American College of Medical Genetics Laboratory Quality Assurance Committee. Section E9 of the American College of Medical Genetics technical standards and guidelines: Fluorescence in situ hybridization. *Genet. Med.* **2011**, 13, 667–675.

19. (CLSI) CaLSI. MM07-A2. *Fluorescence In Situ Hybridization Methods for Clinical Laboratories Approved.* Guideline. 2nd Edition. Vol. 33. Clinical and Laboratory Standards Institute, Wayne, PA; Aug. **2013**.

20. Wiktor, A.E.; Van Dyke, D.L.; Stupca, P.J.; Ketterling, R.P.; Thorland, E.C.; Shearer, B.M.; Fink, S.R.; Stockero, K.J.; Majorowicz, J.R.; Dewald, G.W. Preclinical validation of fluorescence in situ hybridization assays for clinical practice. *Genet. Med.* **2006**, 8, 16–23.

21. Zneimer, S.M. Validation of fluorescence in situ hybridization (FISH) for chromosome 5 monosomy and deletion. *Curr. Prot. Hum. Genet.* **2020**, 105, e96.

22. Ciolino, A.L.; Tang, M.E.; Bryant, R. Statistical treatment of fluorescence in situ hybridization validation data to generate normal reference ranges using Excel functions. *J. Mol. Diagn.* **2009**, 11, 330–333.

23. McGowan-Jordan, J.; Hastings, A.; Moore, S. *ISCN 2020 an International System for Human Cytogenomic Nomenclature (2020).* Karger, Basel/Freiburg; **2020**.

24. Mascarello, J.T.; Brothman, A.R.; Davison, K.; Dewald, G.W.; Herrman, M.; McCandless, D.; Park, J.P.; Persons, D.L.; Rao, K.W.; Schneider, N.R.; Vance, G.H.; Cooley, L.D; Cytogenetics Resource Committee of the College of American Pathologists and American College of Medical Genetics. Proficiency testing for laboratories performing fluorescence in situ hybridization with chromosome-specific DNA probes. *Arch. Pathol. Lab. Med.* **2002**, 126, 1458–1462.

25. Chai, H.; DiAdamo, A.; Grommisch, B.; Xu, F.; Zhou, Q.; Wen, J.; Mahoney, M.; Bale, A.; McGrath, J.; Spencer-Manzon, M.; Li, P.; Zhang, H. A retrospective analysis of 10-year data assessed the diagnostic accuracy and efficacy of cytogenomic abnormalities in current prenatal and pediatric settings. *Front. Genet.* **2019**, 10, 1162.

26. Liehr, T.; Ziegler, M. Rapid prenatal diagnostics in the interphase nucleus: Procedure and cut-off rates. *J. Histochem. Cytochem.* **2005**, 53, 289–291.

27. Tepperberg, J.; Pettenati, M.J.; Rao, P.N.; Lese, C.M.; Rita, D.; Wyandt, H.; Gersen, S.; White, B.; Schoonmaker, M.M. Prenatal diagnosis using interphase fluorescence in situ hybridization (FISH): 2-year multi-center retrospective study and review of the literature. *Prenat. Diagn.* **2001**, 21, 293–301.

28. Lall, M.; Mahajan, S.; Saviour, P.; Paliwal, P.; Joshi, A.; Setai, N.; Verma, J.C. FISH is not suitable as a standalone test for detecting fetal chromosomal abnormalities. *J. Fet. Med.* **2015**, 2, 53–59.

29. Chai, H.; DiAdamo, A.; Grommisch, B.; Boyle, J.; Amato, K.; Wang, D.; Wen, J.; Li, P. Integrated FISH, karyotyping and aCGH analyses for effective prenatal diagnosis of common aneuploidies and other cytogenomic abnormalities. *Med. Sci. (Basel).* **2019**, 7, 16.

30. Xie, X.; Tan, W.; Li, F.; Carrano, E.; Ramirez, P.; DiAdamo, A.; Grommisch, B.; Amato, K.; Chai, H.; Wen, J.; Li, P. Diagnostic cytogenetic testing following positive noninvasive prenatal screening results of sex chromosome abnormalities: Report of five cases and systematic review of evidence. *Mol. Genet. Genomic Med.* **2020**, 8, e1297.

31. Schinzel, A. Microdeletion syndromes, balanced translocations, and gene mapping. *J. Med. Genet.* **1988**, 25, 454–462.

32. Kuwano, A.; Ledbetter, S.A.; Dobyns, W.B.; Emanuel, B.S.; Ledbetter, D.H. Detection of deletions and cryptic translocations in Miller-Dieker syndrome by in situ hybridization. *Am. J. Hum. Genet.* **1991**, 49, 707–714.

33. Desmaze, C.; Prieur, M.; Amblard, F.; Aikem, M.; LeDeist, F.; Demczuk, S.; Zucman, J.; Plougastel, B.; Delattre, O.; Croquette, M.F.; Brevière, G.-M.; Huon, C.; Le Merrer, M.; Mathieu, M.; Sidi, D.; Stephan, J.-L.; Aurias, A. Physical mapping by FISH of the DiGeorge critical region (DGCR): Involvement of the region in familial cases. *Am. J. Hum. Genet.* **1993**, 53, 1239–1249.

34. Stankiewicz, P.; Lupski, J.R. Genome architecture, rearrangements and genomic disorders. *Trends Genet.* **2002**, 18, 74–82.

35. Ensenauer, R.E.; Adeyinka, A.; Flynn, H.C.; Michels, V.V.; Lindor, N.M.; Dawson, D.B.; Thorland, E.C.; Lorentz, C.P.; Goldstein, J.L.; McDonald, M.T.; Smith, W.E.; Simon-Fayard, E.; Alexander, A.A.; Kulharya, A.S.; Ketterling, R.P.; Clark, R.D.; Jalal, S.M. Microduplication 22q11.2, an emerging syndrome: Clinical, cytogenetic, and molecular analysis of thirteen patients. *Am. J. Hum. Genet.* **2003**, 73, 1027–1040.

36. Liehr, T.; Claussen, U.; Starke, H. Small supernumerary marker chromosomes (sSMC) in humans. *Cytogenet. Genome Res.* **2004**, 107, 55–67.

37. Henegariu, O.; Bray-Ward, P.; Artan, S.; Vance, G.H.; Qumsyieh, M.; Ward, D.C. Small marker chromosome identification in metaphase and interphase using centromeric multiplex fish (CM-FISH). *Lab. Invest.* **2001**, 81, 475–481.

38. Brecevic, L.; Michel, S.; Starke, H.; Müller, K.; Kosyakova, N.; Mrasek, K.; Weise, A.; Liehr, T. Multicolor FISH used for the characterization of small supernumerary marker chromosomes (sSMC) in commercially available immortalized cell lines. *Cytogenet, Genome Res.* **2006**, 114, 319–324.

39. Manolakos, E.; Kefalas, K.; Neroutsou, R.; Lagou, M.; Kosyakova, N.; Ewers, E.; Ziegler, M.; Weise, A.; Tsoplou, P.; Rapti, S.M.; Papoulidis, I.; Anastasakis, E.; Garas, A.; Sotiriou, S.; Eleftheriades, M.; Peitsidis, P.; Malathrakis, D.; Thomaidis, L.; Kitsos, G.; Orru, S.; Liehr, T.; Petersen, M.B.; Kitsiou-Tzeli, S. Characterization of 23 small supernumerary marker chromosomes detected at pre-natal diagnosis: The value of fluorescence in situ hybridization. *Mol. Med. Rep.* **2010**, 3, 1015–1022.

40. Liehr, T.; Ewers, E.; Hamid, A.B.; Kosyakova, N.; Voigt, M.; Weise, A.; Manvelyan, M. Small supernumerary marker chromosomes and uniparental disomy have a story to tell. *J. Histochem. Cytochem.* **2011**, 59, 842–848.
41. Liehr, T.; Klein, E.; Mrasek, K.; Kosyakova, N.; Guilherme, R.S.; Aust, N.; Venner, C.; Weise, A.; Hamid, A.B. Clinical impact of somatic mosaicism in cases with small supernumerary marker chromosomes. *Cytogenet. Genome Res.* **2013**, 139, 158–163.
42. Armanet, N.; Tosca, L.; Brisset, S.; Liehr, T.; Tachdjian, G. Small supernumerary marker chromosomes in human infertility. *Cytogenet Genome Res.* **2015**, 146, 100–108.
43. Ravnan, J.B.; Tepperberg, J.H.; Papenhausen, P.; Lamb, A.N.; Hedrick, J.; Eash, D.; Ledbetter, D.H.; Martin, C.L. Subtelomere FISH analysis of 11 688 cases: An evaluation of the frequency and pattern of subtelomere rearrangements in individuals with developmental disabilities. *J. Med. Genet.* **2006**, 43, 478–489.
44. Hu, Q.; Chai, H.; Shu, W.; Li, P. Human ring chromosome registry for cases in the Chinese population: Re-emphasizing cytogenomic and clinical heterogeneity and reviewing diagnostic and treatment strategies. *Mol. Cytogenet.* **2018**, 11, 19.
45. Zhang, H.Z.; Li, P.; Wang, D.; Huff, S.; Nimmakayalu, M.; Qumsiyeh, M.; Pober, B.R. FOXC1 gene deletion is associated with eye anomalies in ring chromosome 6. *Am. J. Med. Genet.* **2004**, 124A, 280–287.
46. Zhang, H.Z.; Xu, F.; Seashore, M.; Li, P. Unique genomic structure and distinct mitotic behavior of ring chromosome 21 in two unrelated cases. Cytogenet. *Genome Res.* **2012**, 136, 180–187.
47. Xu, F.; DiAdamo, A.J.; Grommisch, B.; Li, P. Interstitial duplication and distal deletion in a ring chromosome 13 with pulmonary atresia and ventricular septal defect: A case report and review of literature. *N. Am. J. Med. Sci.* **2013**, 6, 208–212.
48. Rossi, M.R.; DiMaio, M.S.; Xiang, B.; Lu, K.; Kaymakcalan, H.; Seashore, M.; Mahoney, M.J.; Li, P. Clinical and genomic characterization of distal duplications and deletions of chromosome 4q: Study of two cases and review of the literature. *Am. J. Med. Genet.* **2009**, 149A, 2788–2794.
49. Khattab, M.; Xu, F.; Li, P.; Bhandari, V. A de novo 3.54 Mb deletion of 17q22-q23.1 associated with hydrocephalus: A case report and review of literature. *Am. J. Med. Genet.* **2011**, 155A, 3082–3086.
50. Li, P.; Zhang, H.Z.; Huff, S.; Nimmakayalu, M.; Qumsiyeh, M.; Yu, J.; Szekely, A.; Xu, T.; Pober, B.R. Karyotype-phenotype insights from 11q14.1-q23.2 interstitial deletions: FZD4 haploinsufficiency and exudative vitreoretinopathy in a patient with a complex chromosome rearrangement. *Am. J. Med. Genet* **2006**, 140A, 2721–2729.
51. Chai, H.; Grommisch, B.; DiAdamo, A.; Wen, J.; Hui, P.; Li, P. Inverted duplication: Triplication and quintuplication through sequential breakage-fusion-bridge events induced by a terminal deletion at 5p in a case of spontaneous abortion. *Mol. Genet. Genomic. Med.* **2019**, 7, e00965.
52. Brownstein, C.A.; Adler, F.; Nelson-Williams, C.; Iijima, J.; Li, P.; Imura, A.; Nabeshima, Y.; Reyes-Mugica, M.; Carpenter, T.O.; Lifton, R.P. A translocation causing increased alpha-klotho level results in hypophosphatemic rickets and hyperparathyroidism. *Proc. Natl. Acad. Sci. U. S. A.* **2008**, 105, 3455–3460.
53. Nowell, P.C.; Hungerford, D.A. Chromosome studies on normal and leukemic human leukocytes. *J. Natl. Cancer Inst.* **1960**, 25, 85–109.
54. Rowley, J.D. Genetics: A story of swapped ends. *Science.* **2013**, 340, 1412–1413.
55. Tkachuk, D.C.; Westbrook, C.A.; Andreeff, M.; Donlon, T.A.; Cleary, M.L.; Suryanarayan, K.; Homge, M.; Redner, A.; Gray, J.; Pinkel, D. Detection of bcr-abl fusion in chronic myelogeneous leukemia by in situ hybridization. *Science.* **1990**, 250, 559–562.

56. Losada, A.P.; Wessman, M.; Tiainen, M.; Hopman, A.H.; Willard, H.F.; Solé, F.; Caballín, M.R.; Woessner, S.; Knuutila, S. Trisomy 12 in chronic lymphocytic leukemia: An interphase cytogenetic study. *Blood.* **1991**, 78, 775–779.
57. Anastasi, J.; Le Beau, M.M.; Vardiman, J.W.; Fernald, A.A.; Larson, R.A.; Rowley, J.D. Detection of trisomy 12 in chronic lymphocytic leukemia by fluorescence in situ hybridization to interphase cells: A simple and sensitive method. *Blood.* **1992**, 79, 1796–1801.
58. Jenkins, R.B.; Le Beau, M.M.; Kraker, W.J.; Borell, T.J.; Stalboerger, P.G.; Davis, E.M.; Penland, L.; Fernald, A.; Espinosa, R 3rd.; Schaid, D.J.; Noel, P.; Dewald, G.W. Fluorescence in situ hybridization: A sensitive method for trisomy 8 detection in bone marrow specimens. *Blood.* **1992**, 79, 3307–3315.
59. Dewald, G.; Stallard, R.; Alsaadi, A.; Arnold, S.; Blough, R.; Ceperich, T.M.; Rafael Elejalde, B.; Fink, J.; Higgins, J.V.; Higgins, R.R.; Hoeltge, G.A.; Hsu, W.T.; Johnson, E.B.; Kronberger, D.; McCorquodale, D.J.; Meisner, L.F.; Micale, M.A.; Oseth, L.; Payne, J.S.; Schwartz, S.; Sheldon, S.; Sophian, A.; Storto, P.; Van Tuinen, P.; Wenger, G.D.; Wiktor, A.; Willis, L.A.; Yung, J.F.; Zenger-Hain, J. A multicenter investigation with D-FISH BCR/ABL1 probes. *Cancer Genet. Cytogenet.* **2000**, 116, 97–104.
60. Wolff, D.J.; Bagg, A.; Cooley, L.D.; Dewald, G.W.; Hirsch, B.A.; Jacky, P.B.; Rao, K.W.; Rao, P.N; Association for Molecular Pathology Clinical Practice Committee; American College of Medical Genetics Laboratory Quality Assurance Committee. Guidance for fluorescence in situ hybridization testing in hematologic disorders. *J. Mol. Diagn.* **2007**, 9, 134–143.
61. Liehr, T.; Othman, M.A.; Rittscher, K.; Alhourani, E. The current state of molecular cytogenetics in cancer diagnosis. *Expert Rev. Mol. Diagn.* **2015**, 15, 517–526.
62. Swerdlow, S.H.; Campo, E.; Pileri, S.A.; Harris, N.L.; Stein, H.; Siebert, R.; Advani, R.; Ghielmini, M.; Salles, G.A.; Zelenetz, A.D.; Jaffe, E.S. The 2016 revision of the World Health Organization classification of lymphoid neoplasms. *Blood.* **2016**, 127, 2375–2390.
63. Li, M.M.; Ewton, A.A.; Smith, J.L. Using cytogenetic rearrangements for cancer prognosis and treatment (Pharmacogenetics). *Curr. Genet. Med. Rep.* **2013**, 1, 99–112.
64. Gorusu, M.; Benn, P.; Li, Z.; Fang, M. On the genesis and prognosis of variant translocations in chronic myeloid leukemia. *Cancer Genet. Cytogenet.* **2007**, 173, 97–106.
65. Savage, N.; George, T.I.; Gotlib, J. Myeloid neoplasms associated with eosinophilia and rearrangement of PDGFRA, PDGFRB, and FGFR1: A review. *Int. J. Lab. Hematol.* **2013**, 35, 491–500.
66. Vance, G.H.; Kim, H.; Hicks, G.A.; Cherry, A.M.; Higgins, R.; Hulshizer, R.L.; Tallman, M.S.; Fernandez, H.F.; Dewald, G.W. Utility of interphase FISH to stratify patients into cytogenetic risk categories at diagnosis of AML in an Eastern Cooperative Oncology Group (ECOG) clinical trial (E1900). *Leuk. Res.* **2007**, 31, 605–609.
67. Moorman, A.V. New and emerging prognostic and predictive genetic biomarkers in B-cell precursor acute lymphoblastic leukemia. *Haematologica.* **2016**, 101, 407–416.
68. Gesk, S.; Martin-Subero, J.I.; Harder, L.; Luhmann, B.; Schlegelberger, B.; Calasanz, M.J.; Grote, W.; Siebert, R. Molecular cytogenetic detection of chromosomal breakpoints in T-cell receptor gene loci. *Leukemia.* **2003**, 17, 738–745.
69. Smoley, S.A.; Van Dyke, D.L.; Kay, N.E.; Heerema, N.A.; Dell' Aquila, M.L.; Dal Cin, P.; Koduru, P.; Aviram, A.; Rassenti, L.; Byrd, J.C.; Rai, K.R.; Brown, J.R.; Greaves, A.W.; Eckel-Passow, J.; Neuberg, D.; Kipps, T.J.; Dewald, G.W. Standardization of fluorescence in situ hybridization studies on chronic lymphocytic leukemia (CLL) blood and marrow cells by the CLL Research Consortium. *Cancer Genet. Cytogenet.* **2010**, 203, 141–148.

70. Döhner, H.; Stilgenbauer, S.; Benner, A.; Leupolt, E.; Kröber, A.; Bullinger, L.; Döhner, K.; Bentz, M.; Lichter, P. Genomic aberrations and survival in chronic lymphocytic leukemia. *N. Engl. J. Med.* **2000**, 343, 1910–1916.

71. Chen, Z.; Issa, B.; Huang, S.; Aston, E.; Xu, J.; Yu, M.; Brothman, A.R.; Glenn, M. A practical approach to the detection of prognostically significant genomic aberrations in multiple myeloma. *J. Mol. Diagn.* **2005**, 7, 560–565.

72. Weed, J.; Gibson, J.; Lewis, J.; Carlson, K.; Foss, F.; Choi, J.; Li, P.; Girardi, M. FISH panel for leukemic CTCL. *J. Invest. Dermatol.* **2017**, 137, 751–753.

73. Van Dekken, H.; Pizzolo, J.G.; Reuter, V.E.; Melamed, M.R. Cytogenetic analysis of human solid tumors by in situ hybridization with a set of 12 chromosome-specific DNA probes. *Cytogenet. Cell Genet.* **1990**, 54, 103–107.

74. Arnoldus, E.P.; Noordermeer, I.A.; Peters, A.C.; Voormolen, J.H.; Bots, G.T.; Raap, A.K.; van der Ploeg, M. Interphase cytogenetics of brain tumors. *Genes Chromosomes Cancer.* **1991**, 3, 101–107.

75. Poetsch, M.; Dittberner, T.; Woenckhaus, C.; Kleist, B. Use of interphase cytogenetics in demonstrating specific chromosomal aberrations in solid tumors: New insights in the pathogenesis of malignant melanoma and head and neck squamous cell carcinoma. *Histol. Histopathol.* **2000**, 15, 1225–1231.

76. Arnoldus, E.P.; Dreef, E.J.; Noordermeer, I.A.; Verheggen, M.M.; Thierry, R.F.; Peters, A.C.; Cornelisse, C.J.; Van der Ploeg, M.; Raap, A.K. Feasibility of in situ hybridisation with chromosome specific DNA probes on paraffin wax embedded tissue. *J. Clin. Pathol.* **1991**, 44, 900–904.

77. El-Naggar, A.K.; van Dekken, H.D.; Ensign, L.G.; Pathak, S. Interphase cytogenetics in paraffin-embedded sections from renal cortical neoplasms: Correlation with cytogenetic and flow cytometric DNA ploidy analyses. *Cancer Genet. Cytogenet.* **1994**, 73, 134–141.

78. Jenderny, J.; Koster, E.; Meyer, A.; Borchers, O.; Grote, W.; Harms, D.; Jänig, U. Detection of chromosome aberrations in paraffin sections of seven gonadal yolk sac tumors of childhood. *Hum. Genet.* **1995**, 96, 644–650.

79. Nanjangud, G.; Amarillo, I.; Rao, P.N. Solid tumor cytogenetics: Current perspectives. *Clin. Lab. Med.* **2011**, 31, 785–811.

80. Marino-Enriquez, A.; Li, P.; Samuelson, J.; Rossi, M.R.; Reyes-Mugica, M. Congenital fibrosarcoma with a novel complex 3-way translocation t(12;15;19) and unusual histologic features. *Hum. Pathol.* **2008**, 39, 1844–1848.

81. Parisi, F.; Ariyan, S.; Narayan, D.; Bacchiocchi, A.; Hoyt, K.; Cheng, E.; Xu, F.; Li, P.; Halaban, R.; Kluger, Y. Detecting copy number status and uncovering subclonal markers in heterogeneous tumor biopsies. *BMC Genomics.* **2011**, 12, 230.

82. Nagai, T.; Naiki, T.; Etani, T.; Iida, K.; Noda, Y.; Shimizu, N.; Isobe, T.; Nozaki, S.; Okamura, T.; Ando, R.; Kawai, N.; Yasui, T. UroVysion fluorescence in situ hybridization in urothelial carcinoma: A narrative review and future perspectives. *Transl. Androl. Urol.* **2021**, 10, 1908–1917.

83. Geiersbach, K.B.; Sill, D.R.; Meyer, R.G.; Yuhas, J.A.; Sukov, W.R.; Mounajjed, T.; Carter, J.M.; Jenkins, R.B.; Chen, B. HER2 testing for breast cancer in the genomics laboratory: A sea change for fluorescence in situ hybridization. *Arch. Pathol. Lab. Med.* **2021**, 145, 883–886.

7 FISH—in Leukemia Diagnostics

*Roberto Matos, Mariana de Souza,
Gerson Ferreira, Amanda Figueiredo,
Marcelo Land and Maria Luiza Silva*

CONTENTS

DOI: 10.1201/9781003223658-7

INTRODUCTION

Leukemia comprises a heterogeneous group of hematological neoplasias character-ized by clonal proliferation of hematopoietic cell lineages that underwent genetic alterations, leading to be expressed in a specific differentiation phase. These altera-tions affect genes with pivotal roles in the control of natural mechanisms of prolif-eration, differentiation, and apoptosis.[1, 2]

The main clinical symptoms result in accumulation of immature precursors in the bone marrow (BM). This can harm or even prevent the production of normal blood cells, such as erythrocytes (causing anemia), or neutrophils (resulting in a suscepti-bility to infections and platelets, causing hemorrhages). With the excess proliferation of BM-blast cells, infiltration of tissues such as tonsils, lymph nodes, skin, spleen, kidneys, central nervous system (CNS), and others may occur.[3]

Leukemia cells may be immunophenotypically similar to mature (as observed in Chronic Lymphocytic Leukemia (CLL)), or precursor cells (originating from several lineages, as in Acute Leukemia (AL)), or still cells similar to precursor and mature ones (as in Chronic Myeloid Leukemia (CML)). Leukemia can affect people of any age. However, there are particular distributions depending on leukemia cell types and patients' ages. In this sense, Acute Lymphoblastic Leukemia (ALL) is more common in early childhood compared with Acute Myeloid Leukemia (AML), which is more frequently observed in older adults. CML is extremely rare in younger children, and CLL is almost exclusive to patients over 40 years old, with a median age of 70 years.[4]

Leukemia are a very heterogeneous and vast group of diseases, covering a wide range of ages.[5] Thus, in this chapter, we will focus on a particular group of leuke-mia, more specifically, childhood ALs, to demonstrate the benefits of fluorescence in situ hybridization (FISH)–based technologies in the daily care of these patients. Nonetheless, we reiterate that the entire study flowchart detailed here can be applied to all patients, regardless of age or leukemia type.

Herein we aim to address the importance of FISH-based technologies to characterize the most prevalent cytogenetic abnormalities in the main forms of leukemia. We also will discuss its usefulness in disease diagnosis, classification and prognostication, and therapeutic implications.

CYTOGENETIC ANALYSIS

The possibility to undertake (banding) cytogenetic analysis of malignant cells represented the beginning of one of the greatest advances in understanding the biological nature of neoplasms. This improvement was particularly important in the leukemogenesis field, shedding light on the prognostic implications of chromosomal abnormalities. Regarding the study of genes involved in disease-specific translocations, cytogenetics is considered one of the most powerful tools, leading to a better understanding of chromosomal rearrangements and the mechanisms underlying leukemic transformation. The precise identification of chromosomal breakpoints enabled the cloning of many genes whose interaction with other genes contributes to the neoplastic process.[6]

For more than 30 years, cytogenetic analysis of patients with leukemia, mainly AL, led to the discovery of several recurrent chromosome abnormalities, guiding more efficient treatment regimens. Nonetheless, the prognostic value of rare abnormalities is still unknown. In this regard, further cytogenetic studies gathering data on patients harboring such rare abnormalities are necessary to define prognosis.[7]

Therefore, an adequate experimental design, individualized for each patient, considering the knowledge on FISH-based technical possibilities, can help characterize chromosomal aberrations that in the past would be characterized only as complex/undefined, thus helping with patient's risk stratification.

G-BANDING KARYOTYPING

Refinements of chromosome banding techniques over the last 40 years allowed for improved detection of chromosomal abnormalities in malignant hemopathies. The advent of these techniques brought the first revolution in cytogenetic analysis. Chromosome banding techniques enabled the clear identification of individual chromosomes and the description of chromosomic abnormalities.[7]

Pioneers in this field, Drs. Nowell and Hungerford (1960), described a minute chromosome in BM samples from patients with CML, which was a game-changer for hematology. This observation later proved to be the translocation t(9;22)(q34;q11), generating the *breakpoint cluster region protein (BCR)—Abelson tyrosine-protein kinase 1 (ABL1)* chimeric fusion gene, the product of which was a constitutively activated tyrosine kinase.[8] Another important step in the field of tumor cytogenetics was the description of the reciprocal translocation t(8;21)(q22;q22) by Dr. Rowley (1973). This translocation is a frequent non-random cytogenetic abnormality in AML. Dr. Rowley described the translocation t(8;21) from her observation of repeating pattern of karyotypes analyzed, using banding techniques.[9, 10] Since then, cytogenetic analysis has provided remarkable knowledge about the heterogeneity of a cell population. This approach enabled the detection of several genetic and chromosomal abnormalities in different malignancies, making it possible to classify the disease based on the alteration observed in the patient's karyotype. For such neoplasms, the presence or absence of specific chromosomal aberrations can be a fundamental prognostic indicator.[11–13]

G-banding cytogenetics can reveal chromosomal markers and complex karyo-types. However, it cannot point to the origin of these rearrangements. On the other hand, FISH-based techniques can pinpoint the genes/gene fusions involved in the abnormalities present in the karyotype.[14] In this context, FISH has expanded the capabilities of cytogenetic analysis not only through its higher sensitivity but also, especially because of its quick detection of alterations. A FISH experiment can be processed in 4–24h, and the analysis of 100–200 cells accomplished in 15–45min. This agility/efficiency enables the information about the panorama of karyotype alterations to be achieved within an appropriate period for treatment strategies.[15]

MOLECULAR CYTOGENETICS

In situ hybridization (ISH) is a cytogenetic method used for high-resolution detection and localization of DNA sequences in tissues and cells.[16] The first description of nucleic acid hybridization was made by Gall and Pardue in 1969[17] when DNA dena-turation was performed in an integrated cytological sample from the ovary of the frog *Xenopus* and hybridized with specific tritium-labeled radioactive RNA. John and colleagues (1969)[18] also reported in situ hybridization of DNA from the Hela cell line and *Xenopus* oocytes with labeled RNA. In 1970, Buongiorno-Nardelli and Amaldi[19] created a hybridization method using paraffin-embedded hamster brain and liver tissues.

Immunofluorescence-based protein detection was described in 1941.[20] Until then, only radioactive-based techniques, which took long exposure times and low resolv-ing power, were used. One of the first attempts was the indirect immunofluorescence procedure using an antiserum against a poly(rA)-poly(dT). This approach was used to detect RNA-DNA hybrids of *Drosophila* chromosomes with a rhodamine-conju-gated secondary antibody.[21] In the early 1980s, Bauman and colleagues[22] devel-oped the direct detections of *Trypanosoma* nucleic acids, adenovirus in human tissue cells, and Polytene chromosomes from the salivary gland of *Drosophila*. The latter became possible by using the covalent binding of the fluorochrome rhodamine mol-ecule to the 3′-terminus of the RNA. The year 1981 was a landmark for molecular cytogenetics due to the introduction of nonradioactive labeling methods. The use of fluorescently labeled probes, such as biotin-labeled nucleotides with avidin-based detection methods, was crucial to improving in situ hybridization technologies.[23]

The advent of FISH-based techniques boosted the diagnosis and study of leuke-mia. The application of new multicolor karyotyping approaches allowed the com-plete characterization of complex chromosome rearrangements and the discovery of new abnormalities and putative genes involved in leukemic transformation.[16]

Therefore, the advent of FISH in the late '80s paved the way to the second revo-lution in cytogenetic analysis. Initially, the use of specific chromosome probes led to the exact identification of chromosomes involved in complex rearrangements. Moreover, the Human Genome Project prompted the development of several locus-specific probes. This technology improved gene mapping strategies and breakpoints identification, besides facilitating the delineation of deleted regions associated with specific disease subtypes. Subsequently, the cloning of these translocation breakpoints made possible the generation of probes specific for the detection of these alterations.[16]

Regarding leukemia diagnostics, the applicability of FISH in clinical and laboratory practices includes the hybridization of a fluorochrome-labeled DNA probe on *in situ* chromosomal targets for the confirmation/characterization of one or more cytogenetic abnormalities.[24] Routinely, FISH is performed on metaphase spreads, but unlike the G-banding technique, it does not rely exclusively on metaphases. FISH can be performed in interphase nuclei, monitoring specific chromosomal and gene rearrangements and numerical abnormalities, maintaining reliability in the experiment.[24] On the other hand, FISH requires a prior conception or knowledge of the abnormality to be screened for selecting the most appropriate probe.[16] Therefore, FISH is a powerful tool combined with cytogenetic and molecular studies in the evaluation of chromosomal abnormalities associated with hematological malignancies and other types of cancers.

CHRONIC LEUKEMIA

CHRONIC MYELOID LEUKEMIA

CML is an extensively described malignant disorder of hematopoietic stem cells (HSCs) that accounts for 15–20% of all adult leukemia cases and about 3% in children.[25, 26] This disease is a chronic myeloproliferative neoplasm originating in a defective pluripotent BM stem cell. The hallmark of this disease is the presence of the Philadelphia (Ph) chromosome, translocation t(9;22)(q34;q11.2), generating the *BCR-ABL1* fusion.[8, 27] Pediatric and adult CML share the same molecular features. However, better outcomes have been observed in pediatric transplant patients due to the lack of other chronic diseases or organ dysfunction associated with increased age and lower incidence of graft versus host disease.[26]

CHRONIC LYMPHOCYTIC LEUKEMIA

CLL is characterized by the clonal proliferation and accumulation of mature and typically CD5-positive B-cells in the peripheral blood (PB), BM, lymph nodes, and spleen. The first hit leading to this disease may involve multipotent and self-renewing HSCs.[28, 29] Genomic studies revealed a model of progression of acquired cytogenetic abnormalities in the course of this disease, making the prognosis worse, including resistance to treatments. In early stages, we can observe deletion of the long arm of chromosome 13 in approximately 55% of cases, in addition to trisomy 12 being present in 10–20% of CLL patients. In later stages, we can observe long arm deletions in chromosome 11 in about 10% of the cases, besides the deletion of the short arm of chromosome 17 (5–8%). The deletion del(17p) has been described as a poor prognostic factor since the *TP53* gene is located in this chromosome region.[30, 31]

Chronic Myelomonocytic Leukemia (CMML)

CMML is a rare clonal disorder originating from HSCs. This disease encompasses features of both myelodysplastic syndromes and myeloproliferative neoplasms. CMML affects mostly (90%) male patients over 60 years of age.[5] According to the WHO, the main diagnostic criteria for CMML are PB monocytosis ($\geq 1 \times 10^9$/l), comprising at least 10% of the white blood cell (WBC) count. Besides, (i) dysplasia involving one or more myeloid lineages or (ii) the presence of acquired clonal genetic/cytogenetic abnormalities in hematopoietic cells are present.[5] About 20–30% of patients with CMML present cytogenetic abnormalities at diagnosis. Risk stratification can be categorized according to the abnormalities observed: (i) low risk when presenting a normal karyotype, or Y chromosome loss as an isolated abnormality; (ii) high risk when presenting trisomy 8, chromosome 7 abnormalities, or complex karyotype; (iii) intermediate-risk when low and high-risk features are not observed.[32, 33] FISH is important to exclude typical (*BCR-ABL1*) and atypical CML (*PDGFRA, PDGFRB, FGFR1*) rearrangements, or *PCM1-JAK2* fusions.[5]

ACUTE LEUKEMIA

ALs are a heterogeneous group of diseases that may originate from the interaction between genetic and epigenetic alterations in hematopoietic progenitors, leading to dysregulation of multiple important signal transduction pathways, resulting in hematopoietic insufficiency due to a clonal proliferation of immature precursors.[34] These neoplastic precursor cells originate in the BM and can infiltrate PB and tissues.[35]

ALs are divided into two main groups, myeloid leukemia, when the myeloblast lineage is affected, or lymphoid leukemia when the lymphoblastic lineage is affected. AL groups differ from each other in specific characteristics regarding cytogenetic and genomic abnormalities, morphological and immunophenotypic profile, response to therapy, and the clinical course of the disease. In general, ALs present a rapid fast clinical course and if not properly treated can lead to death in a few months or even weeks.[36, 37] AL is the most common form of childhood cancer, and its incidence is increasing.[38]

ACUTE MYELOID LEUKEMIA

AML is a molecularly heterogeneous group of diseases affecting individuals of all ages. Despite constituting only 20% of pediatric ALs, AML is overtaking ALL as the leading cause of childhood leukemic mortality.[38] Approximately 80% of children with AML present chromosomal alterations in BM and PB blast cells at the time of diagnosis.[39, 40] This frequency is considerably higher compared to adult patients affected by the disease, in which these alterations are detected in approximately 60% of the cases.[41]

About 80% of children with AML present genomic structural alterations associated with the disease and approximately 76% of patients presenting normal karyotype have at least one known mutation (*FLT3, WT1, NPM1*). Thus, currently, about

90% of pediatric patients with AML have at least one recurrent type I and type II mutation.[42]

In this sense, molecular cytogenetic analyses have conferred substantial evidence with regards to the chromosomal architectures in AML and other cancers. Most importantly, the FISH technique that plays a leading role in diagnostic pathology for its single-cell analysis has provided crucial information regarding the heterogeneity of AML malignant cells.[14]

In AML, cytogenetic alterations of numerical and structural types can be found. Numerical alterations include monosomies and nullisomies, which result in a karyotype with less than 46 chromosomes. Trisomies and tetrasomies, among others, will generate a hyperdiploid karyotype with more than 46 chromosomes. Structural abnormalities are mainly translocations, insertions, and inversions.[40]

The presence of specific cytogenetic abnormalities has important prognostic implications for the disease. Thus, karyotypic alterations in leukemic blasts are considered one of the most significant markers that define the biology of AML. In addition to karyotypic changes, gene mutations with clinical relevance associated with the disease are also part of the classification scheme for AML.[5, 39, 40, 43]

Favorable Prognosis AML

Cytogenetic markers with favorable prognoses in AML are listed in the following.

- t(8;21)(q22;q22): The reciprocal translocation t(8;21)(q22;q22) results in the fusion of the *RUNX1* gene on 21q22 and *RUNX1T1* on 8q22. This translocation presents the highest incidence in childhood AML, occurring in approximately 12% of the cases.[44–46] In adults, the translocation t(8;21) can be found in about 5% of patients with AML.[41, 46] As this abnormality is associated with a favorable prognosis, its identification is highly significant for diagnosis and therapy.[40, 47]
- inv(16)(p13q22): Inversion inv(16)(p13q22) encompasses 3–8% of pediatric cases, and about 4% of adults with AML, resulting in the fusion of the core binding factor b (*CBFb*) gene at region 16q22 with the smooth muscle myosin heavy chain (*MYH11*) at 16p13.[40, 46] Additional chromosomal abnormalities such as +22, +8, and 7q alterations are seen in association with this AML subtype.[40, 46, 48]
- t(15;17)(q22;q21): Acute promyelocytic leukemia (APL) accounting for about 2–10% of pediatric AMLs (7–8% in adults) is usually related with a good prognosis. However, in about 20% of patients whose outcome is not favorable, ~10% pass away due to early complications, and the other 10% experiences disease relapse or drug resistance. These include the APL variants translocation t(11;17)(q23;q21)/*PLZF-RAR*, translocation t(5;17)(q35;q21)/*NPM-RAR*, t(11;17)(q13;q21)/*NUMA-RAR*, and derivative der(17) that generates the *STAT5b-RAR*. Prompt diagnosis, including FISH for a fast initiation of treatment, is critical to lowering the risk of early mortality.[40, 41, 46] Although the hallmark of APL is the *PML-RARA* fusion, generated from translocation t(15;17)(q24;q21), it is common to observe

additional chromosomal abnormalities, with a clinical impact not yet established.[49, 50]

- t(9;11)(p22;q23): The translocation t(9;11)(p22;q23) has been reported in approximately 2% of adult cases and about 6% of pediatric AML.[41, 46] This abnormality is the most commonly observed in infants younger than 12 months old. At the molecular level, the translocation t(9;11) is defined by the involvement of the *KMT2A* gene at 11q23 with the *MLLT3* at 9p22. This aberration presents controversial prognostic implications. However, children and infants with t(9;11) tend to have a more favorable prognosis.[11] Still, karyotypes carrying *KMT2A* gene rearrangements (*KMT2A*-r) are more frequent in infants, presenting an intermediate prognosis or being associated with an adverse risk in AML.[51, 52]

Unfavorable Prognosis AML

Cytogenetic markers with unfavorable prognoses in AML are listed in the following.

- t(9;22)(q34;q11): The reciprocal translocation t(9;22)(q34;q11)/*BCR-ABL1* is a rare event in AML, accounting for 1–2% of total cases.[8, 53] The most common fusion protein in acute leukemia is *BCR-ABL1* p190, but *BCR-ABL1* p210 can also occur, conferring an adverse prognosis. Due to this rarity, more data is still necessary on the characteristics of this AML subtype.[8, 54]
- t(6;9)(p23;q34): Translocation t(6;9)(p23;q34) is often associated with an intermediate or poor outcome. Besides, due to its rarity (1–2%) and association with type I mutations (i.e. *FLT3*/ITD), the prognostication of translocation t(6;9) has not been fully defined in children.[41, 55] In adults, it is even rarer, accounting for <1%.[41, 46] The translocation t(6;9) results in the *DEK-NUP214* chimeric fusion gene, making it possible to be monitored by FISH.[56]
- t(6;11)(q27;q23): Translocation t(6;11)(q27;q23) is observed in about 5% of ALs with *KMT2A-r* and can be found in ALL and AML. It is more frequent in AML (7.8% of AML; 1.8% of ALL). The translocation t(6;11) is rare in infants (0.3%) and more frequent in children (6.6%) and adults (6%).[51] This abnormality can be very subtle, escaping detection by G-banding cytogenetics, and has been associated with a poor prognosis. Additional chromosome abnormalities have been reported in 10% of the cases. Molecularly, the translocation t(6;11)(q27;q23) is characterized by the fusion between the *KMT2A* and the *AFDN* gene.[57, 58]
- t(11;19)(q23; p13.3): The subtle translocation t(11;19)(q23; p13.3), which results in the *KMT2A-MLLT1* gene fusion, is a rare abnormality in AML, accounting for less than 1% of cases.[51, 59] It is important to highlight that this abnormality has a frequency of 4% in infants with AML. The translocation t(11;19) is related to an unfavorable prognosis in AML.[11]
- t(10;11)(p12;q23): A *KMT2A* gene fusion with *MLLT10* at 10p12 is frequently observed in childhood AML as translocation t(10;11)(p12;q23).[58] The translocation t(10;11) is difficult to detect by G-banding since rearrangements

involving *KMT2A* and *MLLT10* often result in cryptic/complex rearrangements.[60, 61] This abnormality is defined as an independent prognostic predictor, so patients with translocation t(10;11)/*KMT2A-MLLT10* should be referred for a high-risk AML treatment protocol.[58]

* t(1;22)(p13;q13): The translocation t(1;22)(p12;q13) juxtaposes the *RBM15* gene located at 1p13 to the *MLK1* at 22q13, generating the *RBM15-MLK1* fusion on derivative chromosome 22.[40, 62] It is a rare translocation in childhood AML (~3%), being practically restricted to acute megakaryoblastic leukemia (AMKL). In this regard, literature data suggest that prenatal genetic factors are involved in the leukemogenesis of this disease. The prognostication is generally poor.[40, 63]

* Chromosome 3 abnormalities: Abnormalities involving the long arm of chromosome 3, including inversion inv(3)(q21q26), translocation t(3;3) (q21;q26), translocation t(3;21), 3q+ and 3q- are considered rare (<1%) and of high risk in AML.[46, 55] Alterations in these chromosomal regions are usually associated with AMKL.[40] Identification of rearrangements in the long arm of chromosome 3 is highly important, especially concerning 3q26.2–26.31, as they can result in rearrangement and overexpression of the *MECOM* gene, conferring a poor prognosis for the patient.[64–66]

ACUTE LYMPHOBLASTIC LEUKEMIA

ALL accounts for 75% of ALs and 25% of all cancers in this age group, with a peak of incidence at 2 to 5 years of age. ALL present a significant heterogeneity at the genetic level but homogeneous morphologic and immunophenotypic aspects. These genetic lesions define disease subsets biologically distinct as well as are used for risk stratification once they lead to different responses to therapy.[66, 67]

In addition, ALL presents distinct subtypes with therapeutic implications, which can be accessed by flow cytometry that enables distinguishing ALL in B-cell precursor ALL, mature (Burkitt) ALL, and T-cell ALL. Cytogenetics, as well as molecular genetic findings at ALL diagnosis represent important prognostic factors in ALL. Therefore, these alterations are mandatory for analyzing outcomes in different clinical trials and detecting specific chromosomal aberrations and their equivalents.[68, 69] Current risk-stratified regiments in childhood ALL have led to long-term, event-free survival rates of more than 80% and approximately 90% of ultimately cured patients.[69]

ALL cytogenetic diagnostics can be impaired by suboptimal cytogenetic preparations most of the time. Poor chromosome morphology and indistinct banding can make the interpretation of karyotypes challenging. In this matter, molecular cytogenetic analysis is an important tool to classify and thus rescue patients in which G-banding is not informative.[70]

Both B and T-cell ALL are associated with characteristic and recurrent cytogenetic alterations. B-cell lineage comprises the majority of ALL cases (80–85%), with well-established cytogenetic subgroups. Hyperdiploidy and translocation t(12;21) confer a favorable prognosis, whereas translocation t(9;22), *KMT2A*-r, intrachromosomal amplification of chromosome 21 (iAMP21), and translocation t(9;22) or

Philadelphia chromosome-like ALL (Ph-like ALL) are known predictors of poor outcomes.[70]

Fifteen percent of childhood ALLs are T-cell lineage, where most commonly T-cell receptor genes (*TCR*) either alpha/delta (14q11.2) or beta (7q34) are involved in translocations with transcription factors that are upregulated by the *TCR* enhancer regions. Risk stratification for T-ALL treatment is not currently related to chromosomal abnormalities. However, there are prognostic implications and potential therapeutic options for some of the recurrent abnormalities.[70]

Favorable/Intermediate Prognosis ALL

Cytogenetic markers with favorable/intermediate prognoses in ALL are listed in the following.

- t(8;14)(q24;q32) and variants: Translocation t(8;14)(q24;q32) juxtaposes *MYC* (8q24) with heavy chain locus (14q32) or light chain *loci* (2p12 and 22q11), leading to *MYC* overexpression, which is believed to play a central role in Burkitt Lymphoma (BL) pathogenesis.[71] Because of the high similarity among mature B cell ALL (FAB L3 subtype) (less than 2% of ALL cases) and BL cells, the World Health Organization (WHO) classification has included this subtype as equivalent to BL in the leukemic phase, under the common denomination of Burkitt lymphoma/leukemia (BL/L).[72, 73]

    Mature B-cell ALL fare poorly with conventional ALL therapy, once it is a high-grade disease, with a duplicating time of 24 hours. Although, it presents an 80% event-free survival rate when treated with three to six months of rapid-sequence, intensive chemotherapy.[74, 75]

- Hyperdiploidy: Hyperdiploidy is the most common recurrent abnormality in childhood B-ALL, accounting for 30% of the patients. This group predominantly has a chromosome count of 51–65, with a nonrandom pattern of gain, leading to one additional copy of chromosomes X, 4, 6, 10, 14, 17, 18, and two additional copies of chromosome 21 as the most frequently observed alterations in hyperdiploid clones. High hyperdiploidy with specific gains of chromosomes 4, 10, 17, and 18 has been associated with an excellent prognosis in pediatric B-ALL. Chromosome counts of 57–60 chromosomes typically include trisomies of chromosomes 5, 8, 11, and 12, while karyotypes with 63–67 chromosomes frequently also have gains of chromosomes 2, 3, 9, 16, and 22.[70, 76] Structural abnormalities associated with hyperdiploid karyotypes generally do not alter favorable prognosis.[70]

- t(1;19)(q23;p13): Translocation t(1;19)(q23;p13) fuses the *TCF3* (*E2A*) gene on band 1q23 to *PBX1* on 19p13.3. It is reported in 5% to 6% pediatric ALL and 1% to 3% adult ALL, and can be either balanced or unbalanced, as der(19)t(1;19) with two normal chromosomes 1. Most patients have pseudodiploid karyotypes, and almost all are diagnosed with pre-B-ALL.[41] Despite being previously considered high risk because of central nervous system involvement and relapse, current intensive protocols lead this rearrangement to be used as a favorable/intermediate-risk mark nowadays.[76]

Unfavorable Prognosis ALL

Cytogenetic markers with unfavorable prognoses in ALL are listed in the following.

- t(9;22)(q34;q11): Rarely seen in children (1–3%), translocation t(9;22) (q34;q11.2) is more common in adults B and T-ALL (11–29%) and leads to *BCR-ABL1* fusion or Ph-chromosome. With rare exceptions, Ph-patients are diagnosed with B-ALL. These patients generally present a poor prognosis, with the only curative therapy strategy being allogeneic hematopoietic stem-cell transplantation (HSCT), although this procedure is associated with increased treatment-related mortality.[70]

- t(v;11q23): Rearrangements involving band 11q23 (*KMT2A* gene) are detected in two-thirds of infants with ALL, representing 80% of this age group. In older children, the incidence is much lower (1–2%),[69, 77] and in adults it is up to 7%.[41] It has already been demonstrated that *KMT2A* may be partner of more than 80 different genes, which distinguish distinct subtypes of leukemia with both lymphoid or myeloid features and poor outcomes.[76] Among 11q23 rearrangements, the most expressively found in ALL are translocation t(4;11)(q21;q23)/*MLL-AFF1(AF4)*, detected in more than 50% of patients, followed by translocation t(11;19)(q23;p13.3)/*MLL-MLLT1(ENL)* and, less commonly, translocation t(9;11)(p22;q23)/*MLL-MLLT3*(AF9), translocation t(10;11)(p13–15;q14–21)/*MLL-MLLT10*(AF10), and others.[77]

- t(4;11)(q21;q23): Most common among ALL *KMT2A*-r is translocation t(4;11) (q21;q23), where *KMT2A* (11q23) gene partners *AF4* (4q21). *KMT2A-AF4* fusion is particularly common in infant ALL and represents 2% of pediatric and 3–7% of adult ALL. The translocation t(4;11) is of general interest once it predicts a poor prognosis.[41]

- Hypodiploidy: Hypodiploid karyotypes present an overall frequency of 5% in childhood ALL. These cases are characterized by a chromosome count of <44 (43 or fewer) chromosomes.[69] There are three well-defined subgroups of hypodiploid to note: near haploidy (24–31 chromosomes), low hypodiploid (32–39 chromosomes), which are associated with an adverse prognosis, and high hypodiploid (40–43 chromosomes). In the near-haploid cases, the chromosomes that usually are kept in two copies are chromosomes 14, 18, 21, and the sex chromosomes. In the low hypodiploid karyotypes, the most frequent monosomy chromosomes are 3, 4, 7, 13, 15, 16, and 17.[70]

- iAMP21: Intrachromosomal amplification of chromosome 21 (iAMP21) was first discovered during a routine screening for *ETV6-RUNX1*. Using the FISH approach, multiple copies of the *RUNX1* gene were observed.[67] This abnormality is characterized by the gain of three or more extra *RUNX1* (21q22) copies and is mostly seen in older children. iAMP21 is usually seen as a sole cytogenetic alteration, occurring in 2% of childhood ALL and associated with older ages at diagnosis,[69, 78] Patients with iAMP21 show a poor outcome with a high rate of relapse when treated under standard risk protocols. However, intensive therapy can greatly improve the outcome.[76, 78]

Uncertain Prognosis ALL

- t(12;21)(p12;q22): Translocation t(12;21)(p12;q22) fuses the *ETV6* gene (12p13) with *RUNX1* gene (21q22) and is observed in up to 25% of childhood pre-B ALL and less than 3% in adult ALL.[79] The translocation t(12;21) is cryptic, highlighting the importance of ISH tools for its detection. Approximately 75% of translocation t(12;21) patients harbor additional genetic changes, most often deletions of 12p with loss of the second copy of *ETV6* gene (55–70% of patients), +21 (15–20%) and an extra der(21) t(12;21) (10–15%).[80] Due to excellent molecular response to treatment and beneficial clinical outcome, it was believed that t(12;21) conferred favorable prognostic, but it was subsequently disrupted as some series found predominantly late relapses occurring in up to 20% of patients.[76, 81]

RECURRENT NUMERICAL ABNORMALITIES IN ACUTE LEUKEMIA

Chromosome 5/del(5q) monosomy is present in <1–2.5% of children with AML, while in adults the percentage is higher, reaching 7% of cases. In ALL, del(5q) is rarer, being observed in 1% of pediatric patients and <2% of adults. This abnormality is associated with a poor prognosis.[41, 55, 82]

Abnormalities of the long arm of chromosome 6 (6q) are seen in about 11% of children with ALL and 3–16% of adults.[41, 83, 84] This subtle alteration has been related to a poor prognosis, since a deletion in this region may lead to the loss of tumor suppressor genes, whose absence may contribute to the genesis of both B-ALL and T-ALL.[11, 84]

Aberrations involving chromosome 7 can occur as a sole abnormality, such as −7/del(7q), or in balanced translocations 7q22/7q32-q35 in AML.[85, 86] Such aberrations are present in 2–7% of pediatric cases and 2.5–8.5% in adults with AML. In ALL, abnormalities involving chromosome 7 have a frequency of 4% in children and 6–11% in adults. For both groups, these changes are more often associated with a poor outcome.[41, 82, 87]

Trisomy of chromosome 8 is seen in 5–10% of pediatric patients with AML and about 9% of adults. It is a frequent alteration in patients with AML and Down syndrome, affecting 27% of cases.[41, 83, 87, 88] In ALL, a higher frequency is reported in adults compared to children, with 10–12% vs. 2%, respectively. This alteration can be observed (i) as an isolated abnormality and thus is commonly associated with a favorable prognosis, and (ii) as a secondary cytogenetic alteration, which may confer a more aggressive prognosis.[83, 89, 90]

Cytogenetic abnormalities in the short arm of chromosome 9 (9p) are present in about 7–9% of cases of childhood ALL.[83] Most of these abnormalities involve the deletion of cell cycle regulatory genes and tumor suppressor genes (*CDKN2*, *CDKN2B*).[11, 91] Del(9q) is a recurrent alteration in AML being present in about 2–4% of children and 2% of adults. They are also usually associated with a lower clinical outcome.[88, 92, 93]

Complex Karyotypes

A complex karyotype (CC) is defined as any karyotype with at least three chromosomal aberrations, regardless of the type of alterations and the chromosomes involved.[94–96] The frequency of CCs increases with the age of the patients.[41, 55, 60, 97] Presence of CCs in the BM blast cells represents a possible indicator of a poor prognosis, which may negatively influence the patient's clinical evolution. The literature has also shown that, in most cases, the disease has been resistant to the initial treatment, with early relapse of patients.[97–99] CCs have been associated with an unfavorable clinical evolution, and it is important to emphasize that these karyotypes can carry, among their alterations, in a cryptic way, chromosomal abnormalities that can be responsible for an adverse prognosis.[61, 100] Since the literature suggests a possible correlation between the combination of aberrations detected in a CC and the clinical response, the importance of refinement in the characterization of this karyotypic profile becomes clear.

Flexibility in the Application of FISH Probes for Leukemia Diagnosis

FISH-based techniques revealed the involvement of several genes and allowed the description of specific abnormalities, uncovering several therapeutic targets related to a specific type of disease.[101, 102] Consequently, treatment has improved with higher cure levels for leukemia patients. Having prior knowledge about the disease and the probes available for testing can provide a vast hall of possibilities to properly and quickly evaluate strategies for risk-adapted therapy.[101, 102]

In this context, the *ETV6-RUNX1* FISH probe, designed for the detection of the cryptic translocation t(12;21)(p13;q22), can be used to detect other cytogenetic abnormalities such as trisomies typical of hyperdiploidy. This probe can also aid in the detection of other molecular alterations involving chromosomes derived from variants of translocation t(12;21)(p13;q22) and *ETV6-RUNX1* gene fusion itself, which have been reported as indicators of different clinical evolutions within this cytogenetic group.[103] Furthermore, in cases where mitoses cannot be obtained and G-banding analysis cannot be performed, this probe allows detection of the iAMP21, a rare alteration of poor prognosis.[104]

Likewise, as the *EVT6-RUNX1* probe can also be used to help detect and characterize other alterations involving chromosomes 12 and 21, the *RUNX1-RUNX1T1* probe can also be used to help uncover abnormalities in chromosome 21, in addition to trisomy 8, a recurrent aberration of important prognosis in leukemia.[83, 90] Hence, the use of gene/gene fusion probes specific to certain diseases and abnormalities can be made flexible to assist in the elucidation of other disease-related alterations. In this context, the application of this knowledge in daily practice can widely contribute to research and diagnostic laboratories.

SAMPLES/TISSUES

For (molecular) cytogenetic analysis, PB and BM samples are processed. The samples are collected in test tubes with anticoagulant heparin and referred to the laboratory by the responsible physicians.

To assist in the risk stratification of patients and in the experimental design based on FISH panels, clinical diagnoses of cytochemistry, immunophenotyping, and morphology are obtained through the study of medical records, slide review, and patient files provided by the responsible physicians in the respective institutions of origin. The clinical-laboratory data of the diagnosis, as well as the clinical information of the follow-up of the patients, are also obtained through the study of the medical records in the institutions of origin.

DESCRIPTION OF METHODS

CELL CULTURE FOR CHROMOSOME PREPARATION

Equipment/Materials
- Class II biological safety cabinet
- Sterile 15ml conic tubes
- Centrifuge
- Sterile transfer pipettes
- Micropipette 50µl (p20–200–1000)
- Pipette tips 50µl (p20–200–1000)
- Neubauer chamber
- Coverslip 24×24mm
- Optic microscope
- Cell culture flasks
- Humidified 37°C, 5% CO_2 incubator

Reagents
- Ready to use growth medium: RPMI 1640 medium at 37°C
- Fetal bovine serum, previously inactivated in a water bath at 56°C for 40 minutes, 12ml of penicillin & streptomycin
- L-glutamine (add in RPMI 1640 medium—1:100 proportion)
- Thomas buffer solution: Dilute 10ml of acetic acid and 10 drops of Giemsa stain in 1l of distilled water. Store in a dark bottle.[105]

Workflow
1. Into the biological safety cabin, transfer the sample (BM/PB) for a 15ml conic tube;
2. Dilute 50µl of the sample in 1ml Thomas buffer solution using a micropipette;
3. Apply this solution to the surface of the Neubauer chamber and place the coverslip;
4. Use the optic microscope (20×) to count the cells in the four chamber's quadrants;
5. The number of cells is obtained by adding the number of counted cells, multiplying the result for 21 and afterwards for 2,500;
6. Culture is established using 80% growth medium, 20% fetal bovine serum, and 5×10^6 cells of the sample in each cell culture flask;

7. Cell culture flasks should stay in the humidified 37°C incubator for 24 hours.

Cell Extraction Protocol

Equipment/Materials

- Syringe and needle
- Disposable Pasteur pipettes
- Benchtop centrifuge
- Laboratory water bath 37°C

Reagents

- Colcemid solution 10μg/ml
- KCl (potassium chloride): Add 0.6g of KCl to 100ml of distilled water. Mix by swirling until dissolved.
- Carnoy's solution/fixative solution (acetic acid and methanol solution—1:3 proportion); combine three parts of methanol to one part of glacial acetic acid (v/v). Use freshly. Store at room temperature (RT).[106]

Workflow

1. After 23 hours of incubation, 50μl of Colcemid solution should be added to each of the cell culture flasks;
2. Return the flasks to the humidified 37°C incubator;
3. One hour later (after completing 24 hours total), transfer the cell culture in the flasks to 15ml conic tubes. The contents in each flask shall be transferred to a different conic tube;
4. Centrifuge the material for 6 minutes at 1,500 rpm;
5. Remove the supernatant from the tubes and homogenize the pellet. To perform the hypotonic shock, add, with the aid of a disposable Pasteur pipette, the KCl solution until completing 8 ml;
6. Accommodate the tubes in a water bath at 37°C for 20 minutes;
7. Centrifuge for 6 minutes at 1,500rpm;
8. Remove the supernatant from the tubes, homogenize the pellet, and slowly add the Carnoy's solution until completing 8ml;
9. Leave the tubes for 20 minutes at RT;
10. Centrifuge for 6 minutes at 1,500rpm;
11. Remove the supernatant, homogenize the pellet, unite the contents of all the tubes in a single one and add Carnoy's solution until completing 8ml;
12. Repeat steps 7 and 8 until the material is clear;
13. Store the material at 4°C.

Slides Preparation Protocol

Equipment/Materials

- Microscope slides

- Disposable Pasteur pipettes
- Glass Coplin jar
- Bunsen burner

Reagents

- Ethanol 70%
- Giemsa solution (Merck): Dilute 2.5ml of Giemsa stain and 40ml of phosphate buffer.[107, 108]
- Phosphate buffer: Dissolve 6.808g of KH_2PO_4 and 0.882g of NaOH in 1l of distilled water. Adjust pH to 6.8.

Workflow

1. Centrifuge the tubes containing the chromosome preparation for 6minutes at 1,500 rpm;
2. Remove the supernatant (until it is twice as large as the pellet) and homogenize the content left in the tube;
3. Clean the slides using ethanol 70%;
4. Puff on the cleaned slides;
5. Add three drops from the material onto the slides. Before adding material, keep the pipette at arm's length from the slides for better chromosome dispersion;
6. Quickly flame the opposite part of the slides over medium heat in a Bunsen burner;
7. Add the slides in a Coplin jar with Giemsa solution for 12 minutes;
8. Observe the slides under an optical microscope (in phase contrast) to check for the presence of metaphases;
9. After confirming the presence of metaphases, follow steps 4–6 again to each new slide, and then store the slides in a microscope slide case for aging.

G-BANDING PROTOCOL

Equipment/Materials

- Spatula
- Precision balance
- Beaker 100ml
- Laboratory water bath
- Disposable Pasteur pipettes
- Glass Coplin jar

Reagents

- Dulbecco's Phosphate Buffered Saline (PBS): Dilute 8.0 g of NaCl, 0.2 g of KCl, 0.2 g of KH_2PO_4, and 1.44g of $Na_2HPO_4H_2O$ in 1l of distilled water. Adjust pH to 7.8.
- Trypsin solution: Dilute 0.1g of Trypsin from porcine pancreas in 100ml of Dulbecco's PBS

- Phosphate buffer: Dissolve 6.808g of KH_2PO_4 and 0.882g of NaOH in 1l of distilled water. Adjust pH to 6.8.
- Giemsa solution (Merck): Dilute approximately 2.5ml of Giemsa stain and 40ml of phosphate buffer.[107, 108]

Workflow

1. Heat the trypsin solution (0.1g of trypsin and 100ml of Dulbecco PBS) in the water bath for 30 minutes;
2. Dip the slide in the trypsin solution for 1–10 seconds;
3. Wash the slide in saline solution;
4. Stain the slide in Giemsa solution for 15 minutes;
5. Wash the slide in saline solution and let it dry at RT;
6. Under the microscope, observe whether the time of exposure to trypsin was sufficient to generate the chromosomal banding pattern; if the chromosomes are digested too much, decrease the exposure time. If the time is not enough for the chromosomes to be banded, increase the time until the desired result;
7. Analyze at least 20 metaphases to confirm the cytogenetic diagnosis.

Fluorescence In Situ Hybridization (FISH) Protocol

Pepsin Treatment

Equipment/Materials

- Glass Coplin jar
- Polypropylene Coplin jar
- Glass microscope slides
- Sterile Pasteur pipettes
- Laboratory water bath
- Thermometer

Reagents

- Ethanol 70%, 85%, and 100%; prepare v/v dilutions of 100% ethanol with distilled water to make stock solutions of 70% and 85% ethanol. Between uses, store covered at ambient temperature. Discard stock solutions after six months.
- 2×SSC (pH 7.0): Dilute stock solution (20×SSC) in distilled water in the proportion 1:10. Adjust pH for 7. 20×SSC (stock solution): Dilute 175.34g of NaCl and 88.24g of $C_6H5Na_3O_7.2H_2O$ in 500ml of distilled water on shaker. Add distilled water to bring final volume of solution to 1l. Adjust pH for 5.3.
- Pepsin 0.005%, HCl 0.001M: Dilute 400µl pepsin 0.5% in 39.6ml HCl 0.01M; pepsin 0.5% HCl 0.001M (Stock solution): Dilute 0.5 g pepsin in 100ml HCl 0.01M. Keep in refrigerator.
- 1×PBS: Dilute stock solution (10×PBS) in distilled water the proportion 1:10. Adjust pH for 7.2. 10×PBS (stock solution): Dilute 80g of NaCl, 2g

KCl, 14.4 g of Na_2HPO_4, and 2.4g of KH_2PO_4 in 800ml of distilled water. Add distilled water to bring final volume of solution to 1l.
- Formaldehyde 1%/1×PBS: Dilute 1.1ml formaldehyde P.A. (37%) in 38.9ml of 1×PBS. Mix well.

Workflow

1. Drop the cell suspension in slides washed in ethanol 70% and leave for drying in a wet plate for 2–3 minutes. It is recommended to observe the dropped slides in the optic microscope to pre-select target areas to apply the probes;
2. Incubate the slides in a 2×SSC solution (pH 7.0) at 37°C for 20 minutes;
3. Incubate the slides in pepsin solution (0.005%, 0.001M HCl) at 37°C for 10 minutes;
4. Wash the slides for 3 minutes in 1×PBS at RT;
5. Incubate the slides in a formaldehyde solution (1% 1×PBS) at RT for 10 minutes;
6. Wash the slides in 1×PBS at RT for 3 minutes;
7. Dehydrate slides in a series of ethanol (70%, 85%, and 100%; 2 minutes each);
8. Leave the slides for drying at RT.

FISH Slides Denaturation and Hybridization

Equipment/Materials
- Micropipette 10µl
- Pipette tips
- Eppendorf tube
- Coverslip 20×20mm
- Rubber cement glue
- Laboratory dry bath/dry block
- Thermometer
- Microscope slides case
- CO_2 incubator

Reagents
- Pre-selected FISH probe
- Hybridization buffer
- MiliQ (purified) water

Workflow

Slides Denaturation

1. Add 8µl of the probe hybridization mix (according to the manufacturer's instructions) in the target slides' areas and cover with the coverslip 20×20mm;
2. Seal the edges of the coverslips with the rubber cement glue;
3. Accommodate the slides in the dry bath at 75°C for 7 minutes.

Hybridization

1. Incubate the slides into a humidified microscope slides case (keep the case closed to simulate a dark chamber environment). Leave the case in a CO_2 incubator at 37°C overnight (12–16 hours).[107, 109]

FISH Slides Post Hybridization
Equipment/Materials

- Clamp
- Micropipette 20μl
- Pipette tips
- Polypropylene Coplin jar
- Laboratory water bath
- Thermometer
- Coverslips 24×32mm
- Dark chamber

Reagents

- 0.4×SSC/3%Tween20 solution (pH=7.0); Dilute 10ml of 20×SSC solution in distilled water in the proportion 1:10. Adjust pH for 7; keep in refrigerator; finally dilute 120μl Tween20 in 40ml 0.4×SSC.
- 2×SSC/0.1%Tween20 solution (pH=7.0): Dilute 40μl Tween20 in 40ml 2×SSC.[109]
- DAPI antifade staining

Workflow

1. Remove the coverslips and wash the slides for 2 minutes in 0.4×SSC, 3% Tween20 solution (pH=7.0) at 72°C;
2. Transfer the slides to a 2×SSC 0.1% Tween20 solution (pH=7.0) for 1 minute at RT;
3. Let the slides dry at RT;
4. Apply 15μl antifade staining e.g. DAPI (Vysis) over the hybridization area and cover with a coverslip 24×32mm;
5. Incubate the slides into a humidified microscope slides case (keep the case closed to simulate a dark chamber environment) for at least 10 minutes;
6. Observe the slides in a fluorescent microscope to detect the FISH signals.

YIELDS

FISH PANEL AND PROBES IN LEUKEMIA DIAGNOSTICS

The WHO recommends an integrated approach for ALs cytogenetic diagnosis. Chromosomal banding cytogenetic analysis and FISH-based techniques should be used in the initial diagnosis and monitoring of clonal abnormalities.[5, 101, 102]

The application of a FISH panel enables a quick assessment of the karyotype and the pattern of abnormalities present in it, including its percentage in BM blast cells.

In this sense, for hematological emergencies such as APL, a rapid FISH result is mandatory for the administration of all-trans retinoic acid (ATRA).[101, 110] For cryptic rearrangements that escape detection by chromosome banding, such as translocation t(12;21)(p13;q22)/*ETV6-RUNX1*, translocation t(10;11)(p12;q23)/*KMT2A-MLLT10*, translocation t(6;11)(q27;q23)/*KMT2A-AFDN*, and abnormalities in the 17p13 (*TP53*) region, FISH can be considered as a stand-alone diagnostic assay.[42, 101, 103] The FISH probes and disease-specific panels most commonly used in clinical cytogenetic laboratories are listed in Table 7.1.

TABLE 7.1
FISH Panel and Probes in Leukemia Diagnostics

Cytogenetic Abnormality	Gene Fusion	Prognosis	Disease Types	Probe Types
t(8;21)(q22;q22)	*RUNX1-RUNX1T1*	favorable	AML	DCDF
inv(16)(p13;q22)	*CBFb-MYH11*	favorable	AML	DCDF
t(15;17) (q22;q21)	*PML-RARA*	favorable	AML	DCDF, DCSF
t(9;11)(p22;q23)	*KMT2A-MLLT3*	favorable in AML patients without additional abnormalities or FAB M5 subtype)	ALL, AML	DCDF
11q23	*KMT2A*-r	intermediate or poor in AML, Unfavorable in ALL	ALL, AML	DCBAP
del(5q)	*EGR1*(5q31)	unfavorable	AML	DCE
t(6;9)(p23;q34)	*DEK-NUP214*	often unfavorable	AML	DCDF
t(6;11)(q27;q23)	*KMT2A-AFDN*	unfavorable in AML	ALL, AML	DCDF
t(11;19) (q23;p13.3)	*KMT2A-MLLT1*	unfavorable in AML	ALL, AML	DCDF
t(10;11)(p12;q23)	*KMT2A-MLLT10*	unfavorable in AML	ALL, AML	DCDF
t(1;22)(p13;q13)	*RBM15-MKL1*	unfavorable	AMKL	DCDF
inv(3)	*MECOM* (3q26)	unfavorable	AML	DCBAP
7q	(*RELN/ORC5*) 7q22.1-q22.2/ *TES*(7q31.2)/*TCRB*(7q34)	unfavorable	AML, T-ALL	DCE, DCBAP
t(9;22)(q43;q11)	*BCR-ABL1*	unfavorable	B-ALL, T-ALL, AML, CML	DCDF
t(12;21) (p13;q22)	*ETV6-RUNX1*	favorable	B-ALL	DCDF
+4/+10/+17	*D4Z1*(4cen),*D10Z1* (10cen),*D17Z1*(17cen)	favorable	B-ALL	TCE

(Continued)

TABLE 7.1 (Continued)
FISH Panel and Probes in Leukemia Diagnostics

Cytogenetic Abnormality	Gene Fusion	Prognosis	Disease Types	Probe Types
t(11;14) (p13;q11)	TCRA/D(14q11)	favorable	T-ALL	DCBAP
t(1;19)(q23;p13)	PBX1-E2A	controversial, improved with therapy	B-ALL	DCDF
t(4;11)(q21;q23)	KMT2A-AFF1	unfavorable	ALL	DCDF
iAMP21	RUNX1(21q22)	unfavorable	B-ALL	DCDF, DCE
6q	MYB(6q23),D6Z1(6cen)	unfavorable	ALL	DCE
del(9p)	CDKN2A(9p21), D9Z3 (9 cen), PAX5(9p13.2)	unfavorable	Pre-B-ALL, T-ALL	DCE, DCBAP
t(8;14)(q24;q32)	cMYC(8q24), IGH(14q32)	favorable	Burkitt lymphoma/ leukemia (BL/L)	DCBAP, DCDF, DCE
del(13q)/+12	DLEU1(13q14), D13S25 (13q34), D12Z3 (12 cen)	favorable	CLL	TCE
del(11q)/ del(17p)	ATM(11q22), TP53(17p13)	unfavorable	CLL	DCE

ALL—acute lymphoblastic leukemia; AML—acute myeloid leukemia; CLL—chronic lymphocytic leukemia; CML—chronic myeloid leukemia; DCDF—dual-color, dual fusion; DCSF—dual-color, single fusion; DCBAP—dual-color, break-apart; DCE—dual-color enumerate; TCE—tricolor enumerate; r—rearrangement; FAB—French-American-British Classification; cen—centromere; del—deletion; inv—inversion; t—translocation; p—short chromosome arm; q—long chromosome arm; (+)—trisomy.

FISH in Favorable Prognosis Leukemia

RUNX1(21q22)-RUNX1T1(8q22) Probe

The test shown in Figure 7.1A is important to reveal the presence of the *RUNX1-RUNX1T1* cryptic fusion on the derivative chromosome 8. In addition, it is possible to observe the presence of a *RUNX1T1* split signal on the derivative chromosome 13, which is a rare variant with only a few cases described in the literature.[47, 111]

CBFb(16q22)-MYH11(16p13) Probe

Application of FISH-probe as shown in Figure 7.1B is important to stratify the patient's risk since inversion inv(16) is related to a good prognosis, with well-defined treatment protocol guidelines.[5, 40, 68]

PML(15q24)-RARA(17q21) Probe

FISH for rapid detection of *PML-RARA* fusion (Figure 7.1C) is mandatory in APL characterization since a prompt diagnosis and initiation of treatment are critical to lowering the risk of early mortality.[110]

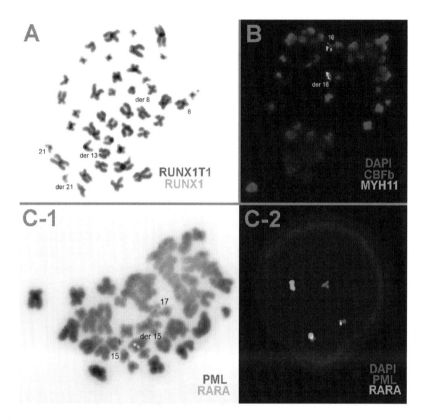

FIGURE 7.1   A) FISH result for the translocation t(8;21)(q22;q22)—*RUNX1* (green signal), *RUNX1T1* (red signal), and *RUNX1-RUNX1T1* fusion gene (yellow signal)—dual-color dual fusion probe (DCDF).

B) FISH dual-color dual fusion assay characterizing a chromosome 16 inversion—*CBFb* (red), *MYH11* (green), *CBFb-MYH11* gene fusion (yellow).

C) FISH test with (1) dual-color single fusion probe on a metaphase spread and (2) dual-color dual fusion in interphase nuclei; *PML* (red), *RARA* (green), *PML-RARA* fusion (yellow).

KMT2A(11q23)-MLLT3(9p22) Probe

With a combination of FISH probes, it can be possible to delineate the patient's karyotypic alterations, confirming, in this case (Figure 7.2A), the recurrent translocation t(9;11) and a trisomy of chromosome 9 with an extra *KMT2A-MLLT3* fusion gene.[11, 77, 112]

cMYC(8q24)-IGH(14q32) Probe

The combination of *IGH-MYC* with a probe for centromere 8 is essential to disclose the recurrent translocation t(8;14), enabling rapid detection and diagnosis of BL (Figure 7.2B). BL is an aggressive disease with a fast clinical course. However,

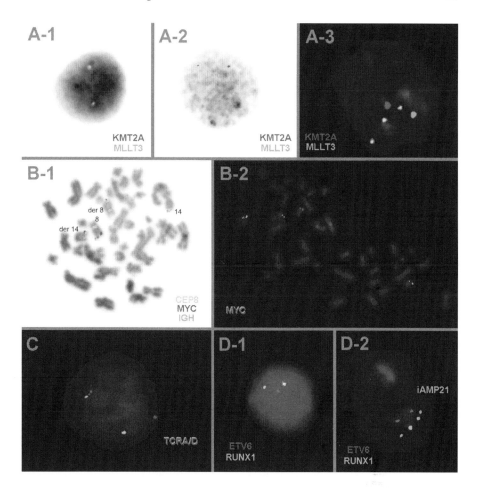

FIGURE 7.2 A) Experiment with FISH in interphase nuclei for the translocation t(9;11) (p22;q23)— 1) Two normal *KMT2A* gene signals (yellow),

2) One standard *KMT2A* (yellow) and *KMT2A* (red and green) split signals with an extra *KMT2A* (red) signal—KMT2A dual-color break-apart probe (DCBAP).

3) Standard *KMT2A* (red) and *MLLT3* (green) signals, besides three copies of *KMT2A-MLLT3* fusion (yellow)—KMT2A-MLLT3 dual color dual fusion.

B) MYC-directed FISH using

1) assay with dual color dual fusion plus centromeric probe (CEP) characterizing the translocation t(8;14)(q24;q32)—*IGH* (green) *cMYC* (red) CEP8 (blue);

2) MYC DCBAP showing a *cMYC* gene rearrangement.

C) FISH results for translocation t(11;14)(p13;q11)/*TCRA/D*—dual-color break-apart probe in interphase nuclei showing a *TCRA/D* gene rearrangement.

D) FISH test with

1) dual color dual fusion on metaphase spread for translocation t(12;21) (p13;q22)—*ETV6* (red) *RUNX1* (green) and *ETV6-RUNX1* fusion gene (yellow);

2) Flexibility of ETV6-RUNX1 DCDF in interphase nuclei for iAMP21 monitoring— two *ETV6* (red) normal signals and *RUNX1* (green) multiple signals revealing the cryptic iAMP21; G-banding showed normal karyotype (result not shown).

excellent clinical outcomes can be achieved provided that it is early diagnosed and properly treated.[71, 74]

TCRA/D(14q11) Probe

Translocation t(11;14)(p13;q11) is associated with an excellent prognosis, so its precise characterization is fundamental for diagnosis and therapy (Figure 7.2C). The translocation t(11;14)(p13;q11) and translocation t(7;11)(q35;p13)/*TCRB* are variant translocations of each other.[41, 113]

ETV6(12p13)-RUNX1(21q22) Probe

The translocation t(12;21) is subtle to the G-banding technique and is associated with a favorable prognosis once adequate treatment is administered.[78, 114] On the other hand, iAMP21 is considered to confer a poor prognosis, including a high risk of relapse. In this example (Figure 7.2D), the *ETV6-RUNX1* probe can be used for a prompt diagnosis of this hematological urgency. Besides, the same probe could help detect the cryptic iAMP21 even with the G-banding karyotype being apparently normal.[104]

FISH IN UNFAVORABLE PROGNOSIS LEUKEMIA

del(5q)/del(7q) Probe

Monosomy 5, 7, or del(5q/7q) as karyotypic changes are associated with a poor outcome in AML. Multiple tumor suppressor genes on the long arms of chromosomes 5 and 7 may play an important role in leukemogenesis.[115–117] They can be detected by FISH as shown in Figure 7.3A.

BCR(22q11)-ABL1(9q34) Probe

The application of a *BCR-ABL1* probe is important to stratify the patient's risk indicating targeted therapy against *BCR-ABL1* fusion (Figure 7.3B) by treatment with tyrosine kinase inhibitors.[5, 101]

KMT2A(11q23)(MLL) Probe

KMT2A—specific probe (11q23) is essential to disclose *KMT2A*-r, a group of aberrations mainly associated with infant leukemia and related to intermediate/adverse prognoses.[51] In this example (Figure 7.3C), the probe helped identify a cryptic insertion in a translocation t(10;11), with chromosomal material between the green and red signals.[61]

TP53 del(17p13) Probe

Loss of the *TP53* gene is an aberration of high prognostic impact (Figure 7.3D), considered to confer the poorest clinical outcome in CLL and an unfavorable prognosis in ALs.[118–120]

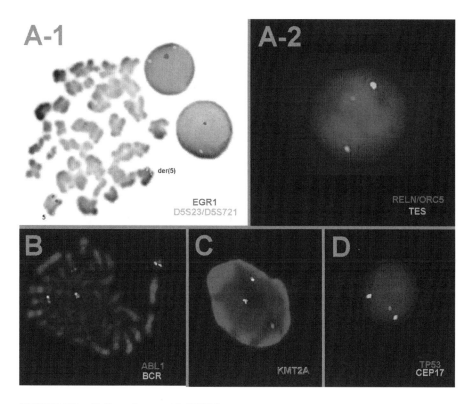

FIGURE 7.3 A) Experiment with FISH in
1) metaphase and interphase nuclei for deletion del(5q)—*EGR1* (5q31) (red), control (5p15) (green) dual-color enumerate probe (DCE);
2) interphase nuclei for del(7q)—(*RELN/ORC5*) 7q22.1-q22.2 (red)/*TES* (7q31.2) (green) DCE.
B) FISH dual color dual fusion result for the translocation t(9;22)(q34;q11)—*BCR* (green), *ABL1* (red), and *BCR-ABL1* fusion genes (yellow signals).
C) FISH assay in metaphase and interphase nuclei with MLL dual-color break-apart probe for *KMT2A*-r monitoring—Normal *KMT2A* (yellow) signal; *KMT2A* split signals (red and green).
D) Experiment with *TP53* deletion dual-color enumerate probe (DCE) in interphase nuclei of CLL BM cells—*TP53* (red) and CEP17 (green), showing the del(17p).

CONCLUSIONS

Hematological neoplasms are classified according to their differences in clinical and biological characteristics. Thus, acknowledging the association between specific cytogenetic abnormalities and clinical-morphological features is fundamental.[121] In this sense, FISH-based techniques can be applied to metaphase chromosomes and interphase nuclei, providing an accurate cell-based diagnosis complementary to genomic tests.[122]

In leukemia, the application of FISH has brought remarkable knowledge in the field of cancer cytogenetics. The characterization of recurrent chromosomal abnormalities and clonal evolution is crucial for classifying the different disease subtypes, selecting treatment strategies, and monitoring minimal residual disease.[123] Besides, at the molecular level, FISH can be the basis for the indication of putative genes with an impact on leukemogenesis.[101] In this context, FISH-based technologies play a major role in the diagnosis, prognosis, and therapeutic choice, facilitating the daily practice of individualized medicine.[121]

Currently, the evolution of FISH-based technologies is moving towards a cell-specific measurement of the genome, transcriptome, and proteome. In this way, a range of possibilities is opened for such techniques to be applied in genetic diagnosis, which can bring important advances for leukemia diagnosis, ranging from the resolution of cytogenetic and genomic abnormalities to epigenetic alterations.

REFERENCES

1. Greaves, M.F. Biological models for leukaemia and lymphoma. *IARC Sci. Publ.* **2004**, 157, 351–372.
2. Brown, C.M.S.; Larsen, S.R.; Iland, H.J.; Joshua, D.E.; Gibson, J. Leukaemias into the 21st century: Part 1: The acute leukaemias. *Intern. Med. J.* **2012**, 42(11), 1179–1186.
3. Bloomfield, C.D., Foon, K.A., Levine, E.G. Leukemias. In: *Medical Oncology*; Calabresi, P., Schein, P.S., Eds. McGraw Hill, Inc, New York, NY, **1993**, pp. 459–501.
4. Juliusson, G.; Hough, R. Leukemia. In: *Tumors in Adolescents and Young Adults*; Stark, D.P., Vassal, G., Eds. Prog. Tumor. Res., Karger, Basel, **2016**, pp. 43, 87–100.
5. Arber, D.A.; Orazi, A.; Hasserjian, R.; Thiele, J.; Borowitz, M.J.; Le Beau, M.M.; Bloomfield, C.D.; Cazzola, M.; Vardiman, J.W. The 2016 revision to the World Health Organization classification of myeloid neoplasms and acute leukemia. *Blood.* **2016**, 127, 2391–2405.
6. Silva, M.L.M.; Land, M. Anomalias Citogenéticas e Moleculares nas Leucemias da Infância. In: *Doenças genéticas em pediatria*; Carakushansky, G., Ed. Guanabara Koogan S.A., Rio de Janeiro, RJ, **2001**, pp. 36, 350–362.
7. Wan, T.S.K. Cancer cytogenetics: Methodology revisited. *Ann. Lab. Med.* **2014**, 34(6), 413–425.
8. Cortes, J.; Pavlovsky, C.; Saußele, S. Chronic myeloid leukaemia. *Lancet Sem.* **2021**, 398, 1914–1926.
9. Udayakumar, A.M.; Alkindi, S.; Pathare, A.V.; Raeburn, J.A. Complex t(8;13;21) (q22;q14;q22)-A novel variant of t(8;21) in a patient with acute myeloid leukemia (AML-M2). *Arch. Med. Res.* **2008**, 39, 252–256.
10. Mangan, J.K.; Speck, N.A. RUNX1 mutations in clonal myeloid disorders: From conventional cytogenetics to next generation sequencing, a story 40 years in the making. *Crit. Rev. Oncog.* **2011**, 16, 77–91.
11. Braoudaki, M.; Tzortzatou-Stathopoulou, F. Clinical cytogenetics in pediatric acute leukemia: An update. *Clin. Lymph., Myel. Leuk.* **2012**, 12, 230–237.
12. Bint, S.; Davies, A.; Ogilvie, C. Multicolor banding remains an important adjunct to array CGH and conventional karyotyping. *Mol. Cytogenet.* **2013**, 6, 55.

13. Bochtler, T.; Stölzel, F.; Heilig, C.E.; Kunz, C.; Mohr, B.; Jauch, A.; Janssen, J.W.; Kramer, M.; Benner, A.; Bornhäuser, M.; Ho, A.D.; Ehninger, G.; Schaich, M.; Krämer, A. Clonal heterogeneity as detected by metaphase karyotyping is an indicator of poor prognosis in acute myeloid leukemia. *J. Clin. Oncol.* **2013**, 31, 3898–3905.

14. Das, K.; Tan, P. Molecular cytogenetics: Recent developments and applications in cancer. *Clin. Genet.* **2013**, 84, 315–325.

15. Gozzetti, A.; Le Beau, M.M. Fluorescence in situ hybridization: Uses and limitations. *Semin. Hematol.* **2000**, 37, 320–333.

16. Wan, T.S.K. Cancer cytogenetics: An introduction. In: *Cancer Cytogenetics Methods and Protocols*; Wan, T.S.K. Ed. Humana Press, New York, NY, **2017**, 5–7.

17. Gall, G.; Pardue, M.L. Formation and detection of RNA-DNA hybrid molecules in cytogenetical preparations. *Proc. Natl. Acad. Sci. USA.* **1969**, 63, 378–381.

18. John, H.A.; Birnstiel, M.L.; Jones, K.W. RNA-DNA hybrids at the cytological level. *Nature.* **1969**, 223, 582–587.

19. Buongiorno-Nardelli, M.; Amaldi, F. Autoradiographic detection of molecular hybrids between rRNA and DNA in tissue sections. *Nature.* **1970**, 225, 946–948.

20. Coons, A.H.; Creech, H.J.; Jones, R.N. Immunological properties of an antibody containing a fluorescent group. *Proc. Soc. Exp. Biol. Med.* **1941**, 47, 200–202.

21. Rudkin, G.T.; Stollar, B.D. High resolution detection of DNA: RNA hybrids in situ by indirect immunofluorescence. *Nature.* **1977**, 265, 472–473.

22. Bauman, J.G.J.; Wiegant, J.; Borst, P.; Van Duijn, P. A new method for fluorescence microscopical localization of specific DNA sequences by in situ hybridization of fluorochrome-labelled RNA. Experimental Cell Research. **1980**, 128, 485–490.

23. Langer, P.R.; Waldrop, A.A.; Ward, D.C. Enzymatic synthesis of biotin-labeled polynucleotides: Novel nucleic acid affinity probes. *Proc. Natl. Acad. Sci USA.* **1981**, 78, 6633–6637.

24. Wolff, D.J.; Bagg, A.; Cooley, L.D.; Dewald, G.W.; Hirsch, B.A.; Jacky, P.B.; Rao, K.W.; Rao, P.N.; Association for Molecular Pathology Clinical Practice Committee; American College of Medical Genetics Laboratory Quality Assurance Committee. Guidance for fluorescence in situ hybridization testing in hematologic disorders. *J. Mol. Diagn.* **2007**, 9, 134–143.

25. Flis, S.; Chojnacki, T. Chronic myelogenous leukemia, a still unsolved problem: Pitfalls and new therapeutic possibilities. *Drug Des. Devel. Ther.* **2019**, 8, 825–843.

26. Andolina, J.R.; Neudorf, S.M.; Corey, S.J. How I treat childhood CML. *Blood.* **2012**, 119, 1821–1830.

27. Khemka, R.; Gupta, M.; Jena, N.K. CML with megakaryocytic blast crisis: Report of 3 cases. Pathol. Oncol. Res. **2018**, 25, 1253–1258.

28. Rozman, C.; Montserrat, E. Chronic lymphocytic leukemia. *N. Engl. J. Med.* **1995**, 333, 1052–1057.

29. Kikushige, Y.; Ishikawa, F.; Miyamoto, T.; Shima, T.; Urata, S.; Yoshimoto, G.; Mori, Y.; Iino, T.; Yamauchi, T.; Eto, T.; Niiro, H.; Iwasaki, H.; Takenaka, K.; Akashi, K. Self-renewing hematopoietic stem cell is the primary target in pathogenesis of human chronic lymphocytic leukemia. *Cancer Cell.* **2011**, 20, 246–259.

30. Calin, G.A.; Dumitru, C.D.; Shimizu, M.; Bichi, R.; Zupo, S.; Noch, E.; Aldler, H.; Rattan, S.; Keating, M.; Rai, K.; Rassenti, L.; Kipps, T.; Negrini, M.; Bullrich, F.; Croce, C.M. Frequent deletions and down-regulation of micro- RNA genes miR15 and miR16 at 13q14 in chronic lymphocytic leukemia. *Proc. Natl. Acad. Sci. USA.* **2002**, 99, 15524–15529.

31. Klein, U.; Lia, M.; Crespo, M.; Siegel, R.; Shen, Q.; Mo, T.; Ambesi-Impiombato, A.; Califano, A.; Migliazza, A.; Bhagat, G.; Dalla-Favera, R. The DLEU2/miR-15a/16-1

cluster controls B cell proliferation and its deletion leads to chronic lymphocytic leukemia. *Cancer Cell.* **2010**, 17, 28–40.

32. Onida, F.; Kantarjian, H.M.; Smith, T.L.; Ball, G.; Keating, M.J.; Estey, E.H.; Glassman, A.B.; Albitar, M.; Kwari, M.I.; Beran, M. Prognostic factors and scoring systems in chronic myelomonocytic leukemia: A retrospective analysis of 213 patients. *Blood.* **2002**, 99, 840–849.

33. Such, E.; Cervera, J.; Costa, D.; Sole, F.; Vallespí, T.; Luño, E.; Collado, R.; Calasanz, M.J.; Hernández-Rivas, J.M.; Cigudosa, J.C.; Nomdedeu, B.; Mallo, M.; Carbonell, F.; Bueno, J.; Ardanaz, M.T.; Ramos, F.; Tormo, M.; Sancho-Tello, R.; del Cañizo, C.; Gómez, V.; Marco, V.; Xicoy, B.; Bonanad, S.; Pedro, C.; Bernal, T.; Sanz, G.F. Cytogenetic risk stratification in chronic myelomonocytic leukemia. *Haematologica.* **2011**, 96, 375–383.

34. Ross, J.A.; Johnson, K.J.; Spector, G.L.; Kersey, J.H. Epidemiology of acute childhood leukemia. In: *Childhood Leukemia: A Practical Handbook*; Reaman, G.H.; Smith, F.O., Eds. Springer, Berlin, Heidelberg, **2011**, pp. 3–21.

35. Gilliland, D.; Jordan, C.; Felix, C. The molecular basis of leukemia. *Am. Soc. Hematol.* **2004**, 80–97.

36. Lo-Coco, F.; Ammatuna, E. The biology of acute promyelocytic leukemia and its impact on diagnosis and treatment. *Am. Soc. Hematol.* **2006**, 514, 156–161.

37. Arceci, R., Aplenc, R. Acute myelogenous leukemia in children. In: *Wintrobe's Clinical Hematology*; Greer, J.P. Ed. Lippincott Williams & Wilkins, Philadelphia, PA, **2009**, pp. 1919–1932.

38. Bolouri, H.; Farrar, J.E.; Triche, T.Jr.; Ries, R.E.; Lim, E.L.; Alonzo, T.A.; Ma, Y.; Moore, R.; Mungall, A.J.; Marra, M.A.; Zhang, J.; Ma, X.; Liu, Y.; Liu, Y.; Auvil, J.M.G.; Davidsen, T.M.; Gesuwan, P.; Hermida, L.C.; Salhia, B.; Capone, S.; Ramsingh, G.; Zwaan, C.M.; Noort, S.; Piccolo, S.R.; Kolb, E.A.; Gamis, A.S.; Smith, M.A.; Gerhard, D.S.; Meshinchi, S. The molecular landscape of pediatric acute myeloid leukemia reveals recurrent structural alterations and age-specific mutational interactions. *Nat. Med.* **2018**, 24, 103–112.

39. Raimondi, S.; Chang, M.N.; Ravindranath, Y.; Behm, F.G.; Gresik, M.V.; Steuber, C.P.; Weinstein, H.J.; Carroll, A.J. Chromosomal abnormalities in 478 children with acute myeloid leukemia: Clinical characteristics and treatment outcome in a cooperative Pediatric Oncology Group Study: POG 8821. *Blood.* **1999**, 94, 3707–3716.

40. Manola, K. Cytogenetics of pediatric acute myeloid leukemia. *Europ J Haematol.* **2009**, 83, 391–405.

41. Mrozek, K.; Heerema, N.; Bloomfield, C. Cytogenetics in acute leukemia. *Blood Reviews.* **2004**, 18, 115–134.

42. Pui, C.H.; Carroll, W.L.; Meshinchi, S.; Arceci, R.J. Biology, risk stratification, and therapy of pediatric acute leukemias: An update. *J Clin Oncol.* **2011**, 29, 551–565.

43. Arceci, R., Meshinchi, S. Biology of acute myeloid Leukemia. In: *Childhood Leukemia: A Practical Handbook*; Reaman, G.H.; Smith, F.O., Eds. Springer, Berlin, Heidelberg, **2011**, pp. 63–74.

44. Creutzig, U.; Zimmermann, M.; Ritter, J.; Reinhardt, D.; Hermann, J.; Henze, G.; Jürgens, H.; Kabisch, H.; Reiter, A.; Riehm, H.; Gadner, H.; Schellong, G. Treatment strategy and long term results in pediatric patients treated in four consecutive AML-BFM trials. *Leukemia.* **2005**, 19, 2030–2042.

45. Gamerdinger, U.; Teigler-Schlegel, A.; Pils, S.; Bruch, J.; Viehmann, S.; Keller, M.; Jauch, A.; Harbott, J. Cryptic chromosomal aberrations leading to an AML1/ETO rearrangement are frequently caused by small insertions. *Genes Chromosomes Cancer.* **2003**, 36, 261–272.

46. Creutzig, U.; Kutny, M.A.; Barr, R.; Schlenk, R.F.; Ribeiro, R.C. Acute myelogenous leukemia in adolescents and young adults. Pediatr. *Blood Cancer.* **2018**, 65, e27089.
47. De Figueiredo, A.F.; Liehr, T.; Bath, S.; Binato, R.; De Souza, M.T.; De Matos, R.R.; Salles, T.de, J.; Jordy, F.C.; Ribeiro, R.C.; Abdelhay, E.; Silva, M.L. A complex karyotype masked a cryptic variant t(8;21)(q22;q22) in a child with acute myeloid leukemia. Leuk. *Lymphoma.* **2011**, 52, 1593–1596.
48. Silva, M.L.M.; Raimondi, S.C.; Abdelhay, E.; Gross, M.; Mkrtchyan, H.; de Figueiredo, A.F.; Ribeiro, R.C.; de Jesus Marques-Salles, T.; Sobral, E.S.; Gerardin Land, M.P.; Liehr, T. Banding and molecular cytogenetic studies detected a CBFB-MYH11 fusion gene that appeared as abnormal chromosomes 1 and 16 in a baby with acute myeloid leukemia FAB M4-Eo. Cancer Genet. *Cytogenet.* **2008**, 182, 56–60.
49. Nakase, K.; Wakita, Y.; Minamikawa, K.; Yamaguchi, T.; Shiku, H. Acute promyelocytic leukemia with del(6)(p23). *Leuk. Res.* **2000**, 24, 79–81.
50. Matos, R.R.C.; Mkrtchyan, H.; Amaral, B.A.S.; Liehr, T.; de Souza, M.T.; Ney-Garcia, D.R.; Santos, N.; Marques-Salles, T.J.; Ribeiro, R.C.; Figueiredo, A.F.; Silva, M.L. An unusual cytogenetic rearrangement originating from two different abnormalities in chromosome 6 in a child with acute promyelocytic leukemia. *Acta Haematol.* **2013**, 130, 23–26.
51. Meyer, C.; Burmeister, T.; Gröger, D.; Tsaur, G.; Fechina, L.; Renneville, A.; Sutton, R.; Venn, N.C.; Emerenciano, M.; Pombo-de-Oliveira, M.S.; Barbieri Blunck, C.; Almeida Lopes, B.; Zuna, J.; Trka, J.; Ballerini, P.; Lapillonne, H.; De Braekeleer, M.; Cazzaniga, G.; Corral Abascal, L.; van der Velden, V.H.J.; Delabesse, E.; Park, T.S.; Oh, S.H.; Silva, M.L.M.; Lund-Aho, T.; Juvonen, V.; Moore, A.S.; Heidenreich, O.; Vormoor, J.; Zerkalenkova, E.; Olshanskaya, Y.; Bueno, C.; Menendez, P.; Teigler-Schlegel, A.; Zur Stadt, U.; Lentes, J.; Göhring, G.; Kustanovich, A.; Aleinikova, O.; Schäfer, B.W.; Kubetzko, S.; Madsen, H.O.; Gruhn, B.; Duarte, X.; Gameiro, P.; Lippert, E.; Bidet, A.; Cayuela, J.M.; Clappier, E.; Alonso, C.N.; Zwaan, C.M.; van den Heuvel-Eibrink, M.M.; Izraeli, S.; Trakhtenbrot, L.; Archer, P.; Hancock, J.; Möricke, A.; Alten, J.; Schrappe, M.; Stanulla, M.; Strehl, S.; Attarbaschi, A.; Dworzak, M.; Haas, O.A.; Panzer-Grümayer, R.; Sedék, L.; Szczepański, T.; Caye, A.; Suarez, L.; Cavé, H.; Marschalek, R. The MLL recombinome of acute leukemias in 2017. *Leukemia.* **2018**, 32, 273–284.
52. De Figueiredo, A.F.; Liehr, T.; Bath, S.; Binato, R.; Ventura, E.M.; de Souza, M.T.; de Matos, R.R.; Ribeiro, R.C.; Abdelhay, E.; Silva, M.L. A new cryptic ins(11;1) (q23;q21q31) detected in a t(1;8;11)(q21;p21;q23) in a baby with acute myeloid leukemia FAB AML-M5. *Blood Cells Mol. Dis.* **2010**, 45, 197–198.
53. Soupir, C.P.; Vergilio, J.A.; Dal Cin, P.; Muzikansky, A.; Kantarjian, H.; Jones, D.; Hasserjian, R.P. Philadelphia chromosome-positive acute myeloid leukemia: A rare aggressive leukemia with clinicopathologic features distinct from chronic myeloid leukemia in myeloid blast crisis. *Am. J. Clin. Pathol.* **2007**, 127, 642–650.
54. Cuneo, A.; Ferrant, A.; Michaux, J.L.; Demuynck, H.; Boogaerts, M.; Louwagie, A.; Doyen, C.; Stul, M.; Cassiman, J.J.; Dal Cin, P.; Castoldi, G.; Van den Berghe, H. Philadelphia chromosome-positive acute myeloid leukemia: Cytoimmunologic and cytogenetic features. *Haematologica.* **1996**, 81, 423–427.
55. Meshinchi, S.; Arceci, R.J. Prognostic factors and risk-based therapy in pediatric acute myeloid leukemia. *The Oncologist.* **2007**, 12, 341–355.
56. Von Lindern, M.; Fornerod, M.; van Baal, S.; Jaegle, M.; de Wit, T.; Buijs, A.; Grosveld, G. The translocation t(6;9), associated with a specific subtype of acute myeloid leukemia, results in the fusion of two genes, DEK and CAN, and the expression of a chimeric, leukemiaspecific DEK-CAN mRNA. *Mol. Cell. Biol.* **1992**, 12, 1687–1697.

57. Martineau, M.; Berger, R.; Lillington, D.M.; Moorman, A.V.; Secker-Walker, L.M. The t(6;11)(q27;q23) translocation in acute leukemia: A laboratory and clinical study of 30 cases: EU concerted action 11q23 Workshop participants. *Leukemia.* **1998**, 12, 788–791.

58. Balgobind, B.V.; Raimondi, S.C.; Harbott, J.; Zimmermann, M.; Alonzo, T.A.; Auvrignon, A.; Beverloo, H.B.; Chang, M.; Creutzig, U.; Dworzak, M.N.; Forestier, E.; Gibson, B.; Hasle, H.; Harrison, C.J.; Heerema, N.A.; Kaspers, G.J.; Leszl, A.; Litvinko, N.; Nigro, L.L.; Morimoto, A.; Perot, C.; Pieters, R.; Reinhardt, D.; Rubnitz, J.E.; Smith, F.O.; Stary, J.; Stasevich, I.; Strehl, S.; Taga, T.; Tomizawa, D.; Webb, D.; Zemanova, Z.; Zwaan, C.M.; van den Heuvel-Eibrink, M.M. Novel prognostic subgroups in childhood 11q23/MLL-rearranged acute myeloid leukemia: Results of an international retrospective study. *Blood.* **2009**, 114, 2489–2496.

59. de Matos, R.R.C.; Ferreira, G.M.; Meyer, C.; Marschalek, R.; Larghero, P.; Ribeiro, R.C.; Liehr, T.; Othman, M.; Bizarro, M.T.S.M.; Sobral da Costa, E.; Land, M.G.P.; Abdelhay, E.; Binato, R.; Silva, M.L.M. KMT2A-MLLT1 and the novel SEC16A-KMT2A in a cryptic 3-way translocation t(9;11;19) present in an infant with acute lymphoblastic leukemia. *J. Pediatr. Hematol. Oncol.* **2022**, 44, e719–e722.

60. De Braekeleer, E.; Meyer, C.; Douet-Guilbert, N.; Morel, F.; Le Bris, M.J.; Berthou, C.; Arnaud, B.; Marschalek, R.; Férec, C.; De Braekeleer, M. Complex and cryptic chromosomal rearrangements involving the MLL gene in acute leukemia: A study of 7 patients and review of the literature. *Blood Cells Mol. Dis.* **2010**, 44, 268–274.

61. De Figueiredo, A.F.; Vieira, T.P.; Liehr, T.; Bhatt, S.; de Souza, M.T.; Binato, R.; Marques-Salles, T.deJ.; Carboni, E.; Ribeiro, R.C.; Silva, M.L.; Abdelhay, E. A rare cryptic and complex rearrangement leading to MLL-MLLT10 gene fusion masked by del(10)(p12) in a child with acute monoblastic leukemia (AML-M5). *Leuk. Res.* **2012**, 36, e74–e77.

62. Ma, Z.; Morris, S.W.; Valentine, V.; Li, M.; Herbrick, J.A.; Cui, X.; Bouman, D.; Li, Y.; Mehta, P.K.; Nizetic, D.; Kaneko, Y.; Chan, G.C.; Chan, L.C.; Squire, J.; Scherer, S.W.; Hitzler, J.K. Fusion of two novel genes, RBM15 and MKL1, in the t(1;22)(p13q13) of acute megakaryoblastic leukemia. *Nat. Genet.* **2001**, 28, 220–221.

63. Carroll, A.; Civin, C.; Schneider, N.; Dahl, G.; Pappo, A.; Bowman, P.; Emami, A.; Gross, S.; Alvarado, C.; Phillips, C.; Krischer, J.; Crist, W.; Head, D.; Gresik, M.; Ravindranath, Y.; Weinstein, H. The t(1;22)(p13;q13) is nonrandom and restricted to infants with acute megakaryoblastic leukemia: A Pediatric Oncology Group study. *Blood.* **1991**, 78, 748–752.

64. Dastugue, N.; Lafage-Pochitaloff, M.; Pagès, M.P.; Radford, I.; Bastard, C.; Talmant, P.; Mozziconacci, M.J.; Léonard, C.; Bilhou-Nabéra, C.; Cabrol, C.; Capodano, A.M.; Cornillet-Lefebvre, P.; Lessard, M.; Mugneret, F.; Pérot, C.; Taviaux, S.; Fenneteaux, O.; Duchayne, E.; Berger, R.; Groupe Français d'Hematologie Cellulaire. Cytogenetic profile of childhood and adult megakaryoblastic leukemia (M7): A study of the Group Francais de Cytogenetique Hematologique (GFCH). *Blood.* **2002**, 100, 618–626.

65. Capela de Matos, R.R.; Othman, M.A.K.; Ferreira, G.M.; Costa, E.S.; Melo, J.B.; Carreira, I.M.; de Souza, M.T.; Lopes, B.A.; Emerenciano, M.; Land, M.G.P.; Liehr, T.; Ribeiro, R.C.; Silva, M.L.M. Molecular approaches identify a cryptic MECOM rearrangement in a child with a rapidly progressive myeloid neoplasm. *Cancer Genet.* **2018**, 221, 25–30.

66. Onciu, M. Acute lymphoblastic leukemia. *Hematol. Oncol. Clin. North Am.* **2009**, 23, 655–674.

67. Bhojwani, D.; Yang, J.J.; Pui, C.H. Biology of childhood acute lymphoblastic leukemia. *Pediatr. Clin. North. Am.* **2015**, 62, 47–60.

68. Pui, C.H. Recent research advances in childhood acute lymphoblastic leukemia. *J. Formos. Med. Assoc.* **2010**, 109, 777–787.

69. Vrooman, L.M.; Silverman, L.B. Treatment of childhood acute lymphoblastic leukemia: Prognostic factors and clinical advances. *Curr. Hematol. Malig. Rep.* **2016**, 11, 385–394.

70. Shago, M. Chromosome preparation for acute lymphoblastic leukemia. In: *Cancer Cytogenetics Methods and Protocols*; Wan, T.S.K., Ed. Humana Press, New York, NY, **2017**, pp. 19–31.

71. Greenought, A.; Dave, S.S. New clues to the molecular pathogenesis of Burkitt lymphoma revealed through next-generation sequencing. *Curr. Opin. Hematol.* **2014**, 21, 326–332.

72. Swerdlow, S.H., Campo, E., Harris, N.L., Jaffe, E.S., Pileri, S.A.; Stein, H.; Thiele, J.; Vardiman, J.W. (eds.). *WHO Classification of Tumours of Haematopoietic and Lymphoid Tissues.* 4th Edition. IARC: Lyon; **2008**.

73. Petit, B.; Mele, L.; Rack, K.; Camera, A.; Vekemans, M.C.; Bassan, R.; Pulsoni, A.; Delannoy, A.; Pagano, L. Characteristics of secondary acute lymphoblastic leukemia with L3 morphology in adult patients. *Leuk Lymphoma.* **2002**, 43, 1599–1604.

74. De Souza, M.T.; Hassan, R.; Liehr, T.; Marques-Salles, T.J.; Boulhosa, A.M.; Abdelhay, E.; Ribeiro, R.C.; Silva, M.L. Conventional and molecular cytogenetic characterization of Burkitt lymphoma with bone marrow involvement in Brazilian children and adolescents. Pediatr. *Blood Cancer.* **2014**, 61, 1422–1426.

75. Chan, K.W. Acute lymphoblastic leukemia. *Curr. Probl. Pediatr. Adolesc. Health Care.* **2002**, 32, 40–49.

76. Kimura, S.; Mullighan, C.H. Molecular markers in ALL: Clinical implications. *Best. Pract. Res. Clin. Haematol.* **2020**, 33, 101193.

77. Ney-Garcia, D.R.; Liehr, T.; Emerenciano, M.; Meyer, C.; Marschalek, R.; Pombo-de-Oliveira, M.do, S.; Ribeiro, R.C.; Poirot Land, M.G.; Macedo Silva, M.L. Molecular studies reveal a MLL-MLLT3 gene fusion displaced in a case of childhood acute lymphoblastic leukemia with complex karyotype. *Cancer Genet.* **2015**, 15, 1–4.

78. Garcia, D.R.; Arancibia, A.M.; Ribeiro, R.C.; Land, M.G.; Silva, M.L. Intrachromosomal amplification of chromosome 21 (iAMP21) detected by ETV6/RUNX1 FISH screening in childhood acute lymphoblastic leukemia: A case report. *Rev. Bras. Hematol. Hemoter.* **2013**, 35, 2–4.

79. Ney-Garcia, D.R.; Liehr, T.; Bhatt, S.; de Souza, M.T.; de Matos, R.R.; Binato, R.; Jordy, F.C.; Abdelhay, E.; Ribeiro, R.C.; Silva, M.L. Molecular cytogenetics studies reveal unexpected chromosomal inversion as variant of t(12;21)(p13;q22) in child with B-cell precursor acute lymphoblastic leukemia. *Leuk Lymphoma.* **2012**, 53, 342–344.

80. Mrozek, K.; Harper, D.P.; Aplan, P.D. Cytogenetics and molecular genetics of acute lymphoblastic leukemia. *Hematol. Oncol. Clin. N. Am.* **2009**, 23, 991–1010.

81. Sundaresh, A.; Williams, O. Mechanism of ETV6-RUNX1 leukemia. *Adv. Exp. Med. Biol.* **2017**, 962, 201–216.

82. Lange, B.J.; Smith, F.O.; Feusner, J.; Barnard, D.R.; Dinndorf, P.; Feig, S.; Heerema, N.A.; Arndt, C.; Arceci, R.J.; Seibel, N.; Weiman, M.; Dusenbery, K.; Shannon, K.; Luna-Fineman, S.; Gerbing, R.B.; Alonzo, T.A. Outcomes in CCG-2961, a Children's Oncology Group Phase 3 trial for untreated pediatric acute myeloid leukemia: A report from the Children's Oncology Group. *Blood.* **2008**, 111, 1044–1053.

83. Martinez-Climent, J.A. Molecular cytogenetics of childhood hematological malignancies. *Leukemia.* **1997**, 11, 1999–2021.

84. Takeuchi, S.; Koike, M.; Seriu, T.; Bartram.; C.R.; Schrappe, M.; Reiter, A.; Park, S.; Taub, H.E.; Kubonishi, I.; Miyoshi, I.; Koeffler, H.P. Frequent loss of heterozygosity on the long arm of chromosome 6: Identification of two distinct regions of deletion in childhood acute lymphoblastic leukemia. *Cancer. Res.* **1998**, 58, 2618–2623.

85. Fischer, K.; Fröhling, S.; Scherer, S.W.; McAllister Brown, J.; Scholl, C.; Stilgenbauer, S.; Tsui, L.C.; Lichter. P.; Döhner, H. Molecular cytogenetic delineation of deletions and translocations involving chromosome band 7q22 in myeloid leukemias. *Blood.* **1997**, 89, 2036–2041.

86. Koike, M.; Tasaka, T.; Spira, S.; Tsuruoka, N.; Koeffler, H.P. Allelotyping of acute myelogenous leukemia: Loss of heterozygosity at 7q31.1 (D7S486) and q33–34 (D7S498, D7S505). *Leuk. Res.* **1999**, 23, 307–310.

87. Betts, D.R.; Ammann, R.A.; Hirt, A.; Hengartner, H.; Beck-Popovic, M.; Kuhne, T.; Nobile, L.; Caflisch, U.; Wacker, P.; Niggli, F.K. The prognostic significance of cytogenetic aberrations in childhood acute myeloid leukaemia: A study of the Swiss Paediatric Oncology Group (SPOG). *European J. Haematology.* **2007**, 78, 468–476.

88. Forestier, E.; Heim, S.; Blennow, E.; Borgström, G.; Holmgren, G.; Heinonen, K.; Johannsson, J.; Kerndrup, G.; Andersen, M.K.; Lundin, C.; Nordgren, A.; Rosenquist, R.; Swolin, B.; Johansson, B.; Nordic Society of Paediatric Haematology and Oncology (NOPHO); Swedish Cytogenetic Leukaemia Study Group (SCLSG); NOPHO Leukaemia Cytogenetic Study Group (NLCSG). Cytogenetic abnormalities in childhood acute myeloid leukaemia: A Nordic series comprising all children enrolled in the NOPHO-93-AML trial between 1993 and 2001. *Br. J. Haematol.* **2003**, 121, 566–577.

89. Hall, G.W. Childhood myeloid leukemias. *Best. Pract. Res. Clin. Haematol.* **2001**, 14, 573–591.

90. Raimondi, S.; Chang, M. N.; Ravindranath, Y.; Behm, F.G.; Gresik, M.V.; Steuber, C.P.; Weinstein, H.J.; Carroll, A.J. Chromosomal abnormalities in 478 children with acute myeloid leukemia: Clinical characteristics and treatment outcome in a cooperative Pediatric Oncology Group Study-POG 8821. *Blood.* **1999**, 94, 3707–3716.

91. Heerema, N.A. 9p rearrangements in ALL. Atlas Genet. Cytogenet. Oncol. Haematol. **1999**. http://AtlasGeneticsOncology.org/Anomalies/9prearrALLID1156.html [accessed on 03/12/2022].

92. Wells, R.J.; Arthur, D.C.; Srivastava, A.; Heerema, N.A.; Le Beau, M.; Alonzo, T.A.; Buxton, A.B.; Woods, W.G.; Howells, W.B.; Benjamin, D.R.; Betcher, D.L.; Buckley, J.D.; Feig, S.A.; Kim, T.; Odom, L.F.; Ruymann, F.B.; Smithson, W.A.; Tannous, R.; Whitt, J.K.; Wolff, L.; Tjoa, T.; Lampkin, B.C. Prognostic variables in newly diagnosed children and adolescents with acute myeloid leukemia: Children's Cancer Group Study 213. *Leukemia.* **2002**, 16, 601–607.

93. Grimwade, D.; Walker, H.; Oliver, F.; Wheatley, K.; Harrison, C.; Harrison, G.; Rees, J.; Hann, I.; Stevens, R.; Burnett, A.; Goldstone, A. The importance of diagnostic cytogenetics on outcome in AML: Analysis of 1,612 patients entered into the MRC AML 10 trial: The medical Research Council Adult and Children's Leukaemia Working Parties. *Blood.* **1998**, 92, 2322–2333.

94. Mrozek, K. Acute myeloid leukemia with a complex karyotype. *Semin. Oncol.* **2008**, 35, 365–377.

95. Orozco, F.; Appelbaum, J. Unfavorable, complex, and monosomal karyotypes: The most challenging forms of acute myeloid leukemia. *Oncology.* **2012**, 26, 1–10.

96. Al-Achkar, W.; Aljapawe, A.; Othman, M.A.K.; Wafa, A. A de novo acute myeloid leukemia (AML-M4) case with a complex karyotype and yet unreported breakpoints. *Mol. Cytogenet.* **2013**, 6, 18.

97. Ney Garcia, D.R.; de Souza, M.T.; de Figueiredo, A.F.; Othman, M.A.K.; Rittscher, K.; Abdelhay, E.; Capela de Matos, R.R.; Meyer, C.; Marschalek, R.; Land, M.G.P.; Liehr, T.; Ribeiro, R.C.; Silva, M.L.M. Molecular characterization of KMT2A fusion partner genes in 13 cases of pediatric leukemia with complex or cryptic karyotypes. *Hematol. Oncol.* **2017**, 35, 760–768.

98. Betts, D.; Ammann, R.A.; Hirt, A.; Hengartner, H.; Beck-Popovic, M.; Kuhne, T.; Nobile, L.; Caflisch, U.; Wacker, P.; Niggli, F.K. The prognostic significance of cytogenetic aberrations in childhood acute myeloid leukaemia: A study of the Swiss Paediatric Oncology Group (SPOG). *European Journal of Haematology.* **2007**, 78, 468–476.

99. Marchesi, F.; Annibali, O.; Cerchiara, E.; Tirindelli, M.C.; Avvisati, G. Cytogenetic abnormalities in adult non-promyelocytic acute myeloid leukemia: A concise review. *Crit. Rev. Oncol. Hemat.* **2011**, 80, 331–346.

100. Ney-Garcia, D.; Vieira, T.P.; Liehr, T.; Bhatt, S.; de Souza, M.T.; de Figueiredo, A.F.; Ribeiro, R.C.; Silva, M.L. A case of childhood T cell acute lymphoblastic leukemia with a complex t(9;9) and homozygous deletion of CDKN2A gene associated with a Philadelphia-positive minor subclone. *Blood Cells, Mol. Dis.* **2012**, 50, 131–133.

101. Cui, C.; Shu, W.; Li, P. Fluorescence in situ hybridization: Cell-based genetic diagnostic and research applications. *Front. Cell. Dev. Biol.* **2016**, 4, 89.

102. Bishop, R. Applications of fluorescence in situ hybridization (FISH) in detecting genetic aberrations of medical significance. Bioscience Horizons: The International Journal of Student Research. **2010**, 3, 85–95.

103. Nordgren, A. Hidden aberrations diagnosed by interphase fluorescence in situ hybridization and spectral karyotyping in childhood acute lymphoblastic leukaemia. *Leuk. Lymphoma.* **2003**, 44, 2039–2053.

104. Capela de Matos, R.R.; Othman, M.A.K.; Ferreira, G.M.; Monteso, K.; de Souza, M.T.; Rouxinol, M.; Melo, J.B.; Carreira, I.M.; Abdelhay, E.; Liehr, T.; Ribeiro, R.C.; Silva, M. Somatic homozygous loss of SH2B3, and a non-Robertsonian translocation t(15;21) (q25.3;q22.1) with NTRK3 rearrangement, in an adolescent with progenitor B-cell lymphoblastic leukemia with the iAMP21. *Cancer Genet.* **2021**, 262–263, 16–22.

105. Testa, J.R.; Misawa, S.; Ogrema, N.; van Slaten, K.; Wiernik, P.H. Chromosomal alterations in acute leukemia patients: Studies with improved culture methods. *Cancer Res.* **1985**, 45, 430–434.

106. Hungerford, D.A. Leukocytes cultures from small in occula of whole blood and the preparation of metaphase chromosomes by treatament with hypotonic (KCl). *Stain Technol.,* **1965**, 40, 333–338.

107. Gisselsson, D. Cytogenetic methods. In: *Cancer Genetics*; Heim, S.; Mitelman, F., Eds. John Wiley & Sons, Inc, New Jersey, NY, **2009**, p. 18.

108. Seabright, M. A rapid banding technique for human chromosomes. *Lancet.* **1971**, 2, 971–972.

109. Guerra, M., Ed. *FISH Conceito e aplicações na citogenética.* Sociedade Brasileira de Genética, Ribeirão Preto, SP; **2004**.

110. Tallman, M.S.; Altman, J.K. How I treat acute promyelocytic leukemia. *Blood.* **2009**, 114, 5126–5135.

111. Capela de Matos, R.R.; de Figueiredo, A.F.; Liehr, T.; Alhourani, E.; De Souza, M.T.; Binato, R.; Ribeiro, R.C.; Silva, M.L. A novel three-way variant t(8;13;21)(q22;q33;q22) in a child with acute myeloid leukemia with RUNX1/RUNX1T1: The contribution of molecular approaches for revealing t(8;21) variants. *Acta Haematol.* **2015**, 134, 243–245.

112. Capela de Matos, R.R.; Ney-Garcia, D.R.; Cifoni, E.; Othman, M.A.K.; Tavares de Souza, M.; Carboni, E.K.; Ferreira, G.M.; Liehr, T.; Ribeiro, R.C.; Macedo Silva, M.L. GAS6 oncogene and reverse MLLT3-KMT2A duplications in an infant with acute

myeloid leukemia and a novel complex hyperdiploid karyotype: Detailed high-resolution molecular cytogenetic studies. *Cytogenet. Genome Res.* **2017**, 152, 33–37.

113. Bilhou-Nabera, C. t(11;14)(p13;q11) TRD/LMO2t(7;11)(q35;p13) TRB/LMO2. *Atlas Genet. Cytogenet. Oncol. Haematol.* **1998**. http://atlasgeneticsoncology.org/ haematological/1070/t(11;14)(p13;q11)-trd-lmo2t(7;11)(q35;p13)-trb-lmo2 [accessed on 03/15/2022].

114. Bacher, U.; Schnittger, S.; Haferlach, C.; Haferlach, T. Molecular diagnostics in acute leukemias. *Clin. Chem. Lab. Med.* **2009**, 47, 1333–1341.

115. Jerez, A.; Sugimoto, Y.; Makishima, H.; Verma, A.; Jankowska, A.M.; Przychodzen, B.; Visconte, V.; Tiu, R.V.; O'Keefe, C.L.; Mohamedali, A.M.; Kulasekararaj, A.G.; Pellagatti, A.; McGraw, K.; Muramatsu, H.; Moliterno, A.R.; Sekeres, M.A.; McDevitt, M.A.; Kojima, S.; List, A.; Boultwood, J.; Mufti, G.J.; Maciejewski, J.P. Loss of heterozygosity in 7q myeloid disorders: Clinical associations and genomic pathogenesis. *Blood.* **2012**, 119, 6109–6118.

116. Joslin, J.M.; Fernald, A.A.; Tennant, T.R.; Davis, E.M.; Kogan, S.C.; Anastasi, J.; Crispino, J.D.; Le Beau, M.M. Haploinsufficiency of EGR1, a candidate gene in the del(5q), leads to the development of myeloid disorders. *Blood.* **2007**, 110, 719–726.

117. De Figueiredo, A.F.; Capela de Matos, R.R.; Moneeb, M.A.K.; Liehr, T.; da Costa, E.S.; Land, M.G.; Ribeiro, R.C.; Abdelhay, E.; Silva, M.L. Molecular cytogenetic studies characterizing a novel complex karyotype with an uncommon 5q22 deletion in childhood acute myeloid leukemia. *Mol. Cytogenet.* **2015**, 8, 62.

118. Baliakas, P.; Hadzidimitriou, A.; Sutton, L.A.; Rossi, D.; Minga, E.; Villamor, N.; Larrayoz, M.; Kminkova, J.; Agathangelidis, A.; Davis, Z.; Tausch, E.; Stalika, E.; Kantorova, B.; Mansouri, L.; Scarfò, L.; Cortese, D.; Navrkalova, V.; Rose-Zerilli, M.J.; Smedby, K.E.; Juliusson, G.; Anagnostopoulos, A.; Makris, A.M.; Navarro, A.; Delgado, J.; Oscier, D.; Belessi, C.; Stilgenbauer, S.; Ghia, P.; Pospisilova, S.; Gaidano, G.; Campo, E.; Strefford, J.C.; Stamatopoulos, K.; Rosenquist, R.; European Research Initiative on CLL (ERIC). Recurrent mutations refine prognosis in chronic lymphocytic leukemia. *Leukemia.* **2015**, 29, 329–336.

119. Seifert, H.; Mohr, B.; Thiede, C.; Oelschlägel, U.; Schäkel, U.; Illmer, T.; Soucek, S.; Ehninger, G.; Schaich, M.; Study Alliance Leukemia (SAL). The prognostic impact of 17p (p53) deletion in 2272 adults with acute myeloid leukemia. *Leukemia.* **2009**, 23, 656–663.

120. Stengel, A.; Schnittger, S.; Weissmann, S.; Kuznia, S.; Kern, W.; Kohlmann, A.; Haferlach, T.; Haferlach, C. TP53 mutations occur in 15.7% of ALL and are associated with MYC-rearrangement, low hypodiploidy, and a poor prognosis. *Blood.* **2014**, 124, 251–258.

121. Wan, T.S.K.; Ma, E.S.K. Molecular cytogenetics: An indispensable tool for cancer diagnosis. *Chang Gung Med. J.* **2012**, 35, 96–110.

122. Xu, F., Li, P. Cytogenomic abnormalities and dosage-sensitive mechanisms for intellectual and developmental disabilities. In: *Developmental Disabilities: Molecules Involved, Diagnosis and Clinical Care*; Salehi, A., Ed. InTech, Rijeka, **2013**, pp. 1–30.

123. Martin, C.L.; Warburton, D. Detection of chromosomal aberrations in clinical practice: From karyotype to genome sequence. *Annu. Rev. Genomics Hum. Genet.* **2015**, 16, 309–326.

8 FISH—in Tissues

Thomas Liehr

CONTENTS

INTRODUCTION

Application of fluorescence in situ hybridization (FISH) in tissues refers here to studies done in interphase cells derived from various kinds of solid tissues from the human body. These tissues are normally fixed in one of two ways: (i) formalin-fixed/paraffin-embedded (FFPE) tissue,[1] or (ii) cryofixed ones.[2] Especially the pathology of solid tumors,[3–5] and more exceptionally postmortem analysis of aborted fetusesare done by FISH.[6] As in FFPE sections embedded after long fixations times in unbuffered formalin may contain highly degraded DNA not suited for FISH in most clinics now FFPE and cryofixation are done in parallel.[7]

Of the four types of FISH-probes, whole chromosome paints, partial chromosome paints, centromeric probes (CPs) and locus-specific probes (LSPs) only the latter two are suited for routine interphase-FISH studies.[8] This is as only CPs and LSPs lead to well-defined and relatively small signals in the interphase nucleus. Thus, such signals can be counted to determine copy numbers of chromosomes or chromosomal regions. Also a smart use of several LSPs or LSPs with CPs labeled by different fluorochromes enables not only proof of deletions, duplications or even amplifications, and translocations or inversions may be picked up on the single-cell level.[9]

Most FISH-studies done on tissues in pathology stick to sectioned and mounted FFPE sections[1–5]; this has the disadvantage of partially overlapping, as well as incomplete, cut nuclei, which cannot be evaluated properly and lead to artificial loss of signals.[10–11] However, it is argued that invasive and small tumor pieces could be identified better on sections.[12] Thus, the alternative approach where a nuclear

DOI: 10.1201/9781003223658-8

extraction technique[1, 2, 13] is applied is not used much in routine pathology, but has been shown to lead to same results with lower cut-off rates; after nuclear extraction, these are non-overlapping, and all incomplete nuclei have been discarded during the extraction procedure.[1–2, 6, 14–16]

Here a commercial standard protocol for handling of FFPE sections is described; also, a procedure how to extract nuclei from a cryosection, which can be used in FISH later. Protocols for nuclear extraction from FFPE and a block of cyrofixed material were previously described.[16]

SAMPLES/TISSUES

Each kind of solid human tissue, which can be either cryo- or FFPE-fixed, is suited for this kind of FISH-study. It is not part of this protocol to describe the production of a cryo- or FFPE-block or how to cut the slices and mount on glass-slides. Slices may be differently thick—normally between from 3 to 8 μm. Also, after cutting and thawing of cryo-sections in most cases a fixation with 3% paraformaldehyde is necessary.[16]

DESCRIPTION OF METHODS

FISH on FFPE-Tissue Sections (No Nuclear Extraction)

(Here a modification of the ZytoVision protocol is described—www.zytovision. com/)

1. Incubate slides with FFPE-sections on slides on a heating plate at 70.5°C for 10 min.
2. Immerse them in NeoClear (Merck, Darmstadt, Germany) two times for 10 min to deparaffinize the sections (room temperature = RT).
3. Rehydrate the sections in an ethanol series 100%, 95%, 70% for 2 min, each; finally put them for 2 min in distilled water (all at RT).
4. Apply "Heat Pretreatment Solution Citric" (ZytoVision, Bremerhaven, Germany) preheated for ~1 h at 99°C. Put max 2–3 slides in a 100 ml cuvette and incubate in a water bath at 99°C for 17 min.
5. Then transfer slides in distilled water (RT) and leave for 2 min.
6. Repeat step 5, remove slides from water, and dry carefully the surface and the underneath of the slide with an absorbent paper, with exception for the sections themselves.
7. Transfer slides in a humid chamber at 37°C and cover sections with a sufficient amount of "Pepsin Solution" (ZytoVision, Bremerhaven, Germany)—incubate for 5 min.
8. Take slides and let solution flow off on an absorbent paper and immediately transfer to "Wash Buffer SSC" (ZytoVision, Bremerhaven, Germany), RT, for 5 min.
9. Transfer to distilled water (at RT) for 1 min and then dehydrate section in an ethanol series 70%, 95%, 100%, 2 min each. Then let slides air dry putting them perpendicular.

10. Adjust FISH probe mix (ZytoVision, Bremerhaven, Germany) according to size of section and size of coverslip. Seal coverslips by rubber cement and let rubber cement dry for ~20 min (best in dark conditions).
11. Co-denature slides and FISH-probe at 74.5°C for 10 min and transfer then to humid chamber at 37°C. Incubate overnight.
12. Mix 10 ml of "Wash Buffer A" (ZytoVision, Bremerhaven, Germany) with 250 ml of distilled water, distribute on three 100 ml cuvettes and heat to 38°C in a water bath.
13. Remove rubber cement from slide by forceps and put in first of three prepared cuvettes; there remove carefully the coverslip.
14. Transfer to second cuvette (38°C) and leave for 5 min.
15. Transfer to third cuvette (38°C) and leave for 5 min.
16. Repeat step 9.
17. Put 1 drop of "DAPI/DuraTect-Solution (ultra)" (ZytoVision, Bremerhaven, Germany) on a 22×22 coverslip wish is laying on the lab table. Approach the coverslip and drop with the slide, which is upside down and let fluid spread between slide and coverslip.
18. Start evaluation about 15 min after coverslip was added.

How to Extract Nuclei from FFPE-Tissue Sections

(a) Perform steps 1 to 6 as described previously.
(b) Latest at this step, but maybe better before starting with step 1 tissue parts not to be analyzed (e.g. normal tissue) is to be removed.
(c) Cover tissue with proteinase K solution (5 mg proteinase K (Roche, Basel, Switzerland, #03115887001), 50 µl 1 M Tris-HCl (pH 7.5), 20 µl 0.5 M EDTA (pH 7.0), 2 µl 5 M NaCl, make up to 1 ml with filtered double-distilled water; make fresh as required).
(d) Incubate for 1 h in a humid chamber (37°C). Do not use a coverslip.
(e) Fluid with disaggregated tissue is collected with a micro-pipette and filtered via a 55 µm nylon mesh (Nytal 55, SEFAR-AG, Heiden, Switzerland #3A07-0049-102-00). Fluid and nuclei pass through the mesh by gravity; wash out the mesh 4 ml 1×PBS; collect all with a 15 ml plastic tube.
(f) Centrifuge the nuclei down at 850×g for 8 min and discard supernatant apart from ~300µl.
(g) Add 4 ml 1×PBS and repeat step f.
(h) Dilute nuclei in remainder 300 µl of 1×PBS.
(i) Put a suspension-drop of on a dry and clean slide; evaporate the fluid from the slide on a 40°C heating plate (~5–10 min), and store suspension at 4°C.
(j) After overnight drying of the slide, fix in 0.1% formalin buffer for 10 min (RT).
(k) Finalize by washing the slides in 1×PBS, distilled water and an ethanol series (70%, 90%, 100%) 1 min each and air dry.
(l) After checking density of nuclei under a phase contrast light microscope, the region with best nuclei density can be marked with a diamond pencil and the slide can be used for FISH. Here a standard FISH protocol can be used (excluding pepsin pretreatment) like e.g. for peripheral blood lymphocyte suspension.

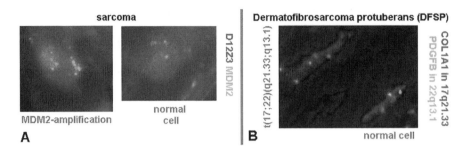

FIGURE 8.1 Typical results as obtained after FISH on FFPE-tissue sections (without nuclear extraction) are shown.

A) A sarcoma cell with MDM2 amplification and a normal cell from same section are shown after application of ZytoLight ® SPEC MDM2/CEN 12 Dual Color Probe (ZytoVision, Bremerhaven, Germany, # Z-2013–50).

B) A Dermatofibrosarcoma protuberans (DFPS) cell with t(17;22)(q21.33;q13.1) (one red, one green and two fusion signals) and a normal cell are visible—probe applied was ZytoLight ® SPEC COL1A1/PDGFB Dual Color Dual Fusion Probe (ZytoVision, Bremerhaven, Germany, # Z-2116–50).

YIELDS

Results of FISH on FFPE sections fixed on slides are shown in Figure 8.1. As standard in interphase FISH analyscs, it is not sufficient to evaluate 1 to 5 nuclei only. In pathology, 50 nuclei of tumor tissue is taken into account.[17–18]

CONCLUSIONS

Even though it has been proven that FISH results from archived, cryo- or FFPE-fixed tissues are at least as reliable, if not even better, when instead of tissue-sections including cut and overlaying nuclei, extracted nuclei are studied, most labs stick to sections. This is most likely due to habit and practice but also the argument that after nuclear extraction it is no longer reproducible where exactly the evaluated nuclei are derived from in the original section.

ACKNOWLEDGMENTS

Figure 8.1 was provided by Stefanie Kankel, Institute of Human Genetics, Jena University Hospital, Jena, Germany; sections were provided by the Institute for Forensic Medicine FSU, Pathology Section, Jena University Hospital, Jena, Germany.

REFERENCES

1. Liehr, T.; Grehl, H.; Rautenstrauss, B. FISH analysis of interphase nuclei extracted from paraffin-embedded tissue. *Trends Genet.* **1995**, 11, 377–378.

2. Liehr, T.; Grehl, H.; Rautenstrauss, B. A rapid method for FISH analysis on interphase nuclei extracted from cryofixed tissue. *Trends Genet.* **1996**, 12, 505–506.

3. Wolfe, K.Q.; Herrington, C.S. Interphase cytogenetics and pathology: A tool for diagnosis and research. *J. Pathol.* **1997**, 181, 359–361.

4. Fuller, C.E.; Perry, A. Fluorescence in situ hybridization (FISH) in diagnostic and investigative neuropathology. *Brain Pathol.* **2002**, 12, 67–86.

5. Lim, A.S.; Lim, T.H. Fluorescence in situ hybridization on tissue sections. *Methods Mol. Biol.* **2017**, 1541, 119–125.

6. Fickelscher, I.; Starke, H.; Schulze, E.; Ernst, G.; Kosyakova, N.; Mkrtchyan, H.; MacDermont, K.; Sebire, N.; Liehr, T. A further case with a small supernumerary marker chromosome (sSMC) derived from chromosome 1: Evidence for high variability in mosaicism in different tissues of sSMC carriers. *Prenat. Diagn.* **2007**, 27, 783–785.

7. Long, A.A.; Komminoth, P.; Lee, E.; Wolfe, H.J. Comparison of indirect and direct in-situ polymerase chain reaction in cell preparations and tissue sections: Detection of viral DNA, gene rearrangements and chromosomal translocations. *Histochemistry.* **1993**, 99, 151–162.

8. Liehr, T. Molecular cytogenetics in the era of chromosomics and cytogenomic approaches. *Front Genet.* **2021**, 12, 720507.

9. Iourov, I.; Vorsanova, S.; Yurov, Y. (Eds.). *Human Interphase Chromosomes: Biomedical Aspects.* Springer, Berlin; **2020**.

10. Dhingra, K.; Sahin, A.; Supak, J.; Kim, S.Y.; Hortobagyi, G.; Hittelman, W.N. Chromosome in situ hybridization on formalin-fixed mammary tissue using non-isotopic: Non-fluorescent probes: Technical considerations and biological implications. *Breast Cancer Res. Treat.* **1992**, 23, 201–210.

11. Liehr, T.; Stübinger, A.; Thoma, K.; Tulusan, H.A.; Gebhart, E. Comparative interphase cytogenetics using FISH on human ovarian carcinomas. *Anticancer Res.* **1994**, 14, 183–188.

12. Köpf, I.; Hanson, C.; Delle, U.; Verbiené, I.; Weimarck, A. A rapid and simplified technique for analysis of archival formalin-fixed, paraffin-embedded tissue by fluorescence in situ hybridization (FISH). *Anticancer Res.* **1996**, 16, 2533–2536.

13. Hedley, D.W.; Friedlander, M.L.; Taylor, I.W.; Rugg, C.A.; Musgrove, E.A. Method for analysis of cellular DNA content of paraffin-embedded pathological material using flow cytometry. J. Histochem. *Cytochem.* **1983**, 31, 1333–1335.

14. Qian, J.; Bostwick, D.G.; Takahashi, S.; Borell, T.J.; Brown, J.A.; Lieber, M.M.; Jenkins, R.B. Comparison of fluorescence in situ hybridization analysis of isolated nuclei and routine histological sections from paraffin-embedded prostatic adenocarcinoma specimens. *Am. J. Pathol.* **1996**, 149, 1193–1199.

15. Köpf, I.; Hanson, C.; Delle, U.; Verbiené, I.; Weimarck, A. A rapid and simplified technique for analysis of archival formalin-fixed, paraffin-embedded tissue by fluorescence in situ hybridization (FISH). *Anticancer Res.* **1996**, 16, 2533–2536.

16. Liehr, T. Characterization of archived formalin-fixed/paraffin-embedded or cryofixed tissue, including nucleus extraction. In: *Fluorescence In Situ Hybridization (FISH): Application Guide.* 2nd Edition. Springer, Berlin, **2017**, pp. 201–208.

17. Wayne, P.A. *Fluorescence In Situ Hybridization Methods for Clinical Laboratories-Aproved Guidelines.* 2nd Edition. Clinical Laboratory Standards Institute MM07-A2; **2013**.

18. Wiktor, A.E.; Van Dyke, D.L.; Stupca, P.J.; Ketterling, R.P.; Thorland, E.C.; Shearer, B.M.; Fink, S.R.; Stockero, K.J.; Majorowicz, J.R.; Dewald, G.W. Preclinical validation of fluorescence in situ hybridization assays for clinical practice. *Genet. Med.* **2006**, 8, 16–23.

9 FISH—in Human Sperm and Infertility

Martina Rincic and Thomas Liehr

CONTENTS

INTRODUCTION

According to a generally agreed definition, infertility is the inability to have off-spring despite active sexual intercourse of a couple for at least one year. This problem affects ~12% of the population in reproductive age; one in six couples needs the help of assisted reproduction techniques (ARTs) to conceive a child.[1] Pre-implantation genetic testing (PGT) as an alternative to prenatal diagnosis was introduced over 30 years ago. PGT performed together with ARTs aims to reduce the transmission of genetic or chromosomal abnormality. Recently, the European Society of Human Reproduction and Embryology (ESHRE) has subdivided PGT into (i) PGT-M (for the monogenic disorder), (ii) PGT-SR (for structural rearrangements), which causes miscarriage, and (iii) PGT-A (for aneuploidy screening).[2] Despite all technical advancements in the field of molecular genetics, according to the latest ESHRE recommendations fluorescence in situ hybridization (FISH) and single nucleotide polymorphism (SNP) array are recommended as techniques of choice for PGT-SR.[3]

Infertility is a complex issue where male infertility factors are relatively common conditions, affecting at least 6% of men of reproductive age. There are many causes known from changes in gross semen parameters (for instance, low sperm counts or compromised motility) to more complex and severe factors. In general, gametes of infertile individuals tend to display higher rates of chromosomal abnormalities than those seen in fertile persons, resulting finally in aneuploidy embryos. Ideally, clinical diagnosis of gamete aneuploidy should include analysis of all

DOI: 10.1201/9781003223658-9

human autosomes and sex chromosomes. On the other hand, that type of analysis is time-consuming and with a high cost, and is often unfeasible. Since most aneuploid embryos do not survive the early stages of development, the aneuploidies that are of major clinical importance are those that can be non-lethal, and thus compatible with survival. They include autosomal trisomies, such as Trisomy 13 (Pätau syndrome), Trisomy 18 (Edwards syndrome), and Trisomy 21 (Down syndrome), and aneuploidies of the sex chromosomes, such as e.g. Turner- and Klinefelter-syndrome (45,X and 47,XXY).

In 1978, direct cytogenetic analysis of the human sperm cells was introduced.[4] Given, that this original technique was difficult to perform, time-consuming, expensive, and above all yields only a small number of karyotypes, interphase FISH was an elegant solution to replace it. Nowadays FISH on human sperm is most commonly used to determine the proportion of aneuploidy present in autosomes and sex chromosomes of infertile men.[1]

INDICATIONS

Generally, FISH analysis of human sperm is used to perform a basic study of aneuploidy. According to literature, a FISH sperm analysis is indicated in cases of (i) recurrent pregnancy loss, (ii) repeated in vitro fertilization failure, (iii) abnormal seminal parameters including count, motility, and morphology, and (iv) genetic aberrations (such as but not limited to, Robertsonian and reciprocal translocations, chromosome inversions, ring chromosomes, and numerical sex chromosome aberrations).[5, 6] Additionally, FISH analysis of human sperm is indicated in men exposed to potential mutagens, and to evaluate the impact of lifestyle factors on sperm aneuploidy rate.[7–9]

DESCRIPTION OF METHODS

Considering the diverse causes of man infertility and referral reasons starting sample can be ejaculated, epididymal, or testicular-derived sperm. As already mentioned, the chromosomes that are generally considered for analyses are 13, 18, 21, X, and Y, given that those aneuploidies are compatible with life. Nevertheless, all chromosomes can be studied by us of probes specific for each individual chromosome. The technical procedure of interphase FISH on human sperm nuclei has several phases: A) sample preparation, B) hybridization, C) post-hybridization washes, and D) visualization.[10, 11] As steps B–D are following standard conditions (as e.g. described by the provider of corresponding FISH probes), they are not further detailed here.

SAMPLE PREPARATION

As sperms (including their DNA) are extremely condensed, being caused by extensive intermolecular disulfide cross-links, they must be decondensed prior to the FISH procedure. To allow DNA probes access to the chromatin Otto's solution is applied {100 ml Tris-HCl buffer 10 mmol/l, pH 7.4, 220 mg KCl (30 mmol/l), 100 mg MgCl$_2$×6H$_2$O (5 mmol/l), and 46 mg dithioerythritol (3 mmol/l)}.

1. Suspend ejaculate/sperm sample in 8 ml PBS; use appropriate centrifuge tubes. Centrifuge at 1,250 rpm for 10 min. Remove supernatant carefully and repeat washing two times.
2. After resuspending the pellet in ~1–2 ml of PBS vortex, add 20 drops of Otto's solution (drop by drop); then suspend in 8 ml Otto's solution and incubate at 37°C for 30 min.
3. Centrifuge at 1,250 rpm for 10 min and remove supernatant carefully.
4. After resuspending the pellet in ~1–2 ml of Otto's solution, add 20 drops of ice-cold, freshly prepared Carnoy's fixative (3:1 methanol: acetic acid—drop by drop) while mixing on a vortexer, then add fixative to obtain a final volume of 8 ml.
5. Repeat step 3*; add 4ml fixative.
6. Repeat step 5 two times.
7. Drop suspension on the slide and let air dry.
* If only a small pellet is visible, drop the sample onto a slide immediately.

In order to achieve good FISH-results, pretreatment of now decondensed cells with pepsin is recommended:

1. Add 50 µl pepsin solution (10%) to a Coplin jar with 100 ml of distilled H$_2$O pre-warmed at 37°C in a water bath; mix thoroughly. Place slides and incubate for 8 min in a water bath at 37°C.
2. Incubate 1×2 min in 1×PBS in a Coplin jar.
3. Incubate 10 min in 1% formaldehyde in PBS-MgCl$_2$.
4. Repeat step 2.
5. Dehydrate the slides in ethanol (70%, 90%, 100%) for 2 min each and let slides air-dry.

RESULTS AND INTERPRETATION

As reported in the literature, apparently healthy men exhibit intra-individual and inter-individual disomy and diploidy rates in sperm.[12, 13] Besides needing to determine cut-off rates in sperm derived from fertile males, in general, interphase FISH can also lead to artificially induced signals (see section later). Thus, to obtain reliable results, usually about 1,000 to 10,000 sperm nuclei per patient have to be analyzed and aligned with cut-off rates. Additionally, signal scoring criteria can aid objective analysis. These criteria include: (i) overlapped spermatozoa or sperm heads without a well-defined boundary are not counted, (ii) in cases of disomy or diploidy all signals should have the same intensity and be separated from each other by a distance longer than the diameter of each signal, and (iii) nullisomies are not directly scored.[14]

Interpretation of abnormal findings from FISH sperm analysis can be evaluated using the quantitative or the qualitative approach. The quantitative approach interprets FISH results as a numerical value (a score, essentially) that would indicate the patient's "degree of risk". This approach has some shortcomings that are the features intrinsic to this technique. It is very difficult (if not impossible) to study simultaneously all the chromosomes of the karyotype (normally only X, Y, 13, 18, and 21 are analyzed), and also not all possible chromosomal abnormalities can be registered. Accordingly, this method tends to underestimate the extent of anomaly rates. In the qualitative approach, significant increases in aneuploidy count would have to be interpreted as evidence that there are meiotic defects, thus indicating that the quality of the spermatogenesis is not at optimum. Such an approach has the potential to identify most "at-risk" patients and should be favored over the quantitative one.[15]

LIMITATIONS AND PITFALLS

Shortcomings of FISH sperm analysis are (i) the inability to access structural chromosomal aberrations and (ii) the fact that sub-regions of chromosomes are studied, as locus-specific probes are normally used. Besides, interphase FISH analysis can lead to false positive results as decreased or increased numbers of signals. The signal number may be decreased due to (a) insufficient tissue penetrability resulting in reduced hybridization efficiency, (b) a certain number of signals that are physically so close together that they appear as one, and (c) analysis of damaged or overlapping nuclei. On the other side, signal number may be increased by (I) signal splitting due to dispersed or damaged DNA, and (II) signals caused by artifacts such as bacteria or background fluorescence.

TROUBLESHOOTING

The most common technical points that can interfere with the quality of FISH in human sperm are the following: no or weak signals, too much background, and diffuse signals and cells. Reasons for signal loss may be too short denaturation or hybridization time. Another possibility could be too stringent washing conditions after hybridization. An increase in salt concentration or decrease of temperature could resolve this, while higher dextran sulfate concentration and increase in probe concentration could lead to better signals.

Usually, too much background is the result of some kind of "dirtiness" on slide surface and can be reduced during hybridization or post-hybridization washing by increasing the temperature or decreasing salt concentration. Additionally, the application of directly labeled probes reduces background fluorescence. During the washing steps, it is important to prevent the slide surfaces from drying out; otherwise background problems may arise. In addition, slides can be dirty at delivery and may need some cleaning before use in FISH-procedure/before dropping cell suspension on them.

To resolve diffuse signals and cells usually decreasing denaturation time and temperature by 2°C, and/or shortening decondensation time will help to settle this issue. Pepsin pretreatment conditions, as well as the denaturation time of the target DNA, should be tested in each laboratory at the introduction of this kind of test.

If pepsin concentration is too stringent, it can result in clean slides without any remaining nuclei.

CONCLUSIONS

FISH is the fastest and most accurate test to assess the aneuploidy of chromosomes in sperm nuclei on a single-cell level. Clinically, the result of FISH sperm analysis can be used in counseling couples affected by male infertility, to come to an informed and consented choice regarding their further reproductive plans.

REFERENCES

1. Chandra, A.; Martinez, G.M.; Mosher, W.D.; Abma, J.C.; Jones, J. Fertility, family planning, and reproductive health of U.S. women: Data from the 2002 National Survey of Family Growth. *Vital Health Stat.* **2005**, 23, 1–160.
2. Zegers-Hochschild, F.; Adamson, G.D.; Dyer, S.; Racowsky, C.; de Mouzon, J.; Sokol, R.; Rienzi, L.; Sunde, A.; Schmidt, L.; Cooke, I.D.; Simpson, J.L.; van der Poel, S. The international glossary on infertility and fertility care, 2017. *Hum. Reprod.* **2017**, 32, 1786–1801.
3. ESHRE PGT-SR/PGT-A Working Group; Coonen, E.; Rubio, C.; Christopikou, D.; Dimitriadou, E.; Gontar, J.; Goossens, V.; Maurer, M.; Spinella, F.; Vermeulen, N.; De Rycke, M. ESHRE PGT Consortium good practice recommendations for the detection of structural and numerical chromosomal aberrations. *Hum. Reprod. Open.* **2020**, 2020, hoaa017.
4. Rudak, E.; Jacobs, P.A.; Yanagimachi, R. Direct analysis of the chromosome constitution of human spermatozoa. *Nature.* **1978**, 274, 911–913.
5. Ramasamy, R.; Besada, S.; Lamb, D.J. Fluorescent in situ hybridization of human sperm-diagnostics.; indications.; and therapeutic implications. *Fertil. Steril.* **2014**, 102, 1534–1539.
6. Fakhrabadi, M.P.; Kalantar, S.M.; Montazeri, F.; Ashkezari, M.D.; Fakhrabadi, M.P.; Yazd, S.S.N. FISH-based sperm aneuploidy screening in male partner of women with a history of recurrent pregnancy loss. *Middle East Fert. Soc. J.* **2020**, 25, 23.
7. Saad, A.A.; Hussein, T.; El-Sikaily, A.; Abdel-Mohsen, M.A.; Mokhamer, E.H.; Youssef, A.I.; Mohammed, J. Effect of polycyclic aromatic hydrocarbons exposure on sperm DNA in idiopathic male infertility. *J. Health Pollut.* **2019**, 9, 190309.
8. Srám RJ.; Binková, B.; Rössner, P.; Rubes, J.; Topinka, J.; Dejmek, J. Adverse reproductive outcomes from exposure to environmental mutagens. *Mutat. Res.* **1999**, 428, 203–215.
9. Robbins, W.A.; Vine, M.F.; Truong, K.Y.; Everson, R.B. Use of fluorescence in situ hybridization (FISH) to assess effects of smoking, caffeine, and alcohol on aneuploidy load in sperm of healthy men. *Environ. Mol. Mutagen.* **1997**, 30, 175–183.
10. Sarrate, Z.; Anton, E. Fluorescence in situ hybridization (FISH) protocol in human sperm. *J. Vis. Exp.* **2009**, (31), 1405.
11. Rauch, A. Human sperm cells. In: *FISH Technology*; Rautenstrauss, B.W.; Liehr, T. Eds. Springer, Berlin, Heidelberg, **2002**, pp. 127–137.
12. García-Mengual, E.; Triviño, J.C.; Sáez-Cuevas, A.; Bataller, J.; Ruíz-Jorro, M.; Vendrell, X. Male infertility: Establishing sperm aneuploidy thresholds in the laboratory. *J. Assist. Reprod. Genet.* **2019**, 36, 371–381.

13. Schultz, H.; Mennicke, K.; Schlieker, H.; Al-Hasani, S.; Bals-Pratsch, M.; Diedrich, K.; Schwinger, E. Comparative study of disomy and diploidy rates in spermatozoa of fertile and infertile men: A donor-adapted protocol for multi-colour fluorescence in situ hybridization (FISH). *Int. J. Androl.* **2000**, 23, 300–308.

14. Blanco, J.; Egozcue, J.; Vidal, F. Incidence of chromosome 21 disomy in human spermatozoa as determined by fluorescent in-situ hybridization. *Hum. Reprod.* **1996**, 11, 722–726.

15. Sarrate, Z.; Vidal, F.; Blanco, J. Role of sperm fluorescent in situ hybridization studies in infertile patients: Indications.; study approach.; and clinical relevance. *Fertil. Steril.* **2010**, 93, 1892–1902.

10 FISH—in Spontaneously Aborted Products of Conception

Thomas Liehr

CONTENTS

INTRODUCTION

Infertility is a major problem in Western countries, but also—either less expressed or less registered—in all human societies around the world. Besides the inability to conceive, some couples experience (repeated) abortions at early, middle or late pregnancy stages.[1] Such couples should attract gynecologist's attention during routine patient admission, and these couples should be offered (cyto)genetic studies for both partners.[2] In a subset of these cases, a gonosomal mosaic, a small supernumerary marker chromosome or a balanced rearrangement can be traced, as being causative for the problem.[3, 4] Besides, and especially in cases where no such conclusive results can be found, the spontaneously aborted products of conception (PoCs) themselves should be studied genetically.[5] Chromosomal aberrations identified in these PoCs can be important (i) to make such couples better understand why an abortion has happened,[6] and (ii) also as they can be a hint on a cryptic gonadal mosaicism being present in male or female partner.[7]

In some settings, DNA extraction from PoCs and subsequent chromosome-microarray,[8] multiplex ligation-dependent probe amplification (MLPA),[9] or simple STR (= short tandem repeat DNA) -analyses (own unpublished data) may be used to find genomic imbalances, that have being causative for the abortion. Another straightforward method is tissue cultivation from PoCs and subsequent chromosome analyses by GTG-banding[5, 10, 11]; alternatively and if available, also formalin-fixed formalin embedded material of PoCs may be analyzed by molecular cytogenetics.[12] However, the latter is missing the opportunity to possibly get an informative banding cytogenetic result. Furthermore, as PoC-cell-cultivation is prone to culture failure,

DOI: 10.1201/9781003223658-10

it has been shown that interphase molecular cytogenetics may be a way out here to obtain nonetheless informative results.[5]

SAMPLES/TISSUES

Cytogenetic preparations with and without metaphase spreads, but at least containing sufficient interphase nuclei from PoCs are used for the here described analyses. Preparation is done using standard techniques.[13]

DESCRIPTION OF METHODS

The method applied here is a standard fluorescence in situ hybridization (FISH) approach—see e.g.[14] Commercially available probes, either centromere- or locus-specific ones, are applied. Preferentially used are centromeric probes, still, for chromosomes 13 and 21, and 14 and 22 only the probes D13/21Z1 and D14/22Z1 staining both corresponding chromosome pairs are available. In addition, problems may arise with centromeric probe D1Z7 in 1p11.1–1q11 being identical to D5Z2 in 5p11-q11.1 and D19Z3 in 19q11; chromosome 1 may be identified by D1Z5 in 1q11-q12, but additional (derivatives of) chromosomes 5 and 19 cannot be indubitably distinguished in metaphase and not at all in interphase.[15, 16] Accordingly, commercially available locus-specific probes can be selected if these chromosomes shall be studied.

As the available amount of PoC-material normally is restricted, at least two- to multicolor-FISH settings are recommended. According to available commercially available probes applied multicolor-FISH probe sets may be put together (e.g. Table 10.1). In a study from 2017 we had an ~37% detection rate of abortion causing chromosomal aneuploidies when concentrating on chromosomes 13/21, 14/22, 15, 16, 17, 18, X and Y. Suited probe sets which are composed of commercially available probes from Abbott/Vysis and Cytocell are listed in Table 10.1. Probe sets 1 to 3 (Table 10.1) should be applied subsequently to samples to be tested. In case of five signals for D13/21Z1 or D14/22Z1 probes, the result needs to be checked by locus-specific (LSI) probes, as listed in Table 10.1. The latter is necessary, as it can easily be that instead of a trisomy 13, 14, 21 or 22 suited chromosome-specific LSI probes do not reveal an aberrant signal pattern. This may be due to chromosomal heteromorphisms, which can lead to additional signals in the centromeric regions of another chromosome.[17] Additionally, such a result can be due to a partial trisomy of one of the four chromosomes in the sense of a small supernumerary marker chromosome.[15]

For the here-suggested set-up and composition of probe sets (Table 10.1), commercially available probes are suggested. Thus, the established lab internal standard FISH-procedure may be used, or the protocol as suggested by the provider of the commercial probes can be applied. It has to be tested under each individual lab-condition if a pretreatment with RNAse and/or pepsin is necessary or can be skipped; this is dependent on the plasma load of interphase nuclei after cytogenetic work-up of PoCs.

TABLE 10.1
Probe Sets Adapted from[5] Consisting of Commercially Available Probes Are Listed; With This Setting Polyploidies, Gonsomal Aberrations and Numerical Aberrations of Chromosomes 13, 14, 15, 16, 17, 21 and 22 Can Be Accessed.

Probe Set	Probes	Location	Fluorochrome
1	D13/21Z1	13p11.1-q11 and	SpectrumGreen (SG)
	D15Z3	21p11.1-q11.1	TexasRed (TR)
	D18Z1	15q11	Diethylaminocoumarin
		18p11.1-q11.1	(DEAC)
2	D14/22Z1	14p11.1-q11.1 and	SG
	D16Z2	22p11.1-q11.1	TR
		16p11.1-q11.1	
3	D17Z1	17p11.1-q11.1	SpectrumOrange (SO)
	DXZ1	Xp11.1-q11.1	DEAC
	Yq12	Yq12	SG
if D13/21Z1 gives 5 signals	LSI 13/21	13q14 and 21q22.13-q22.2	SG and SO
if D14/22Z1 gives 5 signals	LSI D22S75/ ARSA	22q11.2 and 22q13.3	SO and SG

YIELDS

Studying PoCs with the suggested probe sets (Table 10.1) identified in a pilot study of 286 PoCs chromosomal abnormalities in 106 cases. Polyploidy was present in 23/106 cases, sex-chromosome abnormalities in 37/106 cases and trisomies in the remainder 46 cases. Twenty-two of the latter 46 cases were trisomy 16.[5]

CONCLUSIONS

Taking advantage of interphase FISH to unravel reasons of (repeated) abortions is still only scarcely used in human genetic labs. The possibilities to obtain a diagnosis on the reasons for an abortion are still by far not exhausted in routine settings; not only the study referred to here,[5] but also other authors highlighted this option,[5, 8–12] which among others provides also the advantage of a single cell specific evaluation, and thus to detect mosaic conditions, too.

REFERENCES

1. Flyckt, R.; Falcone, T. Infertility: A practical framework. *Cleve Clin. J. Med.* **2019**, 86, 473–482.
2. Larsen, E.C.; Christiansen, O.B.; Kolte, A.M.; Macklon, N. New insights into mechanisms behind miscarriage. *BMC Med.* **2013**, 11, 154.
3. Liehr, T. Small supernumerary marker chromosomes detected in connection with infertility. *Zhonghua Nan. Ke. Xue.* **2014**, 20, 771–780.

4. Manvelyan, M.; Schreyer, I.; Höls-Herpertz, I.; Köhler, S.; Niemann, R.; Hehr, U.; Belitz, B.; Bartels, I.; Götz, J.; Huhle, D.; Kossakiewicz, M.; Tittelbach, H.; Neubauer, S.; Polityko, A.; Mazauric, M.L.; Wegner, R.; Stumm, M.; Küpferling, P.; Süss, F.; Kunze, H.; Weise, A.; Liehr, T.; Mrasek, K. Forty-eight new cases with infertility due to balanced chromosomal rearrangements: Detailed molecular cytogenetic analysis of the 90 involved breakpoints. *Int. J. Mol. Med.* **2007**, 19, 855–864.

5. Tkach, I.R.; Huleyuk, N.L.; Zastavna, D.V.; Weise, A.; Liehr, T.; Ciszkowicz, E.; Tyrka, M. Chromosomal aberrations in spontaneously aborted products of conception from Ukraine. *Biopolymers Cell.* **2017**, 33, 424–433.

6. Griebel, C.P.; Halvorsen, J.; Golemon, T.B.; Day, A.A. Management of spontaneous abortion. *Am. Fam. Physician.* **2005**, 72, 1243–1250.

7. Kamel, A.K.; Abd El-Ghany, H.M.; Mekkawy, M.K.; Makhlouf, M.M.; Mazen, I.M.; El Dessouky, N.; Mahmoud, W.; Abd El Kader, S.A. Sex chromosome mosaicism in the gonads of DSD patients: A karyotype/phenotype correlation. *Sex. Dev.* **2015**, 9, 279–288.

8. Gao, J.; Liu, C.; Yao, F.; Hao, N.; Zhou, J.; Zhou, Q.; Zhang, L.; Liu, X.; Bian, X.; Liu, J. Array-based comparative genomic hybridization is more informative than conventional karyotyping and fluorescence in situ hybridization in the analysis of first-trimester spontaneous abortion. *Mol. Cytogenet.* **2012**, 5, 33.

9. Kim, J.W.; Lyu, S.W.; Sung, S.R.; Park, J.E.; Cha, D.H.; Yoon, T.K.; Ko, J.J.; Shim, S.H. Molecular analysis of miscarriage products using multiplex ligation-dependent probe amplification (MLPA): Alternative to conventional karyotype analysis. *Arch. Gynecol. Obstet.* **2015**, 291, 347–354.

10. Vorsanova, S.G.; Kolotii, A.D.; Iourov, I.Y.; Monakhov, V.V.; Kirillova, E.A.; Soloviev, I.V.; Yurov, Y.B. Evidence for high frequency of chromosomal mosaicism in spontaneous abortions revealed by interphase FISH analysis. *J. Histochem. Cytochem.* **2005**, 53, 375–380.

11. Jobanputra, V.; Sobrino, A.; Kinney, A.; Kline, J.; Warburton, D. Multiplex interphase FISH as a screen for common aneuploidies in spontaneous abortions. *Hum. Reprod.* **2002**, 17, 1166–1170.

12. Russo, R.; Sessa, A.M.; Fumo, R.; Gaeta, S. Chromosomal anomalies in early spontaneous abortions: Interphase FISH analysis on 855 FFPE first trimester abortions. *Prenat. Diagn.* **2016**, 36, 186–191.

13. Weise, A.; Liehr, T. Pre- and postnatal diagnostics and research on peripheral blood, bone marrow chorion, amniocytes, and fibroblasts. In: *Fluorescence In Situ Hybridization (FISH): Application Guide*; Liehr, T., Ed. 2nd Edition. Springer, Berlin, **2017**, pp. 171–180.

14. Weise, A.; Liehr, T. Background. In: *Fluorescence In Situ Hybridization (FISH): Application Guide*; Liehr, T., Ed. 2nd Edition. Springer, Berlin, **2017**, pp. 1–14.

15. Nietzel, A.; Rocchi, M.; Starke, H.; Heller, A.; Fiedler, W.; Wlodarska, I.; Loncarevic, I.F.; Beensen, V.; Claussen, U.; Liehr, T. A new multicolor-FISH approach for the characterization of marker chromosomes: Centromere-specific multicolor-FISH (cenM-FISH). *Hum. Genet.* **2001**, 108, 199–204.

16. Liehr, T. *Benign & Pathological Chromosomal Imbalances Microscopic and Submicroscopic Copy Number Variations (CNVs) in Genetics and Counseling.* Academic Press, New York, **2014**.

17. Liehr, T.; Ziegler, M. Rapid prenatal diagnostics in the interphase nucleus: Procedure and cut-off rates. *J. Histochem. Cytochem.* **2005**, 53, 289–291.

11 FISH—Characterization of Chromosomal Alterations, Recombination, and Outcomes after Segregation

Ilda Patrícia Ribeiro, Eunice Matoso,
Joana Barbosa Melo, Ana Jardim,
Thomas Liehr and Isabel Marques Carreira

CONTENTS

DOI: 10.1201/9781003223658-11

INTRODUCTION

The process leading to the formation of haploid gametes from diploid cells through one round of DNA replication and two rounds of chromosome segregation is called meiosis, or more precisely Meiosis I (MI) and Meiosis II (MII).[1] Both meiosis and mitosis begin with replication of DNA in order to originate a cell with four chromatids of each type of chromosome, two from the mother and two from the father.[2] During the first meiotic division (MI), the separation of maternal and paternal copies of each chromosome occurs: this is the so-called reductional division. It is in the second meiotic division (MII) that the separation of sister chromatids occurs: this is the equational division (very similar to mitosis). The resulting four chromatids are distributed to four different nuclei, which occurs simultaneously for all chromosomes in the two rounds of chromosome segregation.[2] During the meiosis process, there is not only a reduction of chromosome number to a haploid set but also the aleatoric segregation of maternal and paternal chromosomes, and genetic combinations in the offspring due to exchange of genetic material in maternal and paternal homologues through recombination.[3] The reduction in chromosome number is pivotal to restore diploid chromosome complement in the subsequent generation through the union of two gametes during sexual reproduction, ensuring continuity of the species.[4]

There are three significant deviations in MI in comparison to mitosis:

(i)   homologous chromosomes are physically linked together through chiasmata, the products of homologous recombination that occurred at pachytene;

(ii) sister kinetochores attach to microtubules from opposite spindle pole (mono-oriented); and

(iii) cohesion is lost in a step-wise manner on chromosome arms during the first division but preserved at the pericentromere.[5]

Malfunction in these three building blocks of chromosome segregation presents as consequence in an inequitable distribution of sister chromatids, and thus in numerical alterations.[6] Therefore, the complex behavior of chromosomes during the two meiotic divisions pave the way for errors to occur. Errors in homologous chromosome segregation during either meiosis or mitosis can lead to an aberrant chromosome number, which is termed aneuploidy (occurrence of one or more extra or missing chromosomes leading to an unbalanced chromosome complement); this could occur in the postnatal, prenatal, or preimplantation stage, having significant clinical consequences. The identification of these chromosomal errors and the understanding of the molecular and cellular mechanisms involved in this process is pivotal for prevention strategies, genetic counseling, and personalized clinical treatment.[7] Frequently aneuploidy involves the whole chromosome, but partial aneuploidy is also common due to chromosome breakage and rearrangement.[8] Chromosomal alterations distributed throughout the whole body (constitutional) could occur during gamete formation, at fertilization, or in the embryo before implantation.[8] Usually, aneuploidy is the result of a failure to segregate chromosomes in cell division, MI or MII that includes non-disjunction, premature disjunction, or anaphase lag.[9] Most chromosome-missegregation events in human meiosis occur because of incorrect segregation of homologues during MI (non-disjunction).[10] A non-disjunction during MI of oocyte formation is more frequent than in sperm; however, in the majority of possible sex chromosome aneuploidies, a paternal meiotic non-disjunction accounts for 50% of cases.[11] Human female meiosis is error prone, presenting an aneuploidy rate of about 15% to 30–70%, depending on maternal age, compared with 1–4% in sperm.[12, 13] So, human aneuploidy has often maternal origin, with exception for 47,XYY (presence of an extra Y-chromosome) and 45,X (a missing sex chromosome), which are often due to paternal or postfertilisation errors.[6] These discrepancies could be explained by the timing differences in female and male meiosis. Live cell imaging screening of mammalian oocytes showed that chromosome segregation defects are mostly due to an error-prone chromosome-mediated spindle assembly and kinetochore splitting.[14] It is also important to stress that the consequences are different depending on which meiotic cycle is involved in non-disjunction, leading for example to a disomic gamete. If non-disjunction occurs in MI, the gamete would have 24 chromosomes from both grandparents (paternal and maternal) homologs, but if occurs in MII, the gamete would have 24 chromosomes with both copies exclusively from the paternal or maternal homolog.[9] Accordingly, segregation of chromosomes in human meiosis is often error-prone, having as a consequence a numerically unbalanced chromosome complement. Additionally, chromosome segregation errors could also emerge during mitotic cell divisions after fertilization, having as consequence mosaicism—the presence of both aneuploid and normal diploid cells within one individual.[6] Mosaicism is observed in 1–2% of pregnancies undergoing chorionic villus sampling and in 0.1% of amniocenteses, and it can result in a less severe phenotype.[6] However, this phenotype/genotype interpretation is especially difficult in prenatal diagnosis, since we cannot deduce at which stage of the development the error occurred. Although mosaicism can encompass either autosomes or gonosomes it is more common in sex chromosomes, being observed in about 0.2% of fetuses prenatally.[15] The clinical phenotype of mosaicism varies with the percentage and tissue distribution of the aneuploid cells. A distinct event is the chimerism, where different cell lines are originated from more than one zygote.

CHROMOSOMAL ALTERATIONS

In 1956, a new era of clinical cytogenetics started with the identification of the exact chromosome number in humans, and therefore the description of several chromosomal syndromes with altered chromosome numbers.[15] Chromosomal alterations could be categorized as numerical or structural and could involve more than one chromosome. Numerical chromosomal alterations are more common than structural ones and represent a deviation from the normal diploid number for a given species. This category can be subdivided in two groups: (i) aneuploidy (a group in which individual chromosomes may be either missing or present in one or more additional copies) and (ii) polyploidy (here whole haploid sets of chromosomes may be added or lost, as triploidy (3n) in human with 69 chromosomes, or tetraploidy (4n) in human 92 chromosomes).[16] Triploidy or tetraploidy are lethal disorders often observed in spontaneous abortions and rarely in newborns with a reduced survival time. Triploidy is one of the most common chromosome abnormalities occurring in ~1% of recognized conceptions, and 10% of recognized miscarriages.[17] Triploidy can reflect the double contribution coming from the father (di-*andry*) or the mother (di-*gyny*). *Diandry* is usually the consequence of two sperm simultaneously fertilizing the ovum.[18–20] *Digyny* could be the result of nondisjunction of the entire set at either the MI or MII in oogenesis, or a polar body may fail to be extruded.[21]

Structural rearrangements result from chromosome breakage and reunion within a single chromosome or between two or more non-homologous chromosomes originating either balanced or unbalanced complements.[15] Balanced rearrangement means that there is usually no alteration in chromosome content, while unbalanced ones lead to gains or losses of chromosomal material. At birth, structural rearrangements (balanced and unbalanced) were reported in about 1/400 infants.[22] Significant clinical phenotypes were observed in carriers of unbalanced rearrangements, as a consequence of loss and/or duplication of genetic material. Some examples of unbalanced rearrangements are deletions, duplications, rings and isochromosomes, and of balanced rearrangements are inversions (paracentric and pericentric), insertions (direct or inverted), and translocations (reciprocal and Robertsonian).[15] Additionally, we also have small marker chromosomes (SMCs), which are usually supernumerary (sSMCs).

Balanced as well as unbalanced structural abnormalities can be inherited from a carrier parent or could arise as *de novo*. For inherited balanced structural rearrangements, the risk for a clinical phenotype is low; however, for *de novo* rearrangements, this risk is higher, even if the rearrangement looks balanced: submicroscopic deletions/duplications at the breakpoints and alterations in the genes or regulatory elements near to the breakpoints cannot be excluded.[22] So, the management of balanced rearrangements identified in prenatal cases is difficult, especially with a *de novo* origin. It is important to highlight that a balanced carrier (heterozygote) could present with difficulties in reproduction, namely infertility, spontaneous abortions, or abnormal offspring, which is a result of anomalous pairing and segregation at prophase I.[22]

Moreover, chromosomal alterations can also be categorized as constitutional or acquired. Constitutional chromosomal alterations occur during gametogenesis or early embryogenesis, affecting all, or great fraction of an organism's cells,[11] with an estimated incidence of 20–50% of all human conceptions.[7] Acquired chromosomal abnormalities occur postnatally during adulthood, affecting a single clone of cells with a specific distribution in the body, and are usually associated with cancer.

Human cytogenetic nomenclature describing each type of identified chromosomal alterations was developed by an expert committee and is published in collaboration with Cytogenetic and Genome Research (formerly: Cytogenetics and Cell Genetics) since 1963. The current version was prepared and updated by the International Standing Committee on Human Cytogenomic Nomenclature (ISCN) and adopted in 2020.[23]

Limited due to chromosome resolution at the level of chromosome banding (3–5 Mb), on the basis of recognition and interpretation, masked or cryptic alterations may be difficult if not impossible to ascertain. The advent of molecular cytogenetics, particularly the fluorescence in situ hybridization (FISH) technique, have overcome, in the past 3 decades, the gap between conventional banding cytogenetics and molecular biology. Specific nucleic acid sequences could be fluorescent-labeled and located in interphase cells or metaphase chromosomes, being an important tool in clinical genetic laboratories. The orientation must be given by the clinic or by a previous cytogenetic analysis, which will indicate what probe(s)/sequence(s) to use and screen. Meanwhile also chromosome microarray approaches are an indication for further FISH-tests.

Numerous and diverse DNA sequences have been developed and used as probes: (i) unique sequence; (ii) repetitive sequence (such as α-satellite and telomeric DNA); (iii) locus-specific DNA obtained by PCR amplification, large genomic DNA sequences cloned into cosmids, bacterial artificial chromosomes (BACs), P1-derived artificial chromosomes (PACs), yeast artificial chromosomes (YACs); and (iv) chromosome band or arm specific sequences generated by microdissection or as DNA libraries established by chromosome flow sorting.[24]

The application of diverse strategies, with combination of specific probe sets in multicolor combinations, has increased the characterization of simple to complex chromosomal aberrations. With FISH, genetic alterations can be analyzed at the single-cell level, screening simultaneously different chromosomes and loci, and determine mosaicism or clonal variability. Resolution of FISH largely depends of the target material: at metaphase level varies between 2–5 Mb, but in interphase nuclei varies between 2 Mb and 50 kb. In fiber-FISH, the resolution can achieve 5–500 kb.[24]

NUMERICAL ALTERATIONS

Aneuploidy can affect any one of the 23 pairs of chromosomes. In humans, aneuploidy leads to about 35% of miscarriages and 4% of stillbirths.[10] Aneuploidy is also present in around 30–60% of embryos, in 30–70% of oocytes and in 0.3% of newborns (with the most common abnormalities being trisomy 21 and sex chromosome trisomies), facing developmental disabilities and intellectual disability.[10] Regarding human aneuploidy, examples of trisomic conditions compatible with

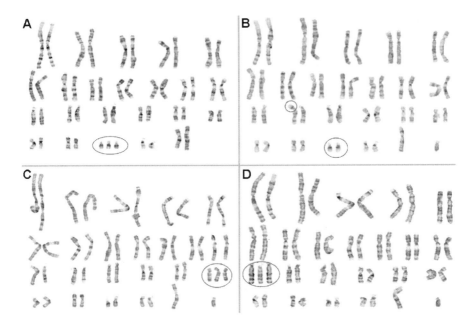

FIGURE 11.1   Representative results of GTG-banding—karyograms of

A) Down syndrome: 47,XX,+21.
B) Down syndrome: 46,XY,der(14;21)(q10;q10),+21.
C) Edwards syndrome: 47,XY,+18.
D) Pätau syndrome: 47,XY,+13.

full-term pregnancies are the Down (trisomy 21), Edwards (trisomy 18), and Pätau (trisomy 13) syndromes (Figure 11.1).

Down syndrome is one of the best-described and most frequent chromosome disorders that causes intellectual disability. The incidence of Down syndrome is approximately 1/600 newborns, being originated in almost 94% of cases from meiotic nondisjunction.[15, 20] In approximately 95% of cases, the extra chromosome 21 is of maternal origin, and around 80% of those are due to an error during MI.[15] Nevertheless, almost 4% of Down syndrome patients present an unbalanced Robertsonian translocation involving chromosome 21 and the long arm of chromosomes 13, 14, 15, 21, or 22.[15] It is important to highlight that if a parent is a balanced 21/21 isochromosome carrier, this represents a 100% risk of having offspring with Down syndrome. Female carriers of balanced 14/21 or 21/22 Robertsonian translocations have a 10–15% risk for unbalanced Down syndrome offspring in comparison to 5% for male carriers.[15] These translocations could be *de novo* or inherited from a balanced carrier parent (usually the mother). Additionally, mosaicism for trisomy 21 is also the cause of Down syndrome in almost 2% of the cases.[15]

Edwards syndrome has a frequency of about 1/6000 live births and a poor post-natal survival estimate (~90% die in the first 6 months).[20] Few cases of trisomy 18 mosaicism were reported, but some patients with unbalanced translocations involving all or most of chromosome 18 long arm were also described presenting with trisomy 18 features.[15]

Pätau syndrome has an estimated incidence of about 1/12,000 live births, and 90% of these patients rarely survive the newborn period.[20] The origin of the extra chromosome 13 is frequently from a maternal meiotic non-disjunction error, but may also arise from an unbalanced Robertsonian translocation involving chromosome 13 (in ~20% of the cases) or yet from trisomy 13 mosaicism, which is a rare event and presents a less severe clinical phenotype.[15]

Likewise, aneuploidies involving sex chromosomes are also a possible consequence of errors in chromosome segregation but present less severe phenotypes in comparison with somatic trisomies. The gene dosage imbalances for genes encoded in sex chromosomes are lower than for the genes encoded on somatic chromosomes, and besides that there is the inactivation process of the extra X chromosome(s), if there is more than one present per cell.[25] Examples of sex chromosome abnormalities are Turner (45,X), triple X (XXX), Klinefelter (47,XXY) and 47,XYY syndromes (Figure 11.2).[25]

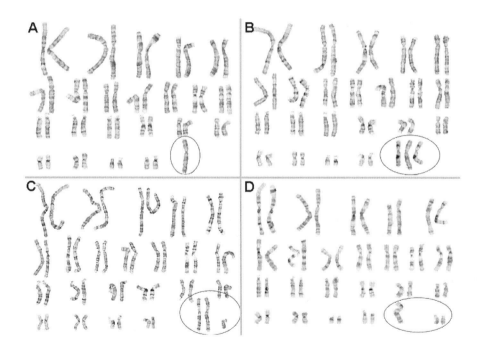

FIGURE 11.2   Representative results of GTG-banding—karyograms of

A) Turner syndrome: 45,X.
B) Triple X syndrome: 47,XXX.
C) Klinefelter syndrome: 47,XXY.
D) 'XYY' syndrome: 47,XYY.

Turner syndrome frequency is around 1/1800 newborn females, with approximately 50% of patients a having 45,X karyotype.[15, 20] Other carriers of this syndrome can have isochromosomes, short-arm deletions or rings involving X-chromosome.[6] Mosaicism is also a possible cause for this syndrome in 30% of patients, which present both a 45,X cell lineage and a 46,XX cell lineage and/or one containing a rearranged X-chromosome. [26] The origin of monosomy X seems to be primarily an error in male meiosis, as the single X-chromosome derives from the mother in 80% of 45,X cases.[6] Although monosomy X is associated with infertility some carriers have healthy babies, supporting the theory of mosaicism and the presence of a normal cell line in the gonads.

Klinefelter syndrome has a frequency of almost 1/500 newborn males and the most common karyotype is 47,XXY.[15, 20] Mosaicism is also detected in few patients, which encompasses normal 46,XY cells combined with cells with two or more X chromosomes. Variants of Klinefelter syndrome are also described, involving more than two X-chromosomes and multiple Y-chromosomes, such as 48,XXYY, 48,XXXY, and 49,XXXXY, that are associated to an increased severity of phenotype.[15] This syndrome can be of maternal or paternal origin resulting from a meiotic non-disjunction error.[6]

Trisomy X syndrome frequency is around 1/900 in newborn females, originating most often from a nondisjunction error in maternal MI.[6, 20] Carriers of this chromosomal alterations are often fertile and with no obvious phenotypic alterations.

All the previously discussed numerical chromosomal alterations are easily diagnosed following conventional banding cytogenetics that can take 3 to 12 days for a final report, depending whether peripheral blood, amniotic fluid or fibroblasts are used as biological material. Every time that a rapid diagnosis of an aneuploidy is needed, a specific FISH analysis can be done in a native, not in vitro cultivated biological material, such as native amniocytes, and a result can be obtained in 24 hours. However, although we can have this rapid diagnosis, for example, for a trisomy 21 (Figure 11.3), we cannot ascertain through this rapid test whether the trisomy 21 is

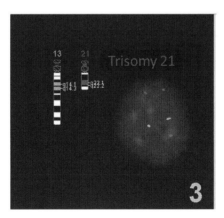

FIGURE 11.3   FISH in native amniocyte-nuclei showing hint of a trisomy 21: 3 SpectrumOrange signals of locus-specific probes for the long arm of chromosome 21 and 2 SpectrumGreen signals (= normal) for locus-specific probes for the long arm of chromosome 13.

associated with a Robertsonian translocation or any other structural rearrangement. In order to have that information, a karyotype has to be done.

UNBALANCED STRUCTURAL REARRANGEMENTS

In cases with unbalanced structural chromosomal rearrangements, there is a disequilibrium compared to normal gene dosage of a particular genomic region that affects the chromosomal structure. Commonly, they have an impact on the phenotype of the carrier and can be detected either by banding cytogenetics (if they are larger than 3–5 Mb) or by FISH analysis (especially, when there is a clinical diagnosis that needs to be confirmed). Unbalanced structural rearrangements include (a) deletions (del), (b) duplications (dup), (c) ring chromosomes (r), (d) isochromosomes (i), or (e) small supernumerary marker chromosomes (sSMCs) (Figure 11.4).

Deletion results from loss of chromosomal material of a specific chromosome. The clinical phenotype is dependent on size, gene content, and location of deleted sequences within the genome. There are two different types of deletions characterized by the number of chromosome breaks involved. (i) A terminal deletion results from a single break in a distal part of one chromosome, and the loss of the distal acentric segment. Cri-du-chat syndrome is an example that is caused by a terminal deletion in the short arm of chromosome 5 (Figure 11.5). (ii) An interstitial deletion results from two breaks in a single chromosome and loss of the acentric fragment between the two breaks.[22] Moreover, a deletion may be due to segregation and recombination of a balanced rearrangement, i.e. translocation, inversion and insertion (see later). If deletions involve segments smaller than 3–5 Mb, they might not be detected by banding cytogenetics. In these cases, clinical information can give a clue towards which FISH probe to use (locus specific), to make, confirm, or dismiss the clinical suspicion of a corresponding diagnosis (Figure 11.6).

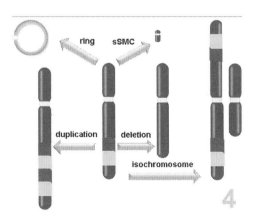

FIGURE 11.4 Schematic representation on formation of duplication, deletion, of isochromosomes of long and short chromosome arms, ring chromosomes, and small supernumerary marker chromosomes (sSMCs).

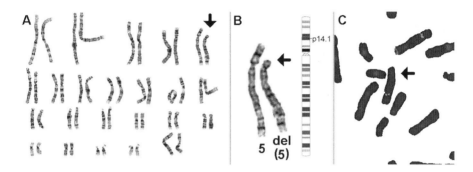

FIGURE 11.5 Cri-du-chat syndrome is characterized by a deletion in the short arm of a chromosome 5.

A) GTG-banding karyogram 46,XX,del(5)(p14.1) of a corresponding patient.
B) GTG-banded pair of chromosomes 5, showing del(5)(p14.1) in higher magnification.
C) After FISH, using a locus-specific probe for 5p15.2 (loci D5S817/D5S2875, Q-Biogene, SpectrumRed) and control probe for 5q31 region (locus D5S89 SpectrumGreen) reveals a heterozygote deletion on one chromosome 5 (arrow).

Duplication results in extra copies of a chromosomal region and consequently in higher copy numbers of corresponding genes, which may affect phenotype by modifying gene dosage. A tandem and an inverted duplication results from duplicated sections adjacent to the original ones. Furthermore, duplications may result from insertion or addition of the duplicated region in a completely different genomic environment.[27] Duplications often originate from unequal crossing over, especially in genomic regions with repetitive DNA sequences. Additionally, duplications arise after recombination and/or segregation of structural balanced rearrangements, namely translocations, inversions, and insertions.[22] Interestingly, almost 5% of the human genome includes duplicated sequences also known as low-copy repeats, having a link with genomic disorders since highly homologous flanking repeats predisposes these regions to recurrent rearrangement by nonallelic homologous recombination, which could lead to deletion, duplication, or inversion.[28] Cat eye syndrome, a common example of tetrasomy of the proximal long arm of chromosome 22 due to an extra dicentric chromosome, is associated with coloboma of the eye, intellectual impairment, and anal atresia (Figure 11.7).[22]

Ring chromosomes usually form by two terminal breaks in both arms of the same chromosome, followed by fusion of the broken ends, with loss of the distal acentric genetic material.[29] Alternative mechanisms were also described, namely the fusion of subtelomeric sequences or telomere–telomere fusion with no deletion, the terminal deletion and a contiguous inverted duplication due to an inv-dup-del rearrangement.[30] There are three most common types of ring chromosomes: (i) large rings with minimal loss from the terminal segments of the short and long arms, (ii) very small rings as extra chromosomes in the karyotype, and (iii) rings of the X-chromosome that are detected in females with Turner syndrome characteristics.[22]

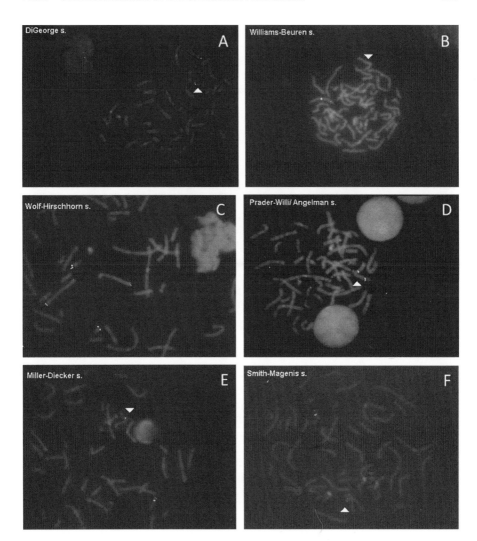

FIGURE 11.6   FISH-detection of microdeletion syndromes with corresponding suited commercially available probes is shown; derivative chromosomes are highlighted by arrowheads.

A) DiGeorge syndrome: specific probe D22S75 (22q11.2 SpectrumRed) and control ARSA (22q13.3 SpectrumGreen).

B) Williams-Beuren syndrome: specific probe ELN (7q11.23 SpectrumGreen) and control TWIST1 (7p21.1 SpectrumRed).

C) Wolf-Hirschhorn syndrome: specific probe (SpectrumRed) and control (SpectrumGreen).

D) Prader-Willi syndrome: specific probe SNRPN (15q11.2 SpectrumRed) and control PML (15q24.1 SpectrumGreen). This probe can also be used for Angelman syndrome patients.

E) Miller-Diecker syndrome: specific probe LIS1 (17p13.3 SpectrumRed) and control RAI1 (17p11.2 SpectrumGreen).

F) Smith-Magenis syndrome: specific probe RAI1 (17p11.2 SpectrumGreen) and control LIS1 (17p13.3 SpectrumRed).

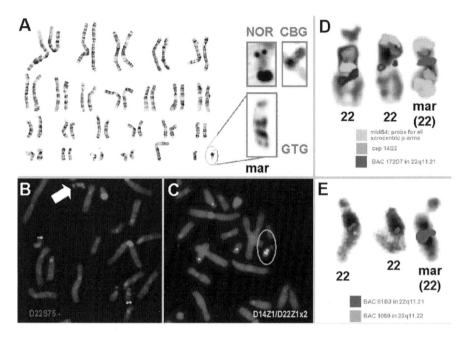

FIGURE 11.7   Characterization of a small supernumerary marker chromosome (sSMC) is depicted:

A) Karyogram of 6 year old boy with cat eye syndrome: GTG-banding revealed a karyotype 47,XY,+mar. The marker has two AgNOR positive and two C-band positive regions.
B) Partial metaphase after FISH highlighting the two normal 22 chromosomes and the sSMC (white arrow). The SMC does not present signals for D22S75 (red), a probe located in the DiGeorge syndrome region.
C) Partial metaphase showing one normal 22 chromosome and the sSMC (green circle) with two signals for the centromeric probe specific for chromosomes 14/22.
D) Normal 22 chromosomes and the sSMC(22) showing the presence of 2 centromeres (cep14/22 in green), signals for BAC RP11–172D7 in 22q11.21 (cat eye syndrome critical region), and two signals corresponding to acrocentric arms (midi54).
E) One signal in the marker for BAC RP11–81B3 in 22q11.21 and absence of signal for BAC RP11–1058B20 in 22q11.22 further confirmed that the sSMC was a typical cat eye syndrome marker.

It may be possible to characterize them by FISH using subtelomeric probes (Figure 11.8). Ring chromosomes have been described for all human chromosomes, although 50% of them arise from acrocentric chromosomes.[31] Ring chromosomes present variable sizes according to the fraction of material lost and are frequently unstable during cell division, being rarely transmitted to offspring.[15] A great majority of these chromosomes arise *de novo* (99%), although a few cases of inherited ring chromosomes have been reported,[32, 33] which are mostly of maternal origin (90%) because ring chromosomes induce (male) infertility.[31] The phenotype of ring chromosome carriers is very variable depending on the chromosome involved, the size

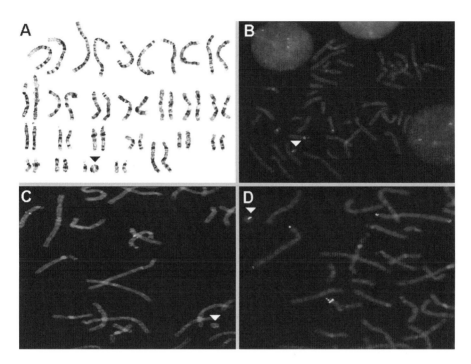

FIGURE 11.8   Characterization of a ring chromosome derived from chromosome 21 is shown:

A) Karyogram of a case with 46,XX,r(21)(p11.2q22.3).
B) Metaphase FISH using a centromeric probe for chromosomes 13/21 (D13Z1/D21Z1) in red confirmed chromosome 13 or 21 origin of the ring.
C) Dual color FISH using probes for RB1 (13q14.2 green) and DSCR1 (21q22 red) suggests a chromosome 21 ring origin.
D) Dual color FISH using probes for GS10K2-T7 (4p16.3 green) and A224XH1 (4q35.2 red) and a probe for 21q22 (blue) confirmed chromosome 21 ring origin.

of the deleted acentric fragment and consequently the gene content and also due to ring chromosome instability, but usually overlaps that of the syndromes associated with deletion of both ends of the corresponding chromosome.[34] So, besides deletions that occur by ring formation, ring instability can also lead to other genomic imbalances with gain or loss of genetic material and consequently to distinct phenotypes. Sister chromatid exchanges happening during mitosis in patients with ring chromosomes result in secondary chromosomal abnormalities like dicentric rings, interlocked rings, and other structural alterations or even ring chromosome loss leading to monosomic cells.[30]

Isochromosomes are composed of two copies of the same arm of one chromosome (mirror image of one of the arms of the chromosome). Some mechanisms have been suggested for their formation, like transversal mis-division of the

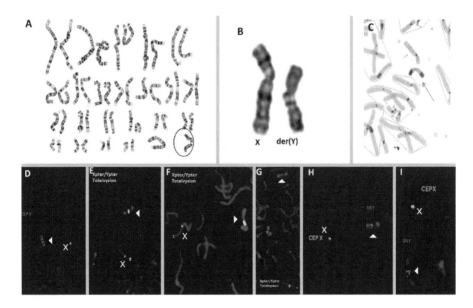

FIGURE 11.9   A case with mosaicism involving a cell line 45,X and five cell lines presenting different dicentric Y chromosomes with the breakpoints in the terminal segment of the short arm, and another cell line with two derivative Y chromosomes is characterized here.

A) Karyogram with a large dicentric Y chromosome (circle).

B) Enlarged sex-chromosomes of the case (GTG-banded).

C) CBG banding staining Yq12 two times on the der(Y)—arrow.

D–I) FISH analysis using different probes as shown in the pictures.

centromere during MII, sister chromatid breakage and reunion near the centromere, or exchange between homologues during meiosis.[15] Isochromosomes X are often observed in Turner syndrome individuals, and represent the most frequent isochromosomes observed, that are, in fact, in the majority of cases dicentric.[15] Isochromosome Yp, that results in duplication of the short arm and loss of the long arm, has been reported to be linked to azoospermia (Figure 11.9).[35] Autosomal isochromosomes occur more often in acrocentric chromosomes with loss of the short arms.

Small supernumerary marker chromosomes (sSMCs) are structurally abnormal chromosomes that cannot be unambiguously identified or characterized by conventional banding cytogenetics alone.[23] These sSMCs are equal to or smaller in size than a chromosome 20 of the same metaphase spread.[36] The origin and composition of an sSMC is recognizable by molecular cytogenetics analysis. They are seen in different shapes like ring, centric minute, and inverted duplication, and could be centric, or acentric, complex (= deriving from more than one chromosome), being discontinuous due to formation via chromothripsis and being present in mosaic.

BALANCED STRUCTURAL REARRANGEMENTS

Up to now, we described chromosomal alterations (numerical and unbalanced) that normally have an impact on the phenotype. Now we will focus on balanced rearrangements, which are often diagnosed after birth of a dysmorphic child, after a prenatal ultrasound of a fetus with morphological alterations, or after studying couples with infertility. The balanced chromosomal alterations comprise three types: inversions (inv), translocations (t), and insertions (ins) (Figure 11.10).

Inversions can be pericentric, when a single break occurs in each arm of the same chromosome, or paracentric, with two breaks in the same arm of a chromosome (Figure 11.11 A). Often, inversion carriers are clinically not affected; however, the offspring has an increased risk of partial duplications and deletions due to the formation of an inversion loop during chromosome pairing in prophase I of meiosis.[15] So, the inversion loop is composed of four chromatids, two normal and two inverted strands, leading to abnormal gametes when crossing-over events happen within this loop.[22] The chromatids resulting from recombination in a paracentric inversion are one normal, one with the same inversion as the progenitor, a dicentric, and an acentric one. For pericentric inversions, recombination also leads to four types of chromatids: one normal, one with the same inversion as the progenitor, and two with unbalances (distal deletion and duplication in a reverse manner in relation to each other).

Translocations are common (1:500) and arise from the exchange of chromosomal segments, usually between two non-homologous chromosomes. Translocations can be grouped into reciprocal and Robertsonian ones (Figure 11.11 B), and if more than two chromosomes are involved we have a complex rearrangement. Reciprocal translocations originate in the exchange of broken-off segments between two different chromosomes; frequently carriers are normal but have an enhanced risk of unbalanced offspring, which depends on the translocation components' segregation, position of breakpoints, and centromere location.[15] Homologues pairing at meiosis is different in translocation carriers, since the two derivative chromosomes and their two normal homologues pair to form a cross-shaped quadrivalent (four chromatids) at pachytene with each homologous segment pairing with its counterpart.[22]

There are five basic segregation patterns from a reciprocal translocation quadrivalent[20, 22]:

- alternate segregation: normal and translocation chromosomes move to opposite poles, originating in balanced gametes.
- adjacent I segregation: adjacent nonhomologous centromeres move to the same pole, leading to a zygote with partial trisomy of one chromosome and partial monosomy of the other, when fertilized by a normal haploid gamete.
- adjacent II segregation: adjacent homologous centromeres move to the same pole, leading to large amounts of unbalanced chromatin, often incompatible with embryonic survival.
- 3: 1 segregation: three of the four chromosomes move to one pole and only one moves to the opposite pole.
- 4: 0 segregation: four chromosomes move together to one pole; this is possible, but not compatible with live.

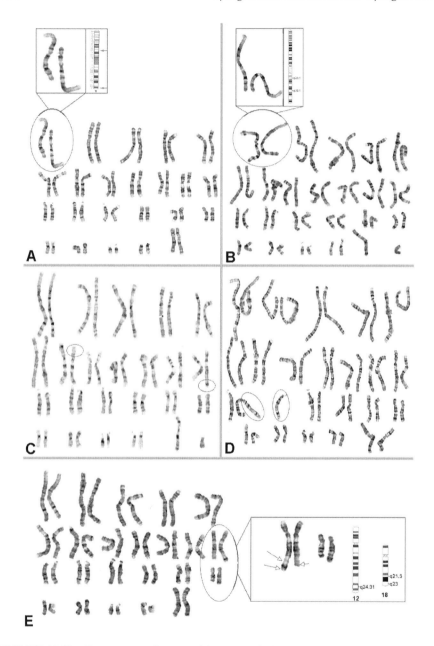

FIGURE 11.10 Karyograms of cases with structural chromosomal aberrations (in parts with schematic depictions):

A) pericentric inversion: 46,XX,inv(1)(p22.1q44).
B) paracentric inversion: 46,XY,inv(1)(q23.1q32.1).
C) reciprocal translocation: 46,XY,t(7;12)(p13;q22).
D) Robertsonian translocation: 45,XX,der(13;14)(q10;q10).
E) Insertion: 46,XX,ins(12;18)(q24.31;q21.3q23).

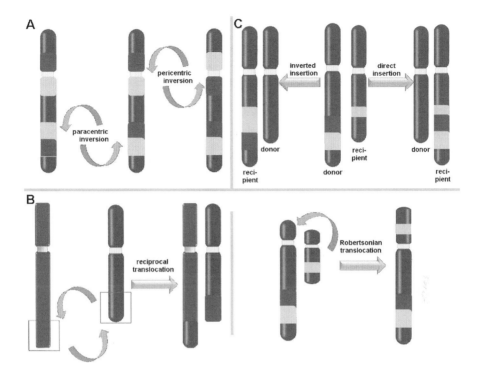

FIGURE 11.11 Schematic representation of

A) paracentric and pericentric inversion.
B) reciprocal and Robertsonian translocation
C) inverted and direct insertions.

Robertsonian translocations affect two acrocentric chromosomes, losing both of their short 'p' arms and joining near their centromeres to originate a single chromosome (carriers with 45 chromosomes); the fused chromosome can also include two centromeres (dicentric chromosome).[15] There seems to be no phenotype in Robertsonian translocation carriers since the genetic material lost is present in each of the short arms of acrocentric chromosomes, and a functional cell only needs six to eight of these regions.

Since in Robertsonian translocations only 'three' chromosomes are implicated, a trivalent is formed at pachytene, resulting in segregation of six types of gametes, i.e. two normal and four with trisomies or monosomies when fertilized by a normal gamete.[20, 22] The risk of unbalanced offspring is higher in female carriers since males often are infertile. The most common Robertsonian translocation involves chromosomes 13 and 14, followed by chromosomes 14 and 21.[26, 37]

Insertions detected by banding cytogenetics are quite rare. They result from three breaks, being the segment from one chromosome inserted into another chromosome, in a directed or inverted orientation (Figure 11.11 C). The incidence of microscopically visible insertions was estimated to be around 1 in 80,000 live births.[38] More recently, using array-comparative genomic hybridization (aCGH) and FISH, the incidence of insertion events was estimated to be about 1 in 500 or 1 in 563 individuals analyzed.[39, 40] In 2012, Nowakowska and colleagues[41] demonstrated that around 2.1% of apparently *de novo*, interstitial copy number variants (CNVs) were in fact imbalances that resulted from parents with balanced insertions. Although insertion carriers are clinically normal, their offspring has an enhanced risk of partial monosomy or partial trisomy of the inserted segment. These cases need to be confirmed by FISH techniques, usually by whole chromosome painting (wcp) probes.

CASE PROBLEMS

Selected clinical cases will be presented as examples of karyotyping and FISH applications in postnatal and in prenatal diagnosis.

For all clinical post and prenatal cases, three questions will be answered:

i) *What kind of chromosomal alteration was found? How to interpret the results? What is the most probable mechanism of origin of the alteration?*
ii) *What is the origin of the chromosomal alteration?*
iii) *What is the future strategy for the family?*

CLINICAL CASE 1

CLINICAL INFORMATION

A 6-year-old girl with microcephaly, moderate to severe psychomotor developmental delay, and prenatal history of intrauterine growth restriction was studied.

TECHNIQUES PERFORMED

G banding cytogenetics;
    FISH for DiGeorge critical region and for the subtelomeric regions.

RESULTS

The combined result for this case was 46,XX.ish del(22)(q13.33)(D22S1726-).

QUESTIONS

i) *What kind of chromosomal alteration was found? How to interpret the results? What is the most probable mechanism of origin of the alteration?*

The high resolution karyotype was normal (Figure 11.12 A) as well as the FISH results for the DiGeorge critical region in 22q11.2 (Figure 11.12 B).

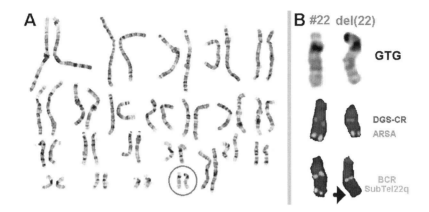

FIGURE 11.12 Example of a cryptic terminal deletion:

A) Seemingly normal karyogram, 46,XX.
B) However, only after FISH, using the depicted probe sets, a subtelomeric deletion is obvious when applying a telomere-near probe for chromosome 22 (arrow).

When the subtelomeric FISH probes—ToTelVysion (Vysis ToTelVysion Multi-Color FISH Probes)—were applied, a subtelomeric deletion on 22q13.33 was observed (Figure 11.12 B).

CURIOSITY

Deletion 22q13 has been accidentally diagnosed by FISH in several individuals primarily analyzed for 22q11.3 deletion syndrome. Numerous commercially available FISH assays for 22q11.3 deletion syndrome use arylsulfatase A (*ARSA*) gene as the control probe. *ARSA* maps to 22q13 and therefore detects deletions of this region. The *ARSA* probe successfully detects the majority of 22q13.3 deletions; however, microdeletions distal to *ARSA* may require FISH analysis using the 22q subtelomere probes,[42] as happened in our case.

ii) What is the origin of the chromosomal alteration?

In order to establish the most probable mechanism of origin of this alteration, both parents were also analyzed by karyotyping and FISH techniques. This did not reveal any numerical or structural alteration (data not shown), and therefore this terminal deletion was classified as *de novo*.

iii) What is the future strategy for the family?

This couple was referred for genetic counselling. Monitoring and testing future pregnancies was recommended.

Conclusion

This terminal deletion of the long arm of chromosome 22 (22q13.33-qter) seems to explain the clinical phenotype of the girl. The deletion 22q13.3 syndrome or Phelan-McDermid syndrome is characterized by neonatal hypotonia, global developmental delay, normal to accelerated growth, absent to severely delayed speech, and minor dysmorphic features. The loss of 22q13.3 can result from simple deletion, translocation, ring chromosome formation, and less commonly structural changes affecting the long arm of chromosome 22, specifically the region containing the *SHANK3* gene.[43] Almost 80% of individuals with Phelan-McDermid syndrome have *de novo*, simple deletions of 22q13,[44] as observed in our case.

CLINICAL CASE 2

Clinical Information

A woman in the 15th week of twin pregnancy did trophoblast biopsy as sonography diagnosed hygroma in one of fetus and a borderline nuchal translucency in the other fetus.

Techniques Performed

QF-PCR for aneuploidy screening;
  G banding cytogenetics;
  FISH technique using subtelomeric probes for chromosomes 11 and 18.

Results Twin 1

QF-PCR for aneuploidy screening was performed with a normal result (data not shown).
  The combined result (Figure 11.13 A and B) for this case was:
  46,XY,der(18)t(11;18)(p15.3;p11.2).ish der(18)t(11;18)(p15.5;p11.32)(D11S2071+,D18S552-)

Results Twin 2

QF-PCR for aneuploidy screening was performed with a normal result (data not shown).
  The combined result (Figure 11.13 C—FISH not shown) for this case was:
46,XX,t(11;18)(p15.5;p11.23).ish t(11;18)(p15.5-,p11.32+;p11.32-,p15.5+)(D11S2071-, D18S552+; D18S552-, D11S2071+)

Questions

i) What kind of chromosomal alteration was found? How to interpret the results? What is the most probable mechanism of origin of the alteration?

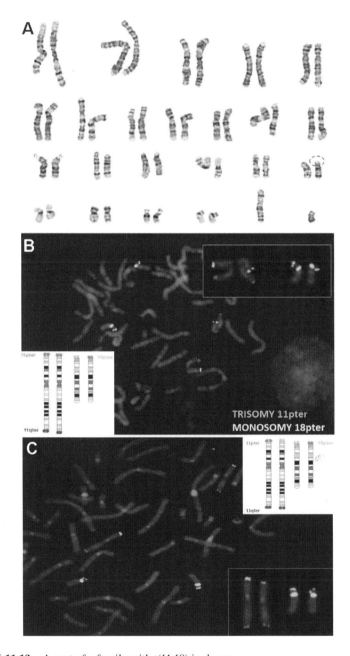

FIGURE 11.13 A part of a family with t(11;18) is shown.

A) karyogram of twin 1: 46,XY,der(18)t(11;18)(p15.3;p11.2).

B) FISH result of twin 1 using subtelomeric probes for chromosomes 11 and 18, confirming presence of the der(18)t(11;18)(p15.3;p11.2).

C) In twin 2, 46,XY,t(11;18)(p15.3;p11.2)mat was found (FISH shown).

The G-banded metaphases of the trophoblast biopsy cells revealed, in twin 1, a male fetus with a chromosome 18 derivative from a translocation involving chromosomes 11 and 18 (Figure 11.13 A) and, in twin 2, a female fetus with an apparently balanced reciprocal translocation involving the same chromosomes (Figure 11.13 C). The structural alterations identified in both twins were confirmed by FISH technique using subtelomeric probes, for 11p15.5 and 18p11.32 (twin 1—Figure 11.13 B; twin 2—not shown). It is important to stress that the result for twin 1 does not contradict that obtained with the screening for aneuploidies by the QF-PCR technique, since this technique is not suitable for the detection of partial aneuploidies in the tested chromosomes, such as this one identified in chromosome 18.

In order to establish the most probable mechanism of origin of the alteration, the study of the parents was recommended.

ii)   What is the origin of the chromosomal alteration?

The study (G-banding from peripheral blood) of both parents revealed a normal karyotype for the mother, and the presence of a reciprocal translocation involving chromosomes 11 and 18 in the father, 46,XY,t(11;18)(p15.3;p11.2) (data not shown). Regarding the most probable biological mechanism, it can be concluded that the male progenitor is a carrier of a structural chromosomic rearrangement, a reciprocal translocation with breakpoints in 11p15.3 and 18p11.2, which has no implication in the phenotype of the carrier. However, during spermatogenesis there is a risk for the offspring to inherited imbalances as a result of anomalous segregations. In this case, the identified structural chromosomal rearrangement in fetus 1 is likely to be due to an adjacent I segregation, having the fetus a derivative chromosome 18 inherited from the paternal translocation. Fetus 2 is the result of an alternate segregation as it carries the same translocation as the father.

iii)   What is the future strategy for the family?

This couple was offered genetic counselling where it was well explained the consequence of this reciprocal translocation in this phenotypically normal father for future pregnancies, and the need of family studies (parents, brothers, sisters, and other family members at risk). To all carriers, invasive prenatal diagnosis should be offered. As chromosome 11 is involved in the rearrangement also the hint should be given, that this imprinting associated chromosome is slightly more likely to be involved in uniparental disomy (UPD) due to the t(11;18). Thus, prenatal UPD test may also be considered in future pregnancies.

CONCLUSION

This clinical case shows the implications for the offspring of a balanced structural rearrangement in one of the progenitors and the need to offer to this couple prenatal diagnosis in future gestations and genetic counselling to explain the implications of these genetic findings, the risk of recurrence, and the need to offer family studies to other relatives. In this clinical case, the male progenitor is a carrier of a reciprocal translocation between chromosomes 11 and 18, and the unbalanced structural chromosomal rearrangement involving partial monosomy of the short arm of chromosome 18 (p11.2-pter) and partial trisomy of the short arm of chromosome 11 (p15.3-pter) observed in twin 1 was originated by an adjacent 1 segregation during spermatogenesis. Twin 2, on the other hand, was a balanced carrier like the male progenitor, and resulted from an alternate type of segregation. The imbalances identified in twin 1 are most probably the cause of the ultrasound alterations detected and consequently are correlated to the clinical phenotype.

CLINICAL CASE 3

CLINICAL INFORMATION

A pregnant woman in 13th week of gestation had a sonographic diagnosis of holoprosencephaly in the fetus, and thus amniotic fluid was collected.

TECHNIQUES PERFORMED

FISH for the rapid detection of chromosome aneuploidies in uncultured amniocytes;
    G banding cytogenetics;
    FISH technique using whole chromosome painting probe for chromosome 21 (WCP21).

RESULTS

FISH for the rapid detection of chromosome aneuploidies in uncultured amniocytes was performed with a normal result (Figure 11.14 A-3). The combined result of G banding and molecular cytogenetics revealed two partial monosomies involving the terminal region of chromosome 13q32-qter and the proximal region of chromosome 21pter-q22.1. Karyotype: 45,XY,der(13)t(13;21)(q32;q22.1),–21.ish t(13;21) (q32;q22.1)(wcp21+;wcp21+).

QUESTIONS

   i) What kind of chromosomal alteration was found? How to interpret the results? What is the most probable mechanism of origin of the alteration?

G-banding in amniotic fluid cells revealed a male fetus with 45 chromosomes due to a monosomy of chromosome 21 and also a structurally altered chromosome 13 (Figure 11.14 A-1). By FISH technique using WCP21, it was possible to characterize the derivative chromosome 13 as an unbalanced structural chromosomal rearrangement involving chromosome 13 and 21 (Figure 11.14 A-1 and 11.14 A-2). This structural rearrangement was also confirmed in fibroblast cells of the fetus (results not shown) and seems to be the origin of the detected ultrasound anomaly, being consequently

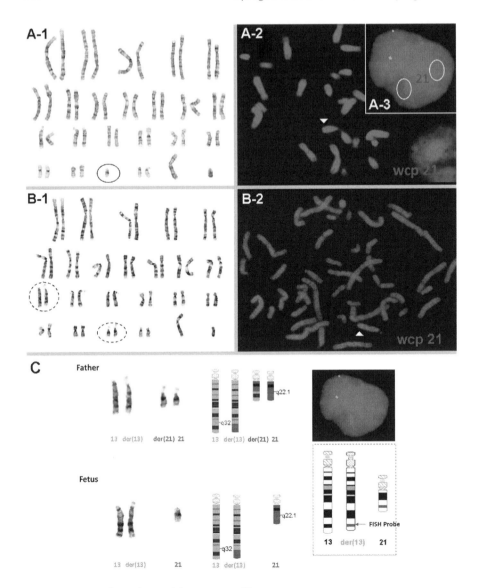

FIGURE 11.14   Case of a partial monosomy 21:

A) 1) Fetal GTG-banding seems to show a monosomy 21 and a derivative chromosome 13 (derivative chromosome 13 appears on the left of chromosome 13 pair); 2) FISH using a whole chromosome paint for chromosome 21 showing one chromosome 21 and also that there is chromosome 21 material on another chromosome; 3) Previous FISH quick test in native amniocyte nuclei gave a normal result for locus specific probes in 21q22.13~22.2.

B) 1) Paternal GTG-banding revealed a karyotype of 46,XY,t(13;21)(q32;q22.1). 2) This result was confirmed by FISH.

C) Schematic depiction of balanced structural rearrangement in the father and the unbalanced structural rearrangement in the fetus and the location of FISH probe used for the rapid detection of aneuploidies.

correlated with the clinical phenotype. In order to ascertain the origin of this structural alteration, both parents were also analyzed by conventional cytogenetics and FISH.

CURIOSITY

Monosomy 21 as seemingly present in GTG-banding is not compatible with life, and such cases abort spontaneously in very early embryogenesis; thus, here a cryptic rearrangement should be suspected by the clinical laboratory geneticists.

ii)   What is the origin of the chromosomal alteration?

Both parents, phenotypically normal, were analyzed by G-banded metaphases obtained from peripheral blood that revealed a normal karyotype for the mother (46,XX), and the presence of a reciprocal translocation involving chromosomes 13 and 21 in the fathers' karyotype, which was also confirmed by FISH: 46,XY,t(13;21) (q32;q22.1).ish t(13;21)(wcp21+;wcp21+) (Figure 11.14 B-1 and 11.14 B-2).

iii)   What is the future strategy for the family?

This couple accepted genetic counselling so that the consequences of this translocation in this phenotypically normal father for future pregnancies and the need for family studies (parents, brothers, sisters, and other family members at risk) were outlined. Also all carriers in the family should be offered invasive prenatal diagnosis, when expecting a child.

CONCLUSION

It was possible to infer that the alteration of the fetus was inherited from the male progenitor (Figure 11.14 C). The explanation for the normal result obtained using FISH for the rapid detection of aneuploidies is because this fetus carries an unbalanced structural chromosomal rearrangement involving chromosomes 13 and 21 and the FISH probe used for rapid detection of aneuploidies is located more distally at 21q22.13-q22.2; and since this region is present in the derivative chromosome 13, the FISH study showed two signals for chromosome 21.

CLINICAL CASE 4

CLINICAL INFORMATION

In a 14-month-old girl, hypotonia due to damage to the 1st neuron, agenesis of the corpus callosum, and macrocephaly were diagnosed. Peripheral blood was taken for cytogenetic analyses.

TECHNIQUES PERFORMED

G banding cytogenetics;

FISH using subtelomeric probes, VIJyRM2000 and MS607 (Vysis-Totelvysion) for 7q36.3 and 22q13.3, respectively.

RESULTS

Karyotyping showed a normal result, 46,XX (data not shown).

FISH using subtelomeric probes for 7q36.3 and 22q13.3 revealed a partial monosomy of the terminal region of the long arm of chromosome 22 and a partial trisomy of the terminal region of the long arm of chromosome 7, 46,XX.ish der(22)t(7;22)(q36.3;q13.3)(MS607-,VIJyRM2000+) (Figure 11.15).

QUESTIONS

i) What kind of chromosomal alteration was found? How to interpret the results? What is the most probable mechanism of origin of the alteration?

The presence of a derivative chromosome 22 from a translocation involving chromosomes 22 and 7 (Figure 11.15 A) seems to be the origin of the phenotype of this girl. Since these chromosomal alterations are submicroscopic, it is not possible to detect them by G banding and, therefore, in these cases, molecular cytogenetics is essential for the diagnosis. This proband had an older brother with severe psychomotor delay, short stature, microcephaly, choanal atresia, and hypogenitalism,

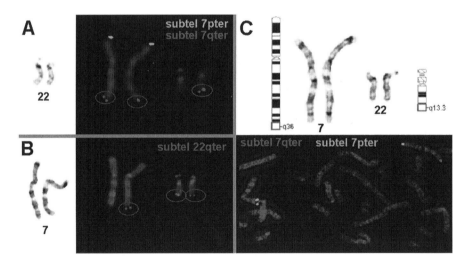

FIGURE 11.15   Family with t(7;22): Banding and molecular cytogenetics results obtained are shown for

A) the affected daughter.
B) the affected brother.
C) the carrier mother.

which was also referred for G banding and molecular cytogenetic study. These studies revealed a partial monosomy of the terminal region of chromosome 7q and a partial trisomy of the terminal region of the chromosome 22q, 46,XY.ish der(7)t(7;22)(q36.3;q13.3)(VIJyRM2000-,MS607+). The presence of this derivative 7 from a translocation involving chromosomes 7 and 22 seems to explain the phenotype of this boy. These alterations are the exact opposite of those observed in the sister (Figure 11.15 B).

ii)   What is the origin of the chromosomal alteration?

Peripheral blood of both parents, phenotypically normal, was analyzed by G-banding and FISH. These studies revealed a normal karyotype in the father (46,XY) and the presence of a reciprocal translocation involving chromosomes 7 and 22 in the mother, 46,XX.ish t(7;22)(q36;q13.3)(MS607+,VIJyRM2000+) without impact on the phenotype. It was possible to infer that the chromosomal alterations in the girl and her brother were inherited from the female progenitor (Figure 11.15 C) due to an anomalous segregation (adjacent segregation) during meiosis.

iii)   What is the future strategy for the family?

Once again, genetic counselling was offered and all implications for the specific situation explained further, including prenatal diagnosis for future pregnancies and the need for additional studies offered to relatives of the carrier. As chromosome 7 is involved in the rearrangement, also the hint should be given that this imprinting associated chromosome is slightly more likely to be involved in UPD due to the t(7;22). Thus, prenatal UPD test may also be considered in future pregnancies.

CONCLUSION

These two different unbalanced chromosomal structural rearrangements in these siblings involving the same two chromosomes (7 and 22) were inherited from the female progenitor and seem to explain their respective phenotypes.

CONCLUSIONS

The complex behavior of chromosomes during the two meiotic divisions is error-prone. Aneuploidy can affect any chromosome; however, only trisomies of the sex chromosomes, of autosomes 13, 18, or 21, and monosomy of the X chromosome, are compatible with survival up to the end of pregnancy, the remainder having been eliminated by natural selection. Structural alterations originated from chromosomal breakage and rejoining could affect the number, distribution, and expression of genes with a significant impact in the offspring phenotype. Couples at particularly increased risk of having offspring with a numerical or structural alteration are those:

- where one partner carries a chromosomal rearrangement such as a translocation or an inversion, as described earlier.

- where one partner has gonadal or germinal mosaic for a trisomic cell line. Couples with various conceptions involving the same trisomy give a clear evidence of a gonadal mosaicism for a trisomic cell line affecting the primordial germ cells.

Such chromosomal alterations and their mechanisms of origin can be identified through banding cytogenetic studies, but in many situations need to be complemented by FISH, depending on the clinical suspicion or on the previous study of the karyotype or from both. It is important to stress that through the diagnosis of both numerical and structural alterations (balanced and unbalanced) by karyotyping, the recurrence risk can be calculated for each specific situation. Genetic counselling must be offered to these couples and, if appropriate, prenatal diagnosis can be recommended ensuring, that a future pregnancy is chromosomally balanced. For couples that presented with repeated early miscarriages or primary infertility, a preimplantation diagnosis with selective transfer of embryos might be appropriate.

DEDICATION

To all our patients and their families, to everyone our laboratory (Cytogenetics and Genomics Laboratory of the Faculty of Medicine of the University of Coimbra) and clinical colleagues, to our students

REFERENCES

1. Pezza, R.J. Mechanisms of chromosome segregation in meiosis: New views on the old problem of aneuploidy. *FEBS J.* **2015**, 282, 2424–2425.
2. Petronczki, M.; Siomos, M.F.; Nasmyth, K. Un ménage à quatre: The molecular biology of chromosome segregation in meiosis. *Cell.* **2003**, 112, 423–440.
3. Greaney, J.; Wei, Z.; Homer, H. Regulation of chromosome segregation in oocytes and the cellular basis for female meiotic errors. *Hum. Reprod. Update.* **2018**, 24, 135–161.
4. Villeneuve, A.M.; Hillers, K.J. Whence meiosis? *Cell.* **2001**, 106, 647–650.
5. Duro, E.; Marston, A.L. From equator to pole: Splitting chromosomes in mitosis and meiosis. *Genes Dev.* **2015**, 29, 109–122.
6. Storchova, Z. Chromosomal numerical aberrations. In: *eLS.* John Wiley & Sons, Ltd, Chichester, **2016**.
7. Queremel Milani, D.A.; Tadi, P. *Genetics, Chromosome Abnormalities.* Treasure Island, FL, StatPearls Publishing; **2022**.
8. Delhanty, J.D. Origins of human aneuploidy. In: *eLS.* John Wiley & Sons, Ltd, Chichester, **2018**.
9. Mahdieh, N.; Rabbani, B. An overview of mutation detection methods in genetic disorders. *Iran J. Pediatr.* **2013**, 23, 375–388.
10. Hassold, T.; Hunt, P. To err (meiotically) is human: The genesis of human aneuploidy. *Nat. Rev. Genet.* **2001**, 2, 280–291.
11. McFadden, D.E.; Friedman, J.M. Chromosome abnormalities in human beings. *Mutat. Res.* **1997**, 396, 129–140.
12. Nagaoka, S.I.; Hassold, T.J.; Hunt, P.A. Human aneuploidy: Mechanisms and new insights into an age-old problem. *Nat. Rev. Genet.* **2012**, 13, 493–504.
13. Ottolini, C.S.; Newnham, L.; Capalbo, A.; Natesan, S.A.; Joshi, H.A.; Cimadomo, D.; Griffin, D.K.; Sage, K.; Summers, M.C.; Thornhill, A.R.; Housworth, E.; Herbert, A.D.;

Rienzi, L.; Ubaldi, F.M.; Handyside, A.H.; Hoffmann, E.R. Genome-wide maps of recombination and chromosome segregation in human oocytes and embryos show selection for maternal recombination rates. *Nat. Genet.* **2015**, 47, 727–735.

14. Holubcová, Z.; Blayney, M.; Elder, K.; Schuh, M. Human oocytes: Error-prone chromosome-mediated spindle assembly favors chromosome segregation defects in human oocytes. *Science.* **2015**, 348, 1143–1147.

15. Luthardt, F.W.; Keitges, E. Chromosomal syndromes and genetic disease. In: *eLS*. John Wiley & Sons, Ltd, Chichester, **2001**.

16. McFeely, R.A. Chromosome abnormalities. *Vet. Clin. North Am. Food Anim. Pract.* **1993**, 9, 11–22.

17. McFadden, D.E.; Robinson, W.P. Phenotype of triploid embryos. *J. Med. Genet.* **2006**, 43, 609–612.

18. Zaragoza, M.V.; Surti, U.; Redline, R.W.; Millie, E.; Chakravarti, A.; Hassold, T.J. Parental origin and phenotype of triploidy in spontaneous abortions: Predominance of diandry and association with the partial hydatidiform mole. *Am. J. Hum. Genet.* **2000**, 66, 1807–1820.

19. McFadden, D.E.; Langlois, S. Parental and meiotic origin of triploidy in the embryonic and fetal periods. *Clin. Genet.* **2000**, 58, 192–200.

20. Gardner, R.J.M.; Amor, D.J. *Gardner and Sutherland's Chromosome Abnormalities and Genetic Counseling.* 5th Edition. Oxford University Press, Oxford; **2018**.

21. Filges, I.; Manokhina, I.; Peñaherrera, M.S.; McFadden, D.E.; Louie, K.; Nosova, E.; Friedman, J.M.; Robinson, W.P. Recurrent triploidy due to a failure to complete maternal meiosis II: Whole-exome sequencing reveals candidate variants. *Mol. Hum. Reprod.* **2015**, 21, 339–346.

22. Moore, C.M.; Best, R.G. Chromosomal genetic disease: Structural aberrations. In *eLS*. John Wiley & Sons, Ltd, Chichester, **2001**.

23. ISCN 2020. *An International System for Human Cytogenomic Nomenclature (2020)*; McGowan-Jordan J.; Hastings, R.J.; Moore, S., Eds. Karger, Basel; **2020**.

24. Speicher, M.R.; Carter, N.P. The new cytogenetics: Blurring the boundaries with molecular biology. *Nat. Rev. Genet.* **2005**, 6, 782–792.

25. Potapova, T.; Gorbsky, G.J. The consequences of chromosome segregation errors in mitosis and meiosis. *Biology (Basel).* **2017**, 6, 12.

26. Hook, E.B.; Cross, P.K. Rates of mutant and inherited structural cytogenetic abnormalities detected at amniocentesis: Results on about 63,000 fetuses. *Ann. Hum. Genet.* **1987**, 51, 27–55.

27. Clancy, S.; Shaw, K. DNA deletion and duplication and the associated genetic disorders. *Nat. Educ* **2008**, 1, 23.

28. Sharp, A.J.; Locke, D.P.; McGrath, S.D.; Cheng, Z.; Bailey, J.A.; Vallente, R.U.; Pertz, L.M.; Clark, R.A.; Schwartz, S.; Segraves, R.; Oseroff, V.V.; Albertson, D.G.; Pinkel, D.; Eichler, E.E. Segmental duplications and copy-number variation in the human genome. *Am. J. Hum Genet.* **2005**, 77, 78–88.

29. Sigurdardottir, S.; Goodman, B.K.; Rutberg, J.; Thomas, G.H.; Jabs, E.W.; Geraghty, M.T. Clinical, cytogenetic, and fluorescence in situ hybridization findings in two cases of "complete ring" syndrome. *Am. J. Med. Genet.* **1999**, 87, 384–390.

30. Guilherme, R.S.; Meloni, V.F.; Kim, C.A.; Pellegrino, R.; Takeno, S.S.; Spinner, N.B.; Conlin, L.K.; Christofolini, D.M.; Kulikowski, L.D.; Melaragno, M.I. Mechanisms of ring chromosome formation, ring instability and clinical consequences. *BMC Med. Genet.* **2011**, 12, 171.

31. Caba, L.; Rusu, C.; Plăiaşu, 5th.; Gug, G.; Grămescu, M.; Bujoran, C.; Ochiană, D.; Voloşciuc, M.; Popescu, R.; Braha, E.; Pânzaru, M.; Butnariu, L.; Sireteanu, A.; Covic,

M.; Gorduza, E. Ring autosomes: Some unexpected findings. *Balkan J. Med. Genet.* **2012**, 15, 35–46.

32. Knijnenburg, J.; van Haeringen, A.; Hansson, K.B.; Lankester, A.; Smit, M.J.; Belfroid, R.D.; Bakker, E.; Rosenberg, C.; Tanke, H.J.; Szuhai, K. Ring chromosome formation as a novel escape mechanism in patients with inverted duplication and terminal deletion. *Eur. J. Hum. Genet.* **2007**, 15, 548–545.

33. Yip, M.Y. Autosomal ring chromosomes in human genetic disorders. *Transl Pediatr.* **2015**, 4, 164–174.

34. Schinzel, A. *Catalogue of Unbalanced Chromosome Aberrations in Man Berlin.* Walter de Gruyter. Basel; **2001**.

35. Hemmat, M.; Hemmat, O.; Boyar, F.Z. Isochromosome Yp and jumping translocation of Yq resulting in five cell lines in an infertile male: A case report and review of the literature. *Mol. Cytogenet.* **2013**, 6, 36.

36. Liehr, T.; Claussen, U.; Starke, H. Small supernumerary marker chromosomes (sSMC) in humans. *Cytogenet Genome Res.* **2004**, 107, 55–67.

37. Therman, E.; Susman, B.; Denniston, C. The nonrandom participation of human acrocentric chromosomes in Robertsonian translocations. *Ann. Hum. Genet.* **1989**, 53, 49–65.

38. Van Hemel, J.O.; Eussen, H.J. Interchromosomal insertions: Identification of five cases and a review. *Hum Genet.* **2000**, 107, 415–432.

39. Neill, N.J.; Ballif, B.C.; Lamb, A.N.; Parikh, S.; Ravnan, J.B.; Schultz, R.A.; Torchia, B.S.; Rosenfeld, J.A.; Shaffer, L.G. Recurrence, submicroscopic complexity, and potential clinical relevance of copy gains detected by array CGH that are shown to be unbalanced insertions by FISH. *Genome Res.* **2011**, 21, 535–544.

40. Kang, S.H.; Shaw, C.; Ou, Z.; Eng, P.A.; Cooper, M.L.; Pursley, A.N.; Sahoo, T.; Bacino, C.A.; Chinault, A.C.; Stankiewicz, P.; Patel, A.; Lupski, J.R.; Cheung, S.W. Insertional translocation detected using FISH confirmation of array-comparative genomic hybridization (aCGH) results. *Am. J. Med. Genet. A.* **2010**, 152, 1111–1126.

41. Nowakowska, B.A.; de Leeuw, N.; Ruivenkamp, C.A.; Sikkema-Raddatz, B.; Crolla, J.A.; Thoelen, R.; Koopmans, M.; den Hollander, N.; van Haeringen, A.; van der Kevie-Kersemaekers, A.M.; Pfundt, R.; Mieloo, H.; van Essen, T.; de Vries, B.B.; Green, A.; Reardon, W.; Fryns, J.P.; Vermeesch, J.R. Parental insertional balanced translocations are an important cause of apparently de novo CNVs in patients with developmental anomalies. *Eur. J. Hum. Genet.* **2012**, 20, 166–170.

42. Durand, C.M.; Betancur, C.; Boeckers, T.M.; Bockmann, J.; Chaste, P.; Fauchereau, F.; Nygren, G.; Rastam, M.; Gillberg, I.C.; Anckarsäter, H.; Sponheim, E.; Goubran-Botros, H.; Delorme, R.; Chabane, N.; Mouren-Simeoni, M.C.; de Mas, P.; Bieth, E.; Rogé, B.; Héron, D.; Burglen, L.; Gillberg, C.; Leboyer, M.; Bourgeron, T. Mutations in the gene encoding the synaptic scaffolding protein SHANK3 are associated with autism spectrum disorders. *Nat. Genet.* **2007**, 39, 25–27.

43. Phelan, M.C. Deletion 22q13.3 syndrome. Orphanet. *J. Rare Dis.* **2008**, 3, 14.

44. Cammarata-Scalisi, F.; Callea, M.; Martinelli, D.; Willoughby, C.E.; Tadich, A.C.; Araya Castillo, M.; Lacruz-Rengel, M.A.; Medina, M.; Grimaldi, P.; Bertini, E.; Nevado, J. Clinical and genetic aspects of Phelan-McDermid syndrome: An interdisciplinary approach to management. Genes (Basel). **2022**, 13, 504.

12 Multicolor-FISH— Methods and Applications

Thomas Liehr

CONTENTS

INTRODUCTION

Since their establishment in routine (cyto-)genetic diagnostics in the mid-1990s, multicolor fluorescence in situ hybridization (mFISH) approaches have become indispensable for the identification and description of chromosomal rearrangements. The subsequent exact characterization of chromosomal breakpoints is of superior clinical impact and is the requisite condition for further molecular investigations aimed at the identification of disease-related genes.[1, 2] Thus, various approaches for a differentiation of chromosomal subbands based on mFISH assays were established.[3] Here an overview is presented on available mFISH approaches, with special focus on mFISH-banding methods including their advantages, limitations, and possible applications.

Although the G-bands by trypsin using Giemsa (GTG) banding technique is still the starting point of most (molecular) cytogenetic test requests, its technical restrictions are well known (e.g., only chromosome morphology combined with a black and white banding pattern can be evaluated). Nonetheless, a quick and not costly overview on eventual gross changes within the human whole genome can be achieved by that cytogenetic banding approach.[2]

mFISH methods using all 24 human whole chromosome painting probes simultaneously, such as multiplex-FISH (M-FISH), spectral karyotyping (SKY), combined ratio labeling FISH (COBRA-FISH), and others,[3] reach their limits in the identification of intrachromosomal rearrangements (such as small interstitial deletions, duplications, and inversions without change of the centromeric index) and when exact characterization of breakpoints is required. These limitations were overcome

DOI: 10.1201/9781003223658-12

by the development of new mFISH probe sets during the last decades. There were many tumor-[3, 4] and clinical-cytogenetics[2, 3]-oriented probe sets established for the human genome, all of them being suited for any kind of imaginative research application,[2, 5] but also mFISH-probe sets for animal,[3] plant,[3] or even bacteria.[3, 5] Here we concentrate on human mFISH probe sets suited for FISH-banding.[3, 6]

FISH banding methods are defined as "any kind of FISH technique, which provides the possibility to characterize simultaneously several chromosomal subregions smaller than a chromosome arm (excluding the short arms of the acrocentric chromosomes)".[6] In contrast to the standard cytogenetic chromosome banding techniques, giving a protein-related banding pattern,[2] the FISH banding techniques are DNA-mediated.

SAMPLES/TISSUES

In routine settings, mFISH-banding is applicable in each cytogenetic preparation derived from dividing cells on metaphase spreads. These are normally tumor cells, cells derived from bone marrow, blood or fibroblasts (normally limited to fetal tissues including chorion, amnion, and abortion material or skin fibroblasts in postnatal cases).[5] In research, also interphase cells from each tissue, including archival formalin-fixed paraffin embedded or cryofixed tissues.[3, 5]

DESCRIPTION OF METHODS

Five different FISH banding methods are available at present, which differ in their probe composition as well as in their banding resolution:

1. The cross-species color banding (Rx-FISH) or Harlequin FISH probe set provides the lowest resolution of 80–90 bands per haploid human karyotype and consists of flow-sorted gibbon chromosomes.[7] A set of 110 human–hamster somatic cell hybrids (split into two pools and labeled with two fluorochromes), when hybridized to human chromosomes, leads to about 100 "bars" along the genome. This pattern has been called "somatic cell hybrid-based chromosome bar code".[8] A combination of the latter and the aforementioned Rx-FISH probe set results in 160-chromosome region-specific DNA-mediated bands in human karyotypes.[8]
2. Spectral color banding (SCAN) was described exemplarily for the two chromosomes 3 and 10. E.g. eight microdissection libraries were created along chromosome 10 with the goal to obtain a banding pattern similar to GTG banding at the 300-band level.[9]
3. A chromosome can be characterized as well by a specific signal pattern produced by locus-specific probes. The first attempts to label each chromosome by subregional DNA probes in different colors were performed by Lichter et al.[10] and Lengauer et al.[11] Such probe sets are called chromosome bar codes (CBCs) and reached yet a resolution of up to 400 bands per haploid karyotype, depending on the number of the applied probes.[12]
4. The Interspersed PCR multiplex FISH (IPM-FISH) approach[13] has an approximate resolution of 400 bands per haploid karyotype, mainly

dependent on chromosome quality. In IPM-FISH, whole chromosome painting probes are used, which are modified by interspersed PCR, leading to a 24-color FISH painting plus an R-band-like pattern.

5. The high-resolution multicolor banding (MCB) technique is based on overlapping microdissection libraries producing fluorescence profiles along the human chromosomes, and was initially described exemplarily for chromosome 5 in 1999 (Figure 12.1).[14] MCB allows the differentiation of chromosome region-specific areas at the band and subband levels at a resolution of 550 bands per haploid karyotype. As the number of pseudocolored bands per chromosome can freely be assigned using the *isis* software (MetaSystems, Altlussheim, Germany), a resolution higher than that of GTG banding of the corresponding chromosome can be achieved (e.g., up to 10 MCB bands for chromosome 22 equal 800 bands per total haploid karyotype). An MCB set of approximately 140 microdissection libraries covering the entire human genome was described in 2002.[15] Also, the simultaneous use of all human MCB libraries in one hybridization step for the characterization of complex karyotypes is possible by mMCB.[16] MCB has been applied in hundreds of studies yet and is also available a murine variant.[17, 18]

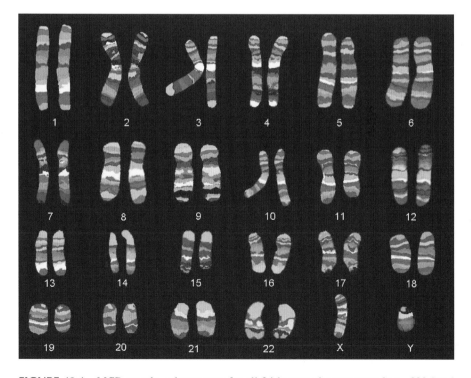

FIGURE 12.1 MCB pseudo-color pattern for all 24 human chromosomes in a ~380-band level. Two homologue autosomes and one gonosome each are presented. The chromosomes depicted here have been put together from 24 different MCB experiments.

FISH banding approaches such as the CBC-technique using locus specific probes, region-specific human–hamster somatic cell hybrids, or nonoverlapping microdissection libraries (SCAN) have, per definition, the disadvantage that unstained and non-informative gaps are left along the chromosome. Such gaps can cause problems, as breakpoints within the unstained gaps cannot exactly be described. Conversely, techniques based on locus-specific probes would theoretically provide the advantage that chromosomal breakpoints could be defined very exactly by the corresponding breakpoint-spanning or flanking clones. The IPM-FISH approach, the Rx-FISH technique, the Rx-FISH combined with somatic cell hybrids, and the MCB methods provide the advantage of leaving no non-informative gaps.

The MCB probe sets are commercially available as m-banding (MetaSystems, Altlussheim, Germany) or are available on request at Thomas.Liehr@med.uni-jena. de. Besides, probes can be prepared as previously described using chromosome microdissection.[15]

YIELDS

Only above mentioned FISH banding probe sets (1), (4), and (5) are finished for the whole human genome and can be applied to achieve comprehensive information in one single hybridization step. Concerning banding resolution, MCB has the highest and most flexible one. Additionally, according to the question to be studied, it can be chosen if only selected chromosomes or the whole genome shall be hybridized.

CONCLUSIONS

The introduction of FISH banding methods was a great step forward for molecular cytogenetic diagnostics. However, none of the mentioned new methods can, for technical reasons, ever be fully informative for itself. The cytogeneticist always has to double check the results obtained in FISH banding with those achieved by other approaches, such as GTG banding, M-FISH or SKY, chromosomal microarray analyses (CMA), sequencing or others. Examples for that modus operandi can be found in the literature.[4] Nevertheless, the goal must be to achieve fully informative cytogenetic results in a minimum of time and with a minimum of FISH experiments. Thus, further developments with respect to probe combinations (such as M-FISH/ SKY with FISH banding methods) will be necessary and are on the way.[19] To perform and evaluate such complex experiments, which then will have to be based on up to nine fluorochromes, further technical developments in microscopy and computer software would be needed.

REFERENCES

1. Liehr, T. Molecular cytogenetics in the era of chromosomics and cytogenomic approaches. *Front. Genet.* **2021**, 12, 720507.
2. Liehr, T. *Cytogenomics.* Academic Press, New York; **2021**.

3. Liehr, T. Basics and literature on multicolor fluorescence in situ hybridization application. http://cs-tl.de/DB/TC/mFISH/5-wcp-oth.html [accessed on 01/12/2022].

4. Liehr, T.; Othman, M.A.; Rittscher, K.; Alhourani, E. The current state of molecular cytogenetics in cancer diagnosis. *Expert Rev. Mol. Diagn.* **2015**, 15, 517–526.

5. Liehr, T. *Fluorescence In Situ Hybridization (FISH): Application Guide.* 2nd Edition. Springer, Berlin; **2017**.

6. Liehr, T.; Starke, H.; Heller, A.; Kosyakova, N.; Mrasek, K.; Gross, M.; Karst, C.; Steinhaeuser, U.; Hunstig, F.; Fickelscher, I.; Kuechler, A.; Trifonov, V.; Romanenko, S.A.; Weise, A. Multicolor fluorescence in situ hybridization (FISH) applied to FISH-banding. *Cytogenet. Genome. Res.* **2006**, 114, 240–244.

7. Müller, S.; Neusser, M.; Wienberg, J. Towards unlimited colors for fluorescence in-situ hybridization (FISH). *Chromosome Res.* **2002**, 10, 223–232.

8. Müller, S.; Rocchi, M.; Ferguson-Smith, M.A.; Wienberg, J. Toward a multicolor chromosome bar code for the entire human karyotype by fluorescence in situ hybridization. *Hum. Genet.* **1997**, 100, 271–278.

9. Liehr, T. About SCAN. http://cs-tl.de/DB/TC/mFISH/6-banding.html#5 [accessed on 01/12/2022].

10. Lichter, P.; Tang, C.J.; Call, K.; Hermanson, G.; Evans, G.A.; Housman, D.; Ward, D.C. High-resolution mapping of human chromosome 11 by in situ hybridization with cosmid clones. *Science* **1990**, 247, 64–69.

11. Lengauer, C.; Speicher, M.R.; Popp, S.; Jauch, A.; Taniwaki, M.; Nagaraja, R.; Riethman, H.C.; Donis-Keller, H.; D'Urso, M.; Schlessinger, D.; Cremer, T. Chromosomal bar codes produced by multicolor fluorescence in situ hybridization with multiple YAC clones and whole chromosome painting probes. *Hum. Mol. Genet.* **1993**, 2, 505–512.

12. Liehr, T. About CBC. http://cs-tl.de/DB/TC/mFISH/6-banding.html#3 [accessed on 01/12/2022].

13. Aurich-Costa, J.; Vannier, A.; Gregoire, E.; Nowak, F.; Cherif, D. IPM-FISH, a new M-FISH approach using IRS-PCR painting probes: Application to the analysis of seven human prostate cell lines. *Genes Chromosomes Cancer.* **2001**, 30, 143–160.

14. Chudoba, I.; Plesch, A.; Lörch, T.; Lemke, J.; Claussen, U.; Senger, G. High resolution multicolor-banding: A new technique for refined FISH analysis of human chromosomes. Cytogenet. *Cell Genet.* **1999**, 84, 156–160.

15. Liehr, T.; Heller, A.; Starke, H.; Rubtsov, N.; Trifonov, V.; Mrasek, K.; Weise, A.; Kuechler, A.; Claussen, U. Microdissection based high resolution multicolor banding for all 24 human chromosomes. *Int. J. Mol. Med.* **2002**, 9, 335–339.

16. Weise, A.; Heller, A.; Starke, H.; Mrasek, K.; Kuechler, A.; Pool-Zobel, B.-L.; Claussen, U.; Liehr, T. Multitude multicolor chromosome banding (mMCB): A comprehensive one-step multicolor FISH banding method. *Cytogenet. Genome Res.* **2003**, 103, 34–39.

17. Leibiger, C.; Kosyakova, N.; Mkrtchyan, H.; Glei, M.; Trifonov, V.; Liehr, T. First molecular cytogenetic high resolution characterization of the NIH 3T3 cell line by murine multicolor banding. *J. Histochem. Cytochem.* **2013**, 61, 306–312.

18. Liehr, T. About MCB. http://cs-tl.de/DB/TC/mFISH/6-banding.html#6 [accessed on 01/12/2022].

19. Zhang, C.; Cerveira, E.; Rens, W.; Yang, F.; Lee, C. Multicolor fluorescence in situ hybridization (FISH) approaches for simultaneous analysis of the entire human genome. *Curr. Protoc. Hum. Genet.* **2018**, 99, e70.

13 FISH—Centromere- and Heterochromatin- Specific Multicolor Probe Sets

Thomas Liehr

CONTENTS

INTRODUCTION

The human genome project has been declared to be finished by 2003.[1] However, in the almost two decades since then, it has been a well-known truism that there was almost 10% of the DNA-sequence still not accessed, yet.[2] This affects especially the highly repetitive DNA-stretches, mainly being concentrated in the highly variant heterochromatic parts of the human genome.[3] This gap was just recently closed by Evan Eichler and colleagues for a single genome.[2] Nonetheless, heterochromatic regions of the human genome are understudied and cannot be accessed by standard molecular genetic approaches, including sequencing, as of now.[4]

Thus, it is also not surprising that since 1996 different fluorescence in situ hybridization (FISH) probe sets were available to characterize euchromatic material along marker or derivative chromosomes.[5] However until 2001, whole genome–oriented heterochromatic FISH-probe sets were not established; the first then available one was specific for all human centromeres.[6] It even lasted until 2012 until a probe set for all large heterochromatic regions in the human genome was developed.[7] The two mentioned probe sets are suited for the one-step-characterization of the chromosomal origin in small supernumerary marker chromosomes (sSMCs),[8, 9] or larger derivative chromosomes with rearrangements including human hererochromatin.[10]

DOI: 10.1201/9781003223658-13

Here, the corresponding centromere-specific multicolor FISH (cenM-FISH),[6] and the heterochromatin-M-FISH (HCM-FISH) probe sets are introduced.[7]

SAMPLES TO BE CHARACTERIZED

In about 50–70% of cases with sSMCs, the derivatives are known to originate from chromosome 15.[8, 9] Among the remaining cases, there is: 1) a great variation in chromosomal and parental origin, 2) possibly a mosaicism, 3) the possibility of genomic imprinting effects, and 4) of homozygosity of autosomal recessively inherited mutations in uniparental isodisomy (UPD). As great dissimilarities in their clinical outcomes are reported, too, the characterization of prenatally detected, particularly de novo sSMCs, is of special interest for genetic counseling and appropriate medical care. Characterization of one or more sSMCs in a patient should be followed by testing for UPD, as the latter can cause clinical signs and symptoms even in heterochromatic sSMC cases, as well.[8, 9, 11, 12]

Many sSMCs consist only of heterochromatic material, and here it is important to distinguish those from such cases with potentially deleterious euchromatic material on them.[8, 9] Besides, there are other incidences where larger derivatives may either include enlarged heterochromatin (chromosomal heteromorphisms), which may be cytogenetically similar to deleterious unbalanced rearrangements.[3, 10]

CENTROMERE- AND HETEROCHROMATIN-SPECIFIC FISH APPROACHES

To characterize an sSMC, whole chromosome painting probe–based FISH approaches, such as multiplex FISH (M-FISH) or spectral karyotyping (SKY),[5] are well suited as long as the marker is larger than the short arm of chromosome 16. If the sSMC is smaller, and mainly heterochromatic, then in that way often no result may be obtained.[8, 9] Thus, for rapid characterization of such sSMCs, a multicolor FISH technique was established, allowing the unambiguous one-step identification of all human centromeric regions, excluding chromosomes 13 and 21 (due to same alphid DNA-sequences on these chromosomes). The so-called cenM-FISH (Figure 13.1) is based on all available centromere-specific DNA probes, labeled in five different fluorochromes.[6]

sSMCs are initially detected by GTG banding analysis and can be characterized for the presence of an acrocentric chromosome-derived short arm by nucleolus organizing region (NOR) staining. According to the NOR staining result (NOR-negative or NOR-positive), the origin of SMCs can be determined by application of all human centromeres in one experiment (i.e., cenM-FISH[6]) or by the acrocentric centromere-specific multicolor FISH (acro-cenM-FISH) probe set,[13] respectively. The latter consists of a probe specific for the acrocentric human p-arms, a NOR-specific probe, a probe specific for Yq12, and all centromere-specific probes for 13/21, 14/22, 15, and 22. These two approaches were successfully applied in about 1,500 cases with SMCs up to present.[8] In general, neocentromeric sSMCs are not stained by any of the two aforementioned approaches; in this case, other

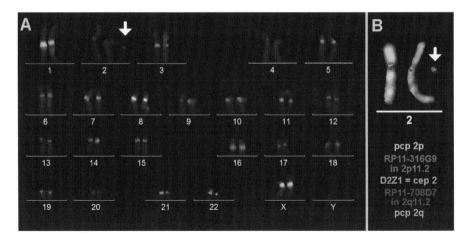

FIGURE 13.1   An sSMC was found in a clinically normal female.

A) CenM-FISH characterized the sSMC (arrow) as chromosome 2 derived.

B) SubcenM-FISH revealed that, irrespective of its small size, the sSMC lead to a partial trisomy 2p11.2 to 2q11.2. According to,[8] there are overall eleven cytogenetically comparable cases without any clinical signs. This case is # 02-O-p11.2/3–2.[8]

techniques for their characterization, such as M-FISH, microdissection of sSMCs, or chromosome microarray (CMA) have to be performed. About 130 cases (~1.8% of all reported sSMCs) with neocentromeres are reported in the literature.[8, 9] After determination of the origin of an sSMC with centromeric DNA, the most important question to address is if there is euchromatic material of the corresponding chromosome on it, or not. This question for partial trisomy can be studied by hybridizing whole chromosome painting or chromosome arm-specific probes. Another possibility is the use of a probe set specific for the pericentric region of all human chromosomes, the centromere-near multicolor FISH (subcenM-FISH) probe set. A chromosome-specific subcenM-FISH probe set consists of a centromere-specific satellite probe, one centromere-near locus-specific probe in the long arm and the short arm (excluding the acrocentric chromosomes), and chromosome arm-specific probes.[14] On the other hand, CMA can also be the initial hint that a patient is not carrier of a simple intra-chromosomal duplication or unbalanced translocation but of an sSMC.

In some cases, the centromeric region of a heterochromatic sSMC can be so small that it hardly stains by a probe contained in the cenM-FISH-probe set. Then the HCM-FISH probe set may be helpful, as it contains probes specific for 1q12, 9p12/q13, 9q12, NOR, 15p11.2, 16q11.2, Yq12, and all acrocentric p-arms.[7]

All mentioned probe sets are available on request at Thomas.Liehr@med.uni-jena.de. Besides, probes can be prepared as previously described,[6, 7, 13, 14] using commercially available locus specific probes and/or chromosome microdissection.[5]

YIELDS

In Figure 13.1, the possible yield of application of (sub-)cenM-FISH probe sets is shown—i.e. the characterization of origin and content of an sSMC. As summarized elsewhere, only molecular cytogenetics can reliably characterized sSMCs, especially in infertile patients.[15]

CONCLUSIONS

In summary, nowadays the detection of an sSMC does not stop after GTG banding, NOR staining, and exclusion of a chromosome 15 origin. It can be exactly described after (acro)cenM-FISH and/or HCM-FISH, subcenM-FISH provides information on additional euchromatin, and molecular genetics uncovers eventually present UPD. CMA can only be applied in euchromatic sSMCs without mosaicism.[15] Meanwhile, a relatively good correlation of sSMC content and clinics is available,[8] being extremely helpful in genetic counseling.

ACKNOWLEDGMENTS

The case shown in Figure 13.1 was provided by Drs. Cramer and Hickmann, Essen, Germany.

REFERENCES

1. https://en.wikipedia.org/wiki/Human_Genome_Project [accessed on 01/12/2022].
2. Nurk, S.; Koren, S.; Rhie, A.; Rautiainen, M.; Bzikadze, A.V.; Mikheenko, A.; Vollger, M.R.; Altemose, N.; Uralsky, L.; Gershman, A.; Aganezov, S.; Hoyt, S.J.; Diekhans, M.; Logsdon, G.A.; Alonge, M.; Antonarakis, S.E.; Borchers, M.; Bouffard, G.G.; Brooks, S.Y.; Caldas, G.V.; Cheng, H.; Chin, C.-S.; Chow, W.; de Lima, L.G.; Dishuck, P.C.; Durbin, R.; Dvorkina, T.; Fiddes, I.T.; Formenti, G.; Fulton, R.S.; Fungtammasan, A.; Garrison, E.; Grady, P.G.S.; Graves-Lindsay, T.-A.; Hall, I.M.; Hansen, N.F.; Hartley, G.A.; Haukness, M.; Howe, K.; Hunkapiller, M.W.; Jain, C.; Jain, M.; Jarvis, E.D.; Kerpedjiev, P.; Kirsche, M.; Kolmogorov, M.; Korlach, J.; Kremitzki, M.; Li, H.; Maduro, V.V.; Marschall, T.; McCartney, A.M.; McDaniel, J.; Miller, D.E.; Mullikin, J.C.; Myers, E.W.; Olson, N.D.; Paten, B.; Peluso, P.; Pevzner, P.A.; Porubsky, D.; Potapova, T.; Rogaev, E.I.; Rosenfeld, J.A.; Salzberg, S.L.; Schneider, V.A.; Sedlazeck, F.J.; Shafin, K.; Shew, C.J.; Shumate, A.; Sims, Y.; Smit, A.F.A.; Soto, D.C.; Sović, I.; Storer, J.M.; Streets, A.; Sullivan, B.A.; Thibaud-Nissen, F.; Torrance, J.; Wagner, J.; Walenz, P.P.; Wenger, A.; Wood, J.M.D.; Xiao, C.; Yan, S.M.; Young, A.C.; Zarate, S.; Surti, U.; McCoy, R.C.; Dennis, M.Y.; Alexandrov, I.A.; Gerton, J.L.; O'Neill, R.J.; Timp, W.; Zook, J.M.; Schatz, M.C.; Eichler, E.E.; Miga, K.H.; Phillippy, A.M. The complete sequence of a human genome. *bioRxiv.* **2021**. preprint doi: https://doi.org/10.1101/2021.05.26.445798.
3. Liehr, T. *Benign & Pathological Chromosomal Imbalances, 1st Edition Microscopic and Submicroscopic Copy Number Variations (CNVs) in Genetics and Counseling.* Academic Press, New York; **2014**.
4. Liehr, T. Repetitive elements in humans. Int. J. Mol. Sci. **2021**, 22, 2072.
5. Liehr, T. Molecular cytogenetics in the era of chromosomics and cytogenomic approaches. *Front. Genet.* **2021**, 12, 720507.

6. Nietzel, A.; Rocchi, M.; Starke, H.; Heller, A.; Fiedler, W.; Wlodarska, I.; Loncarevic, I.F.; Beensen, V.; Claussen, U.; Liehr, T. A new multicolor-FISH approach for the characterization of marker chromosomes: Centromere-specific multicolor-FISH (cenM-FISH). *Hum. Genet.* **2001**, 108, 199–204.

7. Bucksch, M.; Ziegler, M.; Kosayakova, N.; Mulatinho, M.V.; Llerena, J.C. Jr.; Morlot, S.; Fischer, W.; Polityko, A.D.; Kulpanovich, A.I.; Petersen, M.B.; Belitz, B.; Trifonov, V.; Weise, A.; Liehr, T.; Hamid, A.B. A new multicolor fluorescence in situ hybridization probe set directed against human heterochromatin: HCM-FISH. *J. Histochem. Cytochem.* **2012**, 60, 530–536.

8. Liehr, T. Small supernumerary marker chromosomes. http://cs-tl.de/DB/CA/sSMC/0-Start.html [accessed on 01/12/2022].

9. Liehr, T. *Small Supernumerary Marker Chromosomes (sSMC): A Guide for Human Geneticists and Clinicians: With Contributions by UNIQUE (The Rare Chromosome Disorder Support Group).* Springer, Berlin; **2012**.

10. Liehr, T. Cases with heteromorphisms. http://cs-tl.de/DB/CA/HCM/0-Start.html [accessed on 01/12/2022].

11. Liehr, T. *Uniparental Disomy (UPD) in Clinical Genetics: A Guide for Clinicians and Patients.* Springer, Berlin; **2014**.

12. Liehr, T. Cases with uniparental disomy. http://cs-tl.de/DB/CA/UPD/0-Start.html [accessed on 01/12/2022].

13. Trifonov, V.; Seidel, J.; Starke, H.; Prechtel, M.; Beensen, V.; Ziegler, M.; Hartmann, I.; Heller, A.; Nietzel, A.; Claussen, U.; Liehr, T. Enlarged chromosome 13 p-arm hiding a cryptic partial trisomy 6p22.2-pter. *Prenat. Diagn.* **2003**, 23, 427–430.

14. Weise, A.; Liehr, T. Subtelomeric and/or subcentromeric probe sets. In: *Fluorescence In Situ Hybridization (FISH): Application Guide*; Liehr, T., Ed. 2nd Edition. Springer, Berlin, **2017**, pp. 261–269.

15. Liehr, T.; Hamid Al-Rikabi, A.B. Impaired spermatogenesis due to small supernumerary marker chromosomes: The reason for infertility is only reliably ascertainable by cytogenetics. *Sex. Dev.* **2018**, 12, 281–287.

14 FISH—Detection of Individual Radio Sensitivity

Thomas Liehr

CONTENTS

INTRODUCTION

In radiotherapy, alongside tumor control, the prevention of severe treatment-related side effects is a major concern. Although the majority of radio-oncological patients tolerate a standard treatment protocol, toxic side effects can be detected in 0.2–10% of the cases.[1, 2] The latter can be a result of increased individual radio sensitivity, caused by a combination of exogenous and endogenous factors including genetic reasons. However, most of the exact underlying still mechanisms remain unknown. Only in few cases is increased radio sensitivity a result of an identified single gene mutation, e.g., in ataxia telangiectasia (AT) or Nijmegen breakage syndrome (NBS) patients, OMIM #208900 and #251260,[3] but in the majority, it is supposed to be modulated by a mixture of different genes.[4]

To have a chance to detect a possible overreaction prior to therapy, the availability of a predictive test system has been the aim for some decades. Although different approaches were tested (e.g., clonogenic cell survival,[5] G2-assay,[6] comet assay[7]), there is still no reliable routine assay to identify hypersensitive or less sensitive patients to adjust their therapeutic dose. It is known that chromosomal aberrations 1) are indicators of a previous exposure to irradiation and 2) can be used to estimate radio sensitivity. Previous studies demonstrated that molecular cytogenetic methods are superior to conventional cytogenetic analysis in detection and characterization of aberrations. At

DOI: 10.1201/9781003223658-14

this, the frequency of breaks and the occurrence of specific aberration types reflect individual sensitivity to radiation. Especially complex chromosomal aberrations (CCR) were identified as indicators for increased individual radio sensitivity.

To detect individual radio sensitivity by fluorescence in situ hybridization (FISH), three approaches are presently available: i) painting of one up to three different chromosomes in one or different colors, ii) painting of all 24 human chromosomes in different colors (24-color FISH), and iii) FISH-based chromosome banding for the detection of intra-chromosomal rearrangements.

SAMPLES/TISSUES

Our own studies have focused on the analysis of in vitro-induced chromosomal aberrations in NBS and AT homozygote and heterozygote individuals compared with normal reacting controls (Figures 14.1 and 14.2).[8, unpublished data] The aim was to

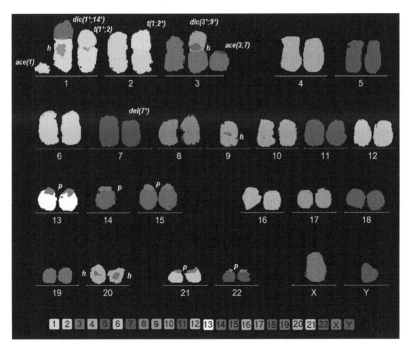

FIGURE 14.1 Example of a 24-color FISH result: karyogram of an NBS-heterozygote person (pseudocolor representation) after in vitro irradiation with 2.0 Gy and scheme of pseudocolors for each individual chromosome. The aberration can be described (nomenclature modified according to the ISCN 2020[29]) as follows: t(1;2),dic(1;14),ace(1),dic(3;9), ace t(3;7),del(7). This result can be summarized as seven breaks per mitosis. The translocation between chromosomes 3, 7, and 9 forms a complex chromosomal rearrangement (CCR).

Abbreviations: p=short arms of the acrocentric chromosomes #13, #14, #15, #21, and #22 consisting of repetitive DNA, pseudocolored in a different paint; h = heterochromatic DNA, which is polymorphic and can be present at #1, #9, and #16. (Stars indicate centromere localization.) Images were captured with the ISIS3 digital FISH imaging system (MetaSystems, Altlussheim, Germany) using a PCO VC45 CCD camera (PCO, Kehl, Germany) on an Axioplan 2 microscope (Zeiss, Jena, Germany).

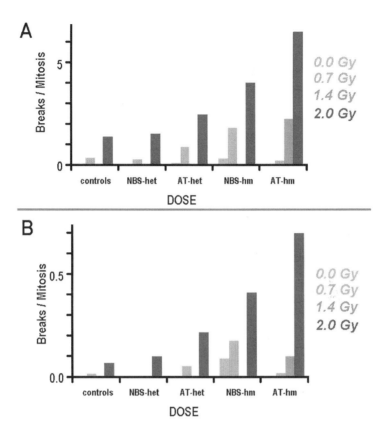

FIGURE 14.2 For the graphic depicted here, four normal controls, two Nijmegen breakage syndrome (NBS), and two ataxia telangiectasia (AT) heterozygotes (NBS-het, AT-het) plus one NBS and one AT homozygote (NBS-hm, AT-hm) were included. (A) shows the average number of breaks per mitosis, while (B) presents the average number of CCR occurring in each mitosis for each of the five groups. Three different in vitro irradiation doses were analyzed per patient. The applied doses were 0.0, 0.7, and 2.0 Gy; in the AT-hm patient, instead of 2.0 Gy, 1.4 Gy was used. A clear differentiation was possible in both diagrams between normal controls and the homozygote patients, as well as between controls and AT-het.

demonstrate that a difference in radio sensitivity which was already known between these individuals is also detectable using the 24-color FISH technique. Thus, the increased radio sensitivity of NBS and AT patients can serve as a positive control in a predictive assay.

DESCRIPTION OF METHODS

ONE- TO THREE-COLOR FISH

FISH applying one up to three whole-chromosome painting (wcp) probes simultaneously was used to detect radiation-induced chromosome instability in peripheral blood lymphocytes and fibroblasts.[9–16] However, in the analyses, the used chromosomes are selected by chance, although a random distribution of chromosomal rearrangements along the chromosomes is up to now still discussed controversially.[17–18] To avoid this problem and to obtain an overview of the whole karyotype, 24-color FISH can be applied using a probe mix of the 24 different human wcp probes.[8, 9, 19, 20]

24-COLOR FISH

This method, first described in 1996,[21, 22] allows the simultaneous visualization of all chromosomes within a metaphase in different specific colors. Nearly all interchromosomal aberrations (as reciprocal and nonreciprocal translocations, complex rearrangements, ring chromosomes, acentric fragments, dicentric fragments, or insertions) can be detected and defined in more detail.

24-color FISH has been used, for example, for studies on normal peripheral blood lymphocytes irradiated in vitro,[8, 19] on radiotherapy-induced residual chromosomal damage in peripheral lymphocytes,[9] on bone marrow cells of Chernobyl victims,[19] and for analysis of chromosomal aberrations after irradiation with ionizing alpha particles. The latter was also used for investigations on the formation of CCR.[23]

MULTICOLOR BANDING

FISH methods using all 24 human whole-chromosome painting probes simultaneously reach their limits in identification of intra-chromosomal rearrangements (such as small interstitial deletions, duplications, and inversions without change of the centromeric index) and when exact characterization of breakpoints is required. These limitations have been overcome by the development of FISH-banding methods during the last decades.[25] In the meantime, one of these approaches, the multicolor banding (MCB) technique (or mBAND),[26] was also used for the analysis of X-ray-induced aberrations.[27, 28] However, only a probe set for chromosome 5 was applied, demonstrating that intra chromosomal aberrations are present in a considerable portion in radiation-induced changes.

SPECIFIC METHODOLOGICAL DETAILS FOR M-FISH EVALUATION

In the following, the criteria used in our own studies are specified; overall, these require an experienced cytogeneticist. After the 24-color FISH procedure including hybridization, posthybridization washes, and detection, at least 100 metaphases per irradiation dose and patient/proband have to be acquired. This can be performed filter-based (m-FISH),[21] or spectracube-based (SKY).[26] All captured metaphases have to be karyotyped. For detailed evaluation of each metaphase, it is not enough to rely only on the pseudocolor functions of the used software, but for unambiguous results, one has to check the different fluorochrome channels or hybridization profiles. Then, aberration types and involved chromosomes have to be registered in detail. All occurring aberrations have to be classified into reciprocal and nonreciprocal translocations, ring chromosomes, acentric fragments, dicentric fragments, inversions, insertions, and complex rearrangements. Translocations, insertions, and complex rearrangements are visible because of a color change along a rearranged chromosome. Dicentrics and ring chromosomes can easily be identified using the inverted DAPI (4′,6-diamidino-2-phenylindole) picture. Complex chromosomal rearrangements (CCR) consist, per definition, of at least two chromosomes with three or more breaks.[24]

For further evaluation, the frequency of break events constituting the observed aberrations was estimated as the minimal number of breaks considered to be necessary for producing the aberrations in each metaphase. The total number of break events in each patient/control and irradiation dose was summed up and divided by the number of metaphases analyzed to obtain the average rate of breaks per mitosis (B/M). The radiosensitivity of lymphocytes is expressed as number of radiation-induced B/M and CCR/M after each irradiation dose, subtracted by the 0.0-Gy control value to correct the influence of spontaneous basic aberration frequencies.[8]

YIELDS

The 24-color FISH is a highly informative approach concerning the characterization of radiation-induced chromosomal rearrangements, but is based on a very sophisticated and time-consuming evaluation procedure. A typical result for one metaphase and a summary of obtained data are shown in Figures 14.1 and 14.2.

CONCLUSIONS

In summary, genetically determined intrinsic radiosensitivity can be detected and quantified by FISH approaches. Twenty-four-color FISH can be used as well as the three-color FISH. FISH is as reliable but more informative than conventional cytogenetic data.[10, 11] As stated before, the 24-color FISH approach requires a very sophisticated evaluation. However, data obtainable by this technique are indispensable to decide whether there exist any hot spots for radiation-induced breakpoints. If so, such hot spots could be characterized in detail by FISH-banding methods,[25] and in the following, locus-specific breakpoint-spanning probes could be applied in the future for quick detection of individual radio sensitivity, possibly even in interphase cytogenetics.

REFERENCES

1. Roberts, S.A.; Spreadborough, A.R.; Bulman, B.; Barber, J.B.; Evans, D.G.; Scott, D. Heritability of cellular radiosensitivity: A marker of low-penetrance predisposition genes in breast cancer? *Am. J. Hum. Genet.* **1999**, 65, 784–794.
2. Rogers, P.B.; Plowman, P.N.; Harris, S.J.; Arlett, C.F. Four radiation hypersensitivity cases and their implications for clinical radiotherapy. *Radiother. Oncol.* **2000**, 57, 143–154.
3. OMIM. www.ncbi.nlm.nih.gov/omim/ [accessed on 01/12/2022].
4. Turesson, I.; Nyman, J.; Holmberg, E.; Oden, A. Prognostic factors for acute and late skin reactions in radiotherapy patients. *Int. J. Radiat. Oncol. Biol. Phys.* **1996**, 36, 1065–1075.

5. Cole, J.; Arlett, C.F.; Green, M.H.; Harcourt, S.A.; Priestley, A.; Henderson, L.; Cole, H.; James, S.E.; Richmond, F. Comparative human cellular radiosensitivity: II. The survival following gamma-irradiation of unstimulated (G0) T-lymphocytes, T-lymphocyte lines, lymphoblastoid cell lines and fibroblasts from normal donors, from ataxia-telangiectasia patients and from ataxia-telangiectasia heterozygotes. *Int. J. Radiat. Biol.* **1988**, 54, 929–943.

6. Sanford, K.K.; Parshad, R.; Price, F.M.; Jones, G.M.; Tarone, R.E.; Eierman, L.; Hale, P.; Waldmann, T.A. Enhanced chromatid damage in blood lymphocytes after G2 phase x irradiation, a marker of the ataxia-telangiectasia gene. *J. Natl. Cancer Inst.* **1990**, 8, 1050–1054.

7. Hovhannisyan, G.G. Fluorescence in situ hybridization in combination with the comet assay and micronucleus test in genetic toxicology. *Mol. Cytogenet.* **2010**, 3, 17.

8. Kuechler, A.; Neubauer, S.; Grabenbauer, G.G.; Claussen, U.; Liehr, T.; Sauer, R.; Wendt, T.G. Is 24-color FISH detection of in-vitro radiation-induced chromosomal aberrations suited to determine individual intrinsic radiosensitivity? Strahlenther. *Onkol.* **2002**, 178, 209–215.

9. Kuechler, A.; Dreidax, M.; Pigorsch, S.U.; Liehr, T.; Claussen, U.; Wendt, T.G.; Dunst, J. Residual chromosomal damage after radiochemotherapy with and without amifostine detected by 24-color FISH. Strahlenther. *Onkol.* **2003**, 179, 493–498.

10. Neubauer, S.; Arutyunyan, R.; Stumm, M.; Dörk, T.; Bendix, R.; Bremer, M.; Varon, R.; Sauer, R.; Gebhart, E. Radiosensitivity of ataxia telangiectasia and Nijmegen breakage syndrome homozygotes and heterozygotes as determined by three-color FISH chromosome painting. *Radiat. Res.* **2002**, 157, 312–321.

11. Gebhart, E.; Neubauer, S.; Schmitt, G.; Birkenhake, S.; Dunst, J. Use of three-color chromosome in situ suppression technique for the detection of past radiation exposure. *Radiat. Res.* **1996**, 145, 47–52.

12. Neubauer, S.; Dunst, J.; Gebhart, E. The impact of complex chromosomal rearrangements on the detection of radiosensitivity in cancer patients. *Radiother. Oncol.* **1997**, 43, 189–195.

13. Stritzelberger, J.; Lainer, J.; Gollwitzer, S.; Graf, W.; Jost, T.; Lang, J.D.; Mueller, T.M.; Schwab, S.; Fietkau, R.; Hamer, H.M.; Distel, L. Ex vivo radiosensitivity is increased in non-cancer patients taking valproate. *BMC Neurol.* **2020**, 20, 390.

14. Schuster, B.; Ellmann, A.; Mayo, T.; Auer, J.; Haas, M.; Hecht, M.; Fietkau, R.; Distel, L.V. Rate of individuals with clearly increased radiosensitivity rise with age both in healthy individuals and in cancer patients. *BMC Geriatr.* **2018**, 18, 105.

15. Gryc, T.; Putz, F.; Goerig, N.; Ziegler, S.; Fietkau, R.; Distel, L.V.; Schuster, B. Idelalisib may have the potential to increase radiotherapy side effects. *Radiat. Oncol.* **2017**, 12, 109.

16. Rave-Fränk, M.; Virsik-Kopp, P.; Pradier, O.; Nitsche, M.; Grunefeld, S.; Schmidberger, H. In vitro response of human dermal fibroblasts to X-irradiation: Relationship between radiation-induced clonogenic cell death, chromosome aberrations and markers of proliferative senescence or differentiation. *Int. J. Radiat. Biol.* **2001**, 77, 1163–1174.

17. Barrios, L.; Miro, R.; Caballin, M.R.; Fuster, C.; Guedea, F.; Subias, A.; Egozcue, J. Cytogenetic effects of radiotherapy. Breakpoint distribution in induced chromosome aberrations. *Cancer Genet. Cytogenet.* **1989**, 41, 61–70.

18. Cigarran, S.; Barrios, L.; Barquinero, J.F.; Caballin, M.R.; Ribas, M.; Egozcue, J. Relationship between the DNA content of human chromosomes and their involvement in radiation-induced structural aberrations, analysed by painting. *Int. J. Radiat. Biol.* **1998**, 74, 449–455.

19. Greulich, K.M.; Kreja, L.; Heinze, B.; Rhein, A.P.; Weier, H.G.; Bruckner, M.; Fuchs, P.; Molls, M. Rapid detection of radiation-induced chromosomal aberrations in lymphocytes and hematopoietic progenitor cells by mFISH. *Mutat. Res.* **2000**, 452, 73–81.

20. Liehr, T. Basics and literature on multicolor fluorescence in situ hybridization application. http://cs-tl.de/DB/TC/mFISH/5-wcp-oth.html#2 [accessed on 01/12/2022].

21. Speicher, M.R.; Gwyn Ballard, S.; Ward, D.C. Karyotyping human chromosomes by combinatorial multi-fluor FISH. *Nat. Genet.* **1996**, 12, 368–375.

22. Schröck, E.; du Manoir, S.; Veldman, T.; Schoell, B.; Wienberg, J.; Ferguson-Smith, M.A.; Ning, Y.; Ledbetter, D.H.; Bar-Am, I.; Soenksen, D.; Garini, Y.; Ried, T. Multicolor spectral karyotyping of human chromosomes. *Science.* **1996**, 273, 494–497.

23. Anderson, R.M.; Stevens, D.L.; Goodhead, D.T. M-FISH analysis shows that complex chromosome aberrations induced by {alpha}-particle tracks are cumulative products of localized rearrangements. *Proc. Natl. Acad. Sci. U. S. A.* **2002**, 99, 12167–12172.

24. Savage, J.R.K.; Simpson, P.J. FISH "painting" pattern resulting from complex exchanges. *Mutat. Res.* **1994**, 312, 51–60.

25. Liehr, T. Molecular cytogenetics in the era of chromosomics and cytogenomic approaches. *Front. Genet.* **2021**, 12, 720507.

26. Liehr, T.; Heller, A.; Starke, H.; Rubtsov, N.; Trifonov, V.; Mrasek, K.; Weise, A.; Kuechler, A.; Claussen, U. Microdissection based high resolution multicolor banding for all 24 human chromosomes. *Int. J. Mol. Med.* **2002**, 9, 335–339.

27. Johannes, C.; Chudoba, I.; Obe, G. Analysis of X-ray-induced aberrations in human chromosome 5 using high-resolution multicolour banding FISH (mBAND). *Chromosom. Res.* **1999**, 7, 625–633.

28. Hande, M.P.; Azizova, T.V.; Geard, C.R.; Burak, L.E.; Mitchell, C.R.; Khokhryakov, V.F.; Vasilenko, E.K.; Brenner, D.J. Past exposure to densely ionizing radiation leaves a unique permanent signature in the genome. *Am. J. Hum. Genet.* **2003**, 72, 1162–1170.

29. ISCN (2020). *An International System for Human Cytogenomic Nomenclature*; McGowan-Jordan, J.; Hastings R.J.; Moore, S., Eds. S Karger, Basel; **1995**.

15 FISH—Detection of CNVs

Tigran Harutyunyan, Anzhela Sargsyan and Rouben Aroutiounian

CONTENTS

INTRODUCTION

DNA copy number variations (CNVs) occur in the genome due to spontaneous and induced gains, losses, and/or insertions of 50 bp (as repeats) to several Mb long DNA segments leading to copy number changes of a particular DNA sequence within the chromosomes.[1] Thus, CNVs are structural changes within chromosomes and are distinguished from whole chromosome gains and losses which are a separate class of cytogenetic alterations (i.e. aneuploidy). When the frequency of a given CNV in a population is greater than 1%, it is referred to as a copy number polymorphism. Mounting evidence demonstrates the importance of de novo CNVs as a source of genomic variation and in development of pathogenic conditions. Therefore, the identification of CNVs in human chromosomes is of crucial task from clinical and research perspectives. However, their detection remains challenging.

DOI: 10.1201/9781003223658-15

Zarrei et al.[2] constructed a high-resolution CNV map of the human genome that includes 2,057,368 variants (195,084 gains and 1,862,284 losses). It was shown that CNVs have unequal frequency among chromosomes. The highest frequencies of gains were observed in chromosomes 22 and Y, followed by chromosomes 16, 9, and 15; however, the highest proportions of variable sequences with losses were detected in chromosomes 19, 22, and Y. In addition, higher proportions of CNVs were identified in pericentromeric and subtelomeric regions of all chromosomes. Moreover, approximately for 100 genes it has been shown that they may be completely deleted without apparent phenotypic consequences.[2] Overall, it was shown that CNVs contribute to 4.8–9.7% of the variability in the human genome.

There are several databases for CNVs of which the "Database of Genomic Variants" is the largest host for data of structural variations in healthy subjects from worldwide populations.[3] The recently developed database CNVIntegrate is hosting data from both healthy persons and cancer patients, and enables comparisons of CNV frequencies within multiple ethnic populations.[4]

Based on differences in breakpoint positions during genomic rearrangements, CNVs are classified as recurrent and non-recurrent. Recurrent CNVs typically occur via non-allelic homologous recombination between region-specific low-copy repeats with >95% nucleotide sequence identity.[5] Therefore, recurrent CNVs have the same DNA sequence at the rearrangement breakpoints and similar size in unrelated individuals. In contrast, non-recurrent CNVs are thought to originate due to inappropriate repair of accidental DNA double-strand breaks in proliferating cells via microhomology-mediated pathways. Thus, non-recurrent CNVs have unique sequences at the breakpoints in each individual.[6] Localization of recurrent and non-recurrent CNVs in the genome is of high importance for distinguishing benign and pathogenic variants, as well as for the identification of environmental factors that potentially are capable of inducing CNVs.

Mounting evidence demonstrates the role of CNVs in various pathological conditions including severe intellectual disability, autism, schizophrenia, cardiovascular diseases, and cancer.[7–10] However, clinical interpretation of CNVs is still a challenging task. Fortin et al.[11] identified CNVs in patients suffering from inherited epilepsy with febrile seizures plus (GEFS+) syndrome. Deletions in 15q11.2, 19p13.3, and 22q11.2 and duplication in 10q11.22 were detected in four patients from 12 families. Nevertheless, the clinical significance of these CNVs is difficult to determine considering the small cohort and the absence of strong evidence demonstrating the causative impact of the identified CNVs. Here intrachromosomal interactions may be suggested as an underlying cause, as recently shown to play a role in 2q37-deletion syndrome.[12]

Very large (> 500 kb) and rare (<1%) CNVs are more likely to be detected in pathological conditions than small and common ones. Nevertheless, the number of genes within the CNV, their dosage sensitivity, and functions (e.g. protein-coding or non-coding) are likely the determinants of the clinical significance of CNVs.[13]

Recently, the American College of Medical Genetics and Genomics and the Clinical Genome Resource have developed guideline for the classification and quantitative interpretation of constitutional CNVs (inherited and de novo). Thus, there are five main categories for postnatal CNVs according to their final point values of evidence scoring metrics: pathogenic (≥0.99), likely pathogenic (0.90–0.98), variant of uncertain significance (−0.89–0.89), likely benign (−0.90 and −0.98), and benign (≤−0.99).[14] The authors developed a publicly available CNV classification calculator (http://cnvcalc.clinicalgenome.org/cnvcalc/) to facilitate the use of this semi-quantitative system. This tool allows applying points for individual evidence categories for a given CNV and automatically determines the CNV category. However, the authors emphasize that these standards cannot be implemented for the acquired CNVs in neoplasia while CNVs are nearly ubiquitous in cancers. Therefore, identification of clinically relevant CNVs and analysis of their frequency are important issues in cancer diagnostics.

Accumulating evidence demonstrates the potential of the environmental mutagens and carcinogens to induce de novo CNVs. In particular, DNA damage and/ or DNA replication stress-inducing agents, such as ionizing radiation, asbestos, benzo(a)pyrene, and hydroxyurea are known to cause non-recurrent CNVs in humans and animal cells.[1] Considering the mechanistic differences between recurrent and non-recurrent CNVs it can be assumed that different groups of environmental mutagens are likely to induce a specific type of CNVs. Therefore, currently used genetic endpoints of mutagenicity testing should also include CNVs as a marker of mutagenic impact. However, the methods of the identification of CNVs and their loci that should be analyzed in the standard genotoxicity testing are frontiers of modern mutagenesis and clinical genetics.

Currently, fluorescence in situ hybridization (FISH) is the most commonly used diagnostic tool and the gold standard for the detection of CNVs of different genes of clinical importance. Namely, in patients with invasive breast cancer, FISH analysis for *HER2/NEU1* typically demonstrates the clear presence or lack of *HER2/NEU1* amplification.[15] However, CNVs in homologous chromosomes can be used to distinguish otherwise cytogenetically similar homologs from each other. Moreover, the application of BAC (bacterial artificial chromosomes) probes specific for CNV loci and allows identification of the parental origin of the homologous chromosomes in the offspring. Accordingly, parental-origin-determination-FISH (pod-FISH) analysis was developed.[16] Analysis of variations in fluorescence intensities between treated and untreated groups enables the detection of alterations in CNV loci. In aflatoxin B1 treated human peripheral blood lymphocytes, application of pod-FISH revealed losses in 8p21.2 and 15q11.2 chromosome loci,[17] while in lymphocytes irradiated with AREAL laser-driven electron accelerator (CANDLE, Armenia), gains were identified in 1p31.1, 7q11.22, 9q21.3, 10q21.1, and 16q23.1 chromosome loci.[18]

In this chapter, the semi-automatic quantitative analysis of CNVs by pod-FISH is described, which can be used for CNV-studies of clinical importance and the investigation of alterations induced by mutagens in by pod-FISH accessible CNV loci.

SAMPLES/TISSUES/MATERIALS

Pod-FISH allows the detection of CNVs on a single-cell level in any cells; however, whole blood lymphocytes are most suitable for the analysis of spontaneous or induced CNVs, as not only interphases but also metaphases can be evaluated. Slides for the pod-FISH analysis can be prepared from heparinized whole blood or blood collected in EDTA containing vacutainers.

HUMAN WHOLE BLOOD CULTURE

- RPMI-1640 medium with L-glutamine (Cat. No. 11875093, Gibco™)
- Fetal bovine serum (Cat. No. F2442, Sigma-Aldrich)
- Penicillin-Streptomycin (Cat. No. P4333, Sigma-Aldrich)
- Phytohemagglutinin (Cat. No. M5030-BC, Merck)

SLIDES WITH METAPHASE CHROMOSOMES

- Colcemid solution (Cat. No. 10295892001, Sigma-Aldrich)
- Potassium chloride (Cat. No. P9541, Sigma-Aldrich)
- Hypotonic solution of KCl: 0.075 M freshly prepared KCl solution used at 37°C
- Methanol 100 % (Cat. No. 34860, Sigma-Aldrich)
- Glacial acetic acid (Cat. No. 1000631011, Sigma-Aldrich)
- Carnoy's fixative: methanol/glacial acetic acid freshly prepared mix in 3:1 (v/v) ratio and used at −20°C

POD-FISH PROBES

- BACs for the specific CNV loci[16]
- Nick translation kit (Cat. No. 11745808910, Roche)
- EDTA 0.5 M (Cat. No. 324506, Sigma-Aldrich)
- Ethanol 100 % (Cat. No. 459836, Sigma-Aldrich)
- tRNA (Cat. No. 10109541001, Sigma-Aldrich)
- Sodium acetate buffer solution 3 M, pH 5.2 (Cat. No. S7899, Sigma-Aldrich)
- Dextran sulfate buffer: 2 g + 10 ml 50 % deionized formamide/2×SSC/50 mM phosphate buffer for 3 h, 70°C. Store at −20°C

PRETREATMENT AND POSTFIXATION

- Ethanol solutions: 70 %, 90 % and 100 %
- PBS (Cat. No. 806544, Sigma-Aldrich)
- Hydrochloric acid solution 0.2 M (Cat. No. 13–1720, Sigma-Aldrich)
- Pepsin (Cat. No. P7012, Sigma-Aldrich)
- Pepsin stock solution: 1 g pepsin + 50 ml double-distilled water. Aliquot in 0.5 ml, store at −20°C
- Pepsin pretreatment solution: 5 ml of 0.2 M HCl + 95 ml of distilled water at 37°C
- Postfix solution: 5 ml 2 % paraformaldehyde + 4.5 ml PBS + 0.5 ml MgCl$_2$ (1 M)

Pod-FISH and Washing

- COT DNA (Cat. No. 11581074001, Sigma-Aldrich)
- Fluoroshield with DAPI (Cat. No. F6057, Sigma-Aldrich)
- 20×SSC (Cat. No. S6639, Sigma-Aldrich)
- Formamide (Cat. No. F9037, Sigma-Aldrich)
- Tween 20 (Cat. No. P9416, Sigma-Aldrich)
- Denaturation buffer: 70 ml formamide + 10 ml 20×SSC + 20 ml distilled water
- 0.4×SSC solution: 10 ml 20×SSC + 490 ml distilled water
- 4×SSC/Tween solution: 100 ml 20×SSC + 400 ml distilled water + 0.25 ml Tween 20

DESCRIPTION OF METHODS

Although pod-FISH is applicable for both interphase nuclei and metaphase chromosomes, we recommend analyzing CNV signals in metaphase chromosomes, which decreases the background noise and provides better resolution for the detection of small changes in fluorescence intensities of FISH signals. For the analysis of radiation-induced CNVs, human whole blood should be exposed to radiation immediately after sampling. For the analysis of chemical mutagen-induced CNVs, whole blood should be cultured for 24 h before exposure.

Human Whole Blood Cultivation

1. Add 1 ml of human whole blood to 9 ml RPMI-1640 medium, containing 10% fetal bovine serum, 1% penicillin/streptomycin, and 10 µg/ml phytohemagglutinin-L, and incubate at 37 °C. The total cultivation period should be 72 h.
2. After 24 or 48 h growth, add mutagen (e.g. aflatoxin B1) to the culture and incubate at 37 °C.

Preparation of Metaphase Spreads

1. 1.5 h before harvesting (at 70.5 h of cultivation), add 0.1 ml colcemid to the culture and incubate at 37 °C to block the cell cycle in the metaphase.
2. Transfer the culture into a 15 ml centrifuge tube.
3. Centrifuge at room temperature for 7 min at 1500 rpm. Discard the supernatant except for about 0.5 mL remaining above the cell pellet, and gently resuspend the pellet.
4. Add 10 ml of prewarmed (37 °C) 0.075 M hypotonic solution of KCl, and incubate for 15 min at 37 °C.
5. Repeat step 3.
6. Quickly add 5 ml of Carnoy's fixative (−20 °C) and vortex immediately to avoid clumps. Add another 5 ml of Carnoy's fixative (−20 °C), and incubate for 10 min at room temperature.
7. Repeat step 3.

8. Resuspend the pellet in 5 ml of Carnoy's fixative (−20°C), and repeat step 3.
9. Repeat step 8 twice.
10. Add 0.5 ml of Carnoy's fixative (−20°C), and store the suspension at −20°C until use.
11. Depending on the density of the suspension, drop (20–50 μl) 2–3 drops onto the precooled (4°C) clean and humid slides, and let the slide dry on the warming plate at 51°C.
12. Keep the slides at room temperature for short-term storage, while slides for long-term storage should be kept at −20°C.

HOMEMADE POD-FISH PROBES

1. The BAC DNA for pod-FISH can be labeled using the corresponding nick translation kit. In particular, nick translation mix (Cat. No. 11745808910, Roche) is designed for direct fluorophore-labeling of FISH probes: dilute 1 μg BAC DNA in double-distilled water to get the final volume of 12 μl, and add 4 μl of 5× fluorophore-labeling mix (contains fluorophore-labeled dUTP) and 4 μl of nick translation mix. The labeled fragments usually have a 200–500 nucleotide length.
2. Mix the solution using the 20 μl pipette tip and incubate at 15°C for 90 min.
3. Terminate the reaction with 1 μl of 0.5 M EDTA, mix vigorously, spin down, and incubate at 65°C for 10 min.
4. Precipitate the labeled probe with 10 μl of tRNA + 5 μl sodium acetate + 100 μl 100 % ethanol at −20°C during the overnight.
5. Centrifuge the mixture at 15,000 rpm at 4°C for 15 min.
6. Discard the supernatant and dry the pellet using a speed vac.
7. Dilute the pellet in 20 μl of dextran sulfate buffer, and mix at 65°C for 5 min.
8. Keep the probe at −20°C until use.

SLIDE PRETREATMENT

1. Dehydrate slides in an ethanol series (70 %, 90 %, 100 %, 3 min each) and air-dry at room temperature.
2. Add 0.5 ml of pepsin to 100 ml of pepsin pretreatment solution, and put slides for 3–5 min in a Coplin jar at 37°C in the water bath.
3. Wash slides in the PBS for 5 min at room temperature.
4. Add 100 μl of postfix solution, and cover with coverslips for 10 min at room temperature.
5. Remove coverslips; repeat step 3 and step 1.

POD-FISH PROCEDURE

1. Mix 1 μl of the probe with 7 μl dextran sulfate in the Eppendorf tube containing COT DNA. The optimal concentration of the COT DNA should be determined experimentally.

2. Prewarm the mixture at 45°C for 20 min.
3. Add 100 µl of denaturation buffer to each slide, cover with coverslip and incubate on a warming plate at 75°C for 3 min, and in parallel begin the denaturation of the probe at 75°C.
4. Remove the coverslip and put slides in a 70 % ethanol at −20°C for 3 min.
5. Dehydrate slides in ethanol series of 90 % and 100 % at room temperature for 3 min each and air-dry.
6. Add 8 µl of the probe onto half of the slide and cover with 24×24 coverslip and seal with rubber cement.
7. Place the slides in the humid chamber and incubate at 37°C overnight.

WASHING AND COUNTERSTAINING OF SLIDES

1. Remove the rubber cement and coverslips, and put the slides in the Coplin jar filled with 0.4×SSC at 65°C for 5 min in the water bath.
2. Put the slides in the new Coplin jar with 4×SSC/Tween on the shaker at room temperature for 5 min.
3. Wash the slides in the PBS for 5 min, then dehydrate in ethanol series (70 %, 90 %, 100 %, 3 min each), and air-dry at room temperature.
4. Add 30 µl of fluoroshield containing DAPI for counterstaining, cover with 24×60 coverslip, and incubate at 4°C for at least 20 min.
5. Analyze the results under a fluorescent microscope.
6. For the analysis of spontaneous CNVs by pod-FISH, at least 20 metaphase spreads should be investigated. However, for the analysis of mutagen-induced CNVs, we recommend capturing at least 50 pairs of CNV signals of homologous chromosomes from each experimental point. This will increase the statistical power of the analysis and will decrease the background noise (and/or false-positive variations).

YIELDS

ANALYSIS OF CNVs USING IMAGEJ PROGRAM

1. The fluorescence intensities of CNV signals can be measured using e.g. the publically available ImageJ program.[19]
2. Capture CNV signals for each fluorochrome and separately save in ".TIF" image.
3. Activate ImageJ and upload each image in the program by pressing "Ctrl+O" and selecting images for evaluation.
4. Zoom in the area containing CNV signals by holding "Ctrl" and scrolling the mouse (Figure 15.1A).
5. Select CNV signals using the "Freehand selection" tool (Figure 15.1B).
6. To obtain a numerical value of the CNV signal, select "Analyze" from the command bar and select "Measure", or simply press "Ctrl+M" after the signal selection step. A "Results table" will occur with numerical values of mean fluorescence intensity or mean area.

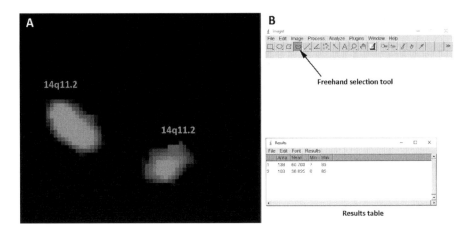

FIGURE 15.1 Example of the analysis of CNVs using pod-FISH. (A) CNV signals in 14q11.2 chromosome loci (Spectrum Green) are detected by pod-FISH. (B) In the ImageJ program, CNV signals are selected using the "Freehand selection tool", and the fluorescence measurement values are presented in the "Results table".

7. Repeat steps 5–7 for the second signal from a homologous chromosome. Thus, the first pair of signals will be measured, and so on (Figure 15.1B).
8. After measuring signals in the selected group of CNVs (e.g. control), go to the "File" in the "Results table" and select "Save as" or press "Ctrl+S".
9. After measuring signals in the control and the treatment groups, statistical analysis can be performed to compare signal intensities within groups and between groups.
10. This approach will allow the detection of even those CNVs that are not detectable by fluorescence microscopy. In addition, the bias by the subjectivity of the researcher visually analyzing the difference between CNV signal intensities will significantly decrease.

CONCLUSIONS

Spontaneous CNVs are an important source of genomic variations and polymorphisms. However, de novo CNVs can have pathogenic consequences such as cancer or neurological disorders. Moreover, induction of DNA strand breaks or inhibition of DNA replication with environmental mutagens can also induce de novo CNVs in human and animal cells. Therefore, we recommend including CNVs as a novel genetic endpoint in mutagenicity testing of environmental factors. Alterations in molecular-cytogenetically visible CNVs can be measured with the application of the ImageJ program. Further studies with the application of standard mutagens are required to establish threshold values of fluorescence intensities of CNVs, as well as to determine CNV loci that can be useful for the determination of recurrent or non-recurrent CNV-inducing agents.

ACKNOWLEDGMENTS

This work was supported by the RA MESCS and BMBF (project #AG-01/20), and RA MESCS (project #21AG-1F068).

REFERENCES

1. Hovhannisyan, G.; Harutyunyan, T.; Aroutiounian, R.; Liehr, T. DNA Copy number variations as markers of mutagenic impact. *Int. J. Mol. Sci.* **2019**, 20, 4723.
2. Zarrei, M.; MacDonald, J.R.; Merico, D.; Scherer, S.W. A copy number variation map of the human genome. *Nat. Rev. Genet.* **2015**, 16, 172–183.
3. MacDonald, J.R.; Ziman, R.; Yuen, R.K.; Feuk, L.; Scherer, S.W. The database of genomic variants: A curated collection of structural variation in the human genome. *Nucleic Acids Res.* **2014**, 42, D986–D992.
4. Chattopadhyay, A.; Teoh, Z.H.; Wu, C.Y.; Juang, J.J.; Lai, L.C.; Tsai, M.H.; Wu, C.H.; Lu, T.P.; Chuang, E.Y. CNVIntegrate: The first multi-ethnic database for identifying copy number variations associated with cancer. Database (Oxford). **2021**, 2021, baab044.
5. Smajlagić, D.; Lavrichenko, K.; Berland, S.; Helgeland, Ø.; Knudsen, G.P.; Vaudel, M.; Haavik, J.; Knappskog, P.M.; Njølstad, P.R.; Houge, G.; Johansson, S. Population prevalence and inheritance pattern of recurrent CNVs associated with neurodevelopmental disorders in 12,252 newborns and their parents. *Eur. J. Hum. Genet.* **2021**, 29, 205–215.
6. Conover, H.N.; Argueso, J.L. Contrasting mechanisms of de novo copy number mutagenesis suggest the existence of different classes of environmental copy number mutagens. *Environ. Mol. Mutagen.* **2016**, 57, 3–9.
7. Lew, A.R.; Kellermayer, T.R.; Sule, B.P.; Szigeti, K. Copy number variations in adult-onset neuropsychiatric diseases. Curr. Genomics. **2018**, 19, 420–430.
8. Velinov, M. Genomic copy number variations in the autism clinic: Work in progress. *Front. Cell Neurosci.* **2019**, 13, 57.
9. Prestes, P.R.; Maier, M.C.; Charchar, F.J. DNA copy number variations: Do these big mutations have a big effect on cardiovascular risk? *Int. J. Cardiol.* **2020**, 298, 116–117.
10. Brezina, S.; Feigl, M.; Gumpenberger, T.; Staudinger, R.; Baierl, A.; Gsur, A. Genome-wide association study of germline copy number variations reveals an association with prostate cancer aggressiveness. *Mutagenesis.* **2020**, 35, 283–290.
11. Fortin, O.; Vincelette, C.; Chénier, S.; Ghais, A.; Shevell, M.I.; Simard-Tremblay, E.; Myers, K.A. Copy number variation in genetic epilepsy with febrile seizures plus. *Eur. J. Paediatr. Neurol.* **2020**, 27, 111–115.
12. Maass, P.G.; Weise, A.; Rittscher, K.; Lichtenwald, J.; Barutcu, A.R.; Liehr, T.; Aydin, A.; Wefeld-Neuenfeld, Y.; Pölsler, L.; Tinschert, S.; Rinn, J.L.; Luft, F.C.; Bähring, S. Reorganization of inter-chromosomal interactions in the 2q37-deletion syndrome. *EMBO J.* **2018**, 37, e96257.
13. Rice, A.M.; McLysaght, A. Dosage sensitivity is a major determinant of human copy number variant pathogenicity. *Nat. Commun.* **2017**, 8, 14366.
14. Riggs, E.R.; Andersen, E.F.; Cherry, A.M.; Kantarci, S.; Kearney, H.; Patel, A.; Raca, G.; Ritter, D.I.; South, S.T.; Thorland, E.C.; Pineda-Alvarez, D.; Aradhya, S.; Martin, C.L. Technical standards for the interpretation and reporting of constitutional copy-number variants: A joint consensus recommendation of the American College of Medical Genetics and Genomics (ACMG) and the Clinical Genome Resource (ClinGen). *Genet Med.* **2020**, 22, 245–257.

15. Liu, Y.; Wu, S.; Shi, X.; Mao, F.; Zeng, X. Cancer with a HER2 IHC2+ and FISH HER2/ CEP17 ratio ≥2.0 and an average HER2 gene copy number <4.0 per tumor cell: HER2 mRNA overexpression is a rare event. *Front. Oncol.* **2020**, 10, 985.
16. Weise, A.; Gross, M.; Mrasek, K.; Mkrtchyan, H.; Horsthemke, B.; Jonsrud, C.; Von Eggeling, F.; Hinreiner, S.; Witthuhn, V.; Claussen, U.; Liehr, T. Parental-origin-determination fluorescence in situ hybridization distinguishes homologous human chromosomes on a single-cell level. *Int. J. Mol. Med.* **2008**, 21, 189–200.
17. Harutyunyan, T.; Hovhannisyan, G.; Babayan, N.; Othman, M.A.; Liehr, T.; Aroutiounian, R. Influence of aflatoxin B1 on copy number variants in human leukocytes in vitro. *Mol. Cytogenet.* **2015**, 8, 25.
18. Harutyunyan, T.; Hovhannisyan, G.; Sargsyan, A.; Grigoryan, B.; Al-Rikabi, A.H.; Weise, A.; Liehr, T.; Aroutiounian, R. Analysis of copy number variations induced by ultrashort electron beam radiation in human leukocytes in vitro. *Mol. Cytogenet.* **2019**, 12, 18.
19. Iourov, I.Y. Quantitative fluorescence in situ hybridization (QFISH). In: *Methods in Molecular Biology: Cancer Cytogenetics Methods and Protocols*; Wan, T.S.K., Ed. Springer, New York, **2017**, pp. 143–149.

16 FISH—Interphase Applications Including Detection of Chromosome Instability (CIN)

*Ivan Y. Iourov, Svetlana G. Vorsanova
and Yuri B. Yurov*

CONTENTS

INTRODUCTION

Fluorescence in situ hybridization (FISH) is a technological basis of interphase molecular cytogenetics. Interphase FISH is found useful for a wide spectrum of applications from molecular cytogenetic diagnosis to interphase chromosome biology.[1–3] Since a kind of a decrease in interest in interphase cytogenetics is observed in the postgenomic era,[4, 5] a brief overview of interphase FISH applications seems to be required.

Molecular cytogenetic analysis is an integral part of the research and medical care in clinical/medical genetics, reproduction, oncology, neurology, psychiatry, pediatrics. The significance of FISH applications in biomedicine has been consistently reported.[2, 3] Interphase FISH techniques are currently suggested to be an important technological element of research in somatic cell genetics/genomics, aging, single-cell biology, chromosome/chromatin biology (i.e. genome organization in interphase).[1, 5]

A more specific application of FISH is the analysis of chromosome instability (CIN) in interphase nuclei. The growing importance of CIN analysis in current biomedicine results from observations on CIN involvement in genetic intercellular/

DOI: 10.1201/9781003223658-16

interindividual diversity and a variety of pathogenic processes including cancerization, neurodegeneration, and tissue dysfunction.[6–9] Accordingly, addressing technological issues of FISH analysis of CIN in interphase nuclei may represent an intriguing part of a description of interphase FISH and related techniques. The present chapter describes interphase FISH applications considering the opportunities in analysis of chromosome abnormalities and instability in interphase nuclei and the place it deserves in current biomedicine.

INTERPHASE FISH AND CIN

Interphase FISH is generally defined as a set of molecular cytogenetic techniques for visualization analysis of genomic loci by means of DNA probes. FISH efficiency and resolution are determined by DNA probe properties. Such probes may be composed of repetitive (e.g. centromeric and telomeric DNA probes), euchromatic (site-specific DNA probes), and microdissected DNA (whole-chromosome painting probes and DNA probes for multicolor banding or partial painting of chromosomes). As in any kind of FISH-based assays, interphase FISH protocols are essentially composed of cell suspension preparation, denaturation of sample/probe DNAs, hybridization, and microscopic evaluation (visual and/or digital) of the results.[10, 11] Since interphase FISH allows visual analysis of DNAs at any cell cycle stage at molecular resolutions, it has become a method of choice for genome analyses at the chromosomal/subchromosomal level in post-mitotic cells or large mitotic cellular populations.[12, 13] Moreover, interphase FISH protocols are applicable for studying genome organization in single cells.[3, 14, 15]

In the postgenomic era, interphase FISH is an efficient technological platform for developing assays to analyze CIN and (re-)arrangements of chromosomal loci in interphase nuclei. Figure 16.1 depicts different interphase FISH assays for studying CIN in interphase (to show the possibilities of interphase FISH assays for studying CIN in interphase nuclei, we provided data on the fetal human brain, inasmuch as this tissue is affected by natural CIN).[16] Thus, FISH analysis with centromeric, chromosome-enumeration, or chromosome-specific DNA probes is used for detecting numerical chromosome abnormalities (aneuploidy and polyploidy) (Figure 16.1A–C), which are probably the most common types of abnormalities resulting in the occurrence and propagation of CIN.[6, 7, 9, 17] FISH assays with site-specific probes for the visualization of genomic loci, usually containing DNA of specific genes, are generally applied for identifying structural chromosome imbalances. This type of interphase FISH techniques is consistently used for specific rearrangements of chromosomes in cancers. The ability of this method to map altered genomic loci provides an opportunity to analyze CIN affecting specific chromosomal regions in interphase.[11, 18–20] Several phenomena of chromosomal arrangement in interphase nuclei may affect the performance of an interphase FISH assay.[1, 3] S phase DNA replication generally results in a double FISH signal. These observations may be used for studying DNA replication in interphase and require researcher's experience to differ replicative signals from signals showing a chromosome rearrangement.[21] Associations or pairing of interphase (homologous) chromosomes, which is common in post-mitotic human cells, may affect the interpretation of FISH results. To

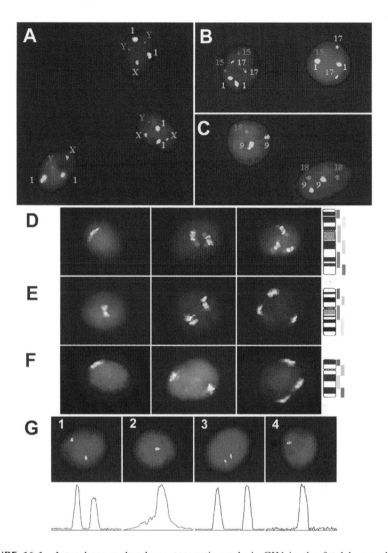

FIGURE 16.1 Interphase molecular cytogenetic analysis CIN in the fetal human brain. Interphase FISH with chromosome-enumeration DNA probes:

(A) two nuclei characterized by additional chromosomes Y and X and a normal nucleus;

(B) a nucleus with monosomy of chromosome 15 and a normal nucleus;

(C) a nucleus with monosomy of chromosome 18 and a normal nucleus.

(D to G) interphase chromosome-specific MCB: nuclei with monosomy, disomy, trisomy, and G-banding ideograms with MCB color-code labeling of a chromosome (from left to right), (D)—chromosome 9, (E)—chromosome 16, and (F)—chromosome 18. (G) interphase QFISH: (1) a nucleus with two signals for chromosomes 18 (relative intensities: 2058 and 1772 pixels), (2) a nucleus with one paired signal mimics monosomy of chromosome 18 (relative intensity: 4012 pixels), (3) a nucleus with two signals for chromosomes 15 (relative intensities: 1562 and 1622 pixels), (4) a nucleus with one signal showing monosomy of chromosome 15 (relative intensity: 1678 pixels).

Source: From Yurov et al.,[16] an open-access article distributed under the terms of the Creative Commons Attribution License.

solve this problem, quantitative FISH may be applied (Figure 16.1G). This technique offers an opportunity to differ between chromosomal associations and chromosome loss (aneuploidy/monosomy).[22–24]

INTERPHASE CHROMOSOME-SPECIFIC MULTICOLOR BANDING

A combination of microdissected DNA probes producing multicolor pseudo-banding of chromosomes by FISH is the basis of multicolor banding (MCB).[25, 26] Using chromosome-specific MCB probes, the technique has been adopted for analysis of interphase chromosomes. The adaptation is the method of interphase chromosome-specific MCB (ICS-MCB), a technique allowing the analysis of interphase chromosomes in their integrity at molecular resolutions (Figure 16.1D–F).[27, 28] ICS-MCB has been repeatedly reported to be an effective technique for analysis of nuclear genome organization at the chromosomal level and CIN in interphase nuclei.[29–34] Indeed, ICS-MCB is actually a unique technique allowing the visualization of a structurally rearranged chromosome in interphase.[32]

DIAGNOSTIC ISSUES

In the diagnostic context, the availability of single-cell molecular cytogenetic analysis via visualization is a striking advantage offered by interphase FISH. Additional advantage offered by FISH are the highest possible cell scoring potential by the visualization of chromosomal changes in interphase nuclei. Since chromosomal imbalances and CIN cause a wide spectrum of human morbid conditions, interphase FISH-based methods are to be recognized as a valuable part of molecular diagnosis.[1–3] It is noteworthy that a diagnosis is referred to as a knowledge about molecular and/or cellular mechanism of a disease. Consequently, interphase FISH-based techniques could be an important part of a diagnostic research (diagnosis and monitoring) of CIN-mediated diseases. Interphase FISH-based study of chromosomal changes may be an addition to pathway-based analysis of genome variations to model the functional consequences of chromosome imbalances and CIN. Chromosomal biomarkers (specific chromosome imbalances or reproducible patterns of CIN) detectable by interphase FISH are to be addressed by systems biology (bioinformatic) methodology for determining causes and consequences of CIN and for developing algorithms of unraveling disease mechanisms.[8, 34–37] Finally, combining immunohistochemical detection of proteins and interphase FISH (Immuno-FISH) offers an opportunity to analyze interphase chromosomes (interphase CIN) in nuclei of specific cell types. This is found useful to detect chromosome instability in post-mitotic tissues and in cancer cells.[31, 32, 38]

CONCLUSIONS

In the postgenomic era, collecting genomic data often lacks chromosomal context. As a result, such important genetic phenomena/processes as chromosomal abnormalities and CIN are still poorly understood.[5, 39] Interphase FISH-based technologies may help to gain further understanding of causes and consequences of CIN and chromosomal rearrangements. Using achievements and developments in genomics and molecular biology, this molecular cytogenetic platform is able to provide new opportunities in basic and diagnostic research. Mosaic and non-mosaic chromosome imbalances and CIN are important targets of research in a significant number of biomedical areas. Interphase FISH-based techniques allow the detection and monitoring of chromosomal changes in large cellular populations at molecular resolutions and at all cell cycle stages. Therefore, interphase molecular cytogenetics using FISH-based approaches represent an important part of studying genetics/genomics, cellular and molecular basis of intercellular or interindividual diversity and diseases. Certainly, interphase FISH should be integrated in sophisticated algorithms including whole-genome scanning methods (optionally, single-cell whole-genome analysis), techniques for visualization of nucleic acid molecules at single-cell level and molecular resolutions, and bioinformatic/systems biology analysis for pathway-based prioritization of chromosomal/genomic variations.

ACKNOWLEDGMENTS

The chapter is dedicated to Dr. Ilia V Soloviev. Interphase FISH studies in authors' labs are supported by the Government Assignment of the Russian Ministry of Science and Higher Education, Assignment no. AAAA-A19–119040490101–6, and by the Government Assignment of the Russian Ministry of Health, Assignment no. 121031000238–1.

REFERENCES

1. Vorsanova, S.G.; Yurov, Y.B.; Iourov, I.Y. Human interphase chromosomes: A review of available molecular cytogenetic technologies. *Mol. Cytogenet.* **2010**, 3, 1.
2. Liehr, T. *Fluorescence In Situ Hybridization (FISH): Application Guide.* Springer, Berlin; **2017**.
3. Iourov, I.Y.; Vorsanova, S.G.; Yurov, Y.B. *Human Interphase Chromosomes: Biomedical Aspects.* 2nd Edition. Springer, Berlin; **2020**.
4. Liehr, T. From human cytogenetics to human chromosomics. *Int. J. Mol. Sci.* **2019**, 20, 826.
5. Iourov, I.Y.; Yurov, Y.B.; Vorsanova, S.G. Chromosome-centric look at the genome. In: *Human Interphase Chromosomes: Biomedical Aspects*; Iourov, I.Y.; Vorsanova, S.G.; Yurov, Y.B., Eds. Springer, Berlin, **2020**, pp. 157–170.
6. Heng, H.H.; Bremer, S.W.; Stevens, J.B.; Horne, S.D.; Liu, G.; Abdallah, B.Y.; Ye, K.J.; Ye, C.J. Chromosomal instability (CIN): What it is and why it is crucial to cancer evolution. *Cancer Metastasis Rev.* **2013**, 32, 325–340.
7. Iourov, I.Y.; Vorsanova, S.G.; Yurov, Y.B.; Kutsev, S.I. Ontogenetic and pathogenetic views on somatic chromosomal mosaicism. *Genes.* **2019**, 10, 379.
8. Iourov, I.Y.; Vorsanova, S.G.; Yurov, Y.B.; Zelenova, M.A.; Kurinnaia, O.S.; Vasin, K.S.; Kutsev, S.I. The cytogenomic "theory of everything": Chromohelkosis may underlie chromosomal instability and mosaicism in disease and aging. *Int. J. Mol. Sci.* **2020**, 21, 8328.
9. Ye, C.J.; Sharpe, Z.; Heng, H.H. Origins and consequences of chromosomal instability: From cellular adaptation to genome chaos-mediated system survival. *Genes.* **2020**, 11, 1162.

10. Iourov, I.Y.; Vorsanova, S.G.; Yurov, Y.B. Recent patents on molecular cytogenetics. *Recent Pat. DNA Gene Seq.* **2008**, 2, 6–15.

11. Liehr, T.; Othman, M.A.; Rittscher, K.; Alhourani, E. The current state of molecular cytogenetics in cancer diagnosis. *Expert Rev. Mol. Diagn.* **2015**, 15, 517–526.

12. Iourov, I.Y.; Vorsanova, S.G.; Yurov, Y.B. Single cell genomics of the brain: Focus on neuronal diversity and neuropsychiatric diseases. *Curr. Genomics.* **2012**, 13, 477–488.

13. Bakker, B.; van den Bos, H.; Lansdorp, P.M.; Foijer, F. How to count chromosomes in a cell: An overview of current and novel technologies. *Bioessays.* **2015**, 37, 570–577.

14. Manvelyan, M.; Hunstig, F.; Bhatt, S.; Mrasek, K.; Pellestor, F.; Weise, A.; Simonyan, I.; Aroutiounian, R.; Liehr, T. Chromosome distribution in human sperm: A 3D multicolor banding-study. *Mol. Cytogenet.* **2008**, 1, 25.

15. McClelland, S.E. Single-cell approaches to understand genome organisation throughout the cell cycle. *Essays Biochem.* **2019**, 63, 209–216.

16. Yurov, Y.B.; Iourov, I.Y.; Vorsanova, S.G.; Liehr, T.; Kolotii, A.D.; Kutsev, S.I.; Pellestor, F.; Beresheva, A.K.; Demidova, I.A.; Kravets, V.S.; Monakhov, V.V.; Soloviev, I.V. Aneuploidy and confined chromosomal mosaicism in the developing human brain. PLoS One. **2007**, 2, e558.

17. Iourov, I.Y.; Yurov, Y.B.; Vorsanova, S.G.; Kutsev, S.I. Chromosome instability, aging and brain diseases. *Cells.* **2021**, 10, 1256.

18. Liehr, T.; Starke, H.; Weise, A.; Lehrer, H.; Claussen, U. Multicolor FISH probe sets and their applications. *Histol. Histopathol.* **2004**, 19, 229–237.

19. Devadhasan, J.P.; Kim, S.; An, J. Fish-on-a-chip: A sensitive detection microfluidic system for Alzheimer's disease. *J. Biomed. Sci.* **2011**, 18, 33.

20. Hu, Q.; Maurais, E.G.; Ly, P. Cellular and genomic approaches for exploring structural chromosomal rearrangements. *Chromosome Res.* **2020**, 28, 19–30.

21. Vorsanova, S.G.; Yurov, Y.B.; Kolotii, A.D.; Soloviev, I.V. FISH analysis of replication and transcription of chromosome X loci: New approach for genetic analysis of Rett syndrome. *Brain Dev.* **2001**, 23, S191–S195.

22. Iourov, I.Y.; Soloviev, I.V.; Vorsanova, S.G.; Monakhov, V.V.; Yurov, Y.B. An approach for quantitative assessment of fluorescence in situ hybridization (FISH) signals for applied human molecular cytogenetics. *J. Histochem. Cytochem.* **2005**, 53, 401–408.

23. Vorsanova, S.G.; Iourov, I.Y.; Beresheva, A.K.; Demidova, I.A.; Monakhov, V.V.; Kravets, V.S.; Bartseva, O.B.; Goyko, E.A.; Soloviev, I.V.; Yurov, Y.B. Non-disjunction of chromosome 21, alphoid DNA variation, and sociogenetic features of Down syndrome. *Tsitol. Genet.* **2005**, 39, 30–36.

24. Iourov, I.Y. Quantitative fluorescence in situ hybridization (QFISH). *Methods Mol. Biol.* **2017**, 1541, 143–149.

25. Liehr, T.; Heller, A.; Starke, H.; Rubtsov, N.; Trifonov, V.; Mrasek, K.; Weise, A.; Kuechler, A.; Claussen, U. Microdissection based high resolution multicolor banding for all 24 human chromosomes. *Int. J. Mol. Med.* **2002**, 9, 335–339.

26. Liehr, T.; Othman, M.A.; Rittscher, K. Multicolor karyotyping and fluorescence in situ hybridization-banding (MCB/mBAND). *Methods Mol. Biol.* **2017**, 1541, 181–187.

27. Iourov, I.Y.; Liehr, T.; Vorsanova, S.G.; Kolotii, A.D.; Yurov, Y.B. Visualization of interphase chromosomes in postmitotic cells of the human brain by multicolour banding (MCB). *Chromosome Res.* **2006**, 14, 223–229.

28. Iourov, I.Y.; Liehr, T.; Vorsanova, S.G.; Yurov, Y.B. Interphase chromosome-specific multicolor banding (ICS-MCB): A new tool for analysis of interphase chromosomes in their integrity. *Biomol. Eng.* **2007**, 24, 415–417.

29. Lemke, J.; Claussen, J.; Michel, S.; Chudoba, I.; Mühlig, P.; Westermann, M.; Sperling, K.; Rubtsov, N.; Grummt, U.W.; Ullmann, P.; Kromeyer-Hauschild, K.; Liehr, T.;

Claussen, U. The DNA-based structure of human chromosome 5 in interphase. *Am. J. Hum. Genet.* **2002**, 71, 1051–1059.

30. Yurov, Y.B.; Iourov, I.Y.; Vorsanova, S.G.; Demidova, I.A.; Kravetz, V.S.; Beresheva, A.K.; Kolotii, A.D.; Monakchov, V.V.; Uranova, N.A.; Vostrikov, V.M.; Soloviev, I.V.; Liehr, T. The schizophrenia brain exhibits low-level aneuploidy involving chromosome 1. *Schizophr. Res.* **2008**, 98, 139–147.

31. Iourov, I.Y.; Vorsanova, S.G.; Liehr, T.; Yurov, Y.B. Aneuploidy in the normal, Alzheimer's disease and ataxia-telangiectasia brain: Differential expression and pathological meaning. *Neurobiol. Dis.* **2009**, 34, 212–220.

32. Iourov, I.Y.; Vorsanova, S.G.; Liehr, T.; Kolotii, A.D.; Yurov, Y.B. Increased chromosome instability dramatically disrupts neural genome integrity and mediates cerebellar degeneration in the ataxia-telangiectasia brain. *Hum. Mol. Genet.* **2009**, 18, 2656–2669.

33. Yurov, Y.B.; Vorsanova, S.G.; Liehr, T.; Kolotii, A.D.; Iourov, I.Y. X chromosome aneuploidy in the Alzheimer's disease brain. *Mol. Cytogenet.* **2014**, 7, 20.

34. Iourov, I.Y.; Vorsanova, S.G.; Kurinnaia, O.S.; Zelenova, M.A.; Vasin, K.S.; Yurov, Y.B. Causes and consequences of genome instability in psychiatric and neurodegenerative diseases. *Mol. Biol.* **2021**, 55, 37–46.

35. Iourov, I.Y.; Vorsanova, S.G.; Yurov, Y.B. In silico molecular cytogenetics: A bioinformatic approach to prioritization of candidate genes and copy number variations for basic and clinical genome research. *Mol. Cytogenet.* **2014**, 7, 98.

36. Liehr, T. Cytogenetically visible copy number variations (CG-CNVs) in banding and molecular cytogenetics of human; about heteromorphisms and euchromatic variants. *Mol. Cytogenet.* **2016**, 9, 5.

37. Iourov, I.Y.; Vorsanova, S.G.; Yurov, Y.B. The variome concept: Focus on CNVariome. *Mol. Cytogenet.* **2019**, 12, 52.

38. Lv, Y.; Mu, N.; Ma, C.; Jiang, R.; Wu, Q.; Li, J.; Wang, B.; Sun, L. Detection value of tumor cells in cerebrospinal fluid in the diagnosis of meningeal metastasis from lung cancer by immuno-FISH technology. *Oncol. Lett.* **2016**, 12, 5080–5084.

39. Heng, H.H. New data collection priority: Focusing on genome-based bioinformation. *Res. Results Biomed.* **2020**, 6, 5–8.

17 FISH—Determination of Telomere Length (Q-FISH/CO-FISH)

Gordana Joksić, Jelena Filipović Tričković
and Ivana Joksić

CONTENTS

INTRODUCTION

Telomeres are specialized nucleoprotein structures at the ends of linear chromosomes. They consist of a shelterin protein complex and a long assembly of hexameric repeats (TTAGGG) oriented in 5′ to 3′ direction, ending as a single strand 3′ overhang. This 3′ overhang invades 5′ double-stranded telomeric duplex and is housed deep inside by clipping with shelterin heteroduplex POT1/TTP1. Shelterin subunits TRF1 and TRF2 (= telomere binding factors 1 and 2) directly bind the double stranded TTAGGG, while TIN2 (= TERF1 interacting nuclear factor) links TPP1/POT1 (= tripeptidyl peptidase 1/protection of telomeres 1) heterodimer and stabilizes its association with chromosome ends. Shelterin assembly creates a telomere-specific structure T-loop (Figure 17.1) that caps the chromosomes and protects their termini from being recognized as breakpoints by DNA repair mechanisms.[1]

DOI: 10.1201/9781003223658-17

189

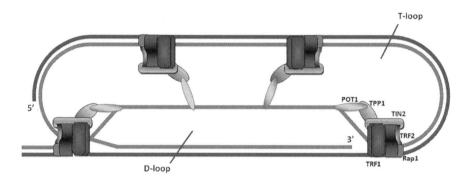

FIGURE 17.1 Telomere T-loop formation.

By possessing at least two active domains in each subunit, shelterins directly inhibit homologous recombination (HR) and non-homologous end-joining (NHEJ).[2, 3] For inhibition of other DNA repair pathways, they use accessory proteins that already act in genome maintenance.[4] Telomeres are transcribed into a class of long noncoding RNA called telomeric repeat-containing RNAs (TERRA),[5] that actively participates in the regulation of telomere homeostasis. TERRAs are involved in the formation of telomeric heterochromatin and act as negative regulators of telomerase.[6, 7] Telomeres shorten through each round of replication, and their length is one of the most important indicators of their function. Telomere shortening is a part of tumor suppressor mechanisms, whereas telomere length homeostasis acts as a mitotic clock regulating cellular life-span.[8] Telomere loss can be reversed by telomerase (in tissues where telomerase is normally active) or by DNA polymerase (in somatic cells), forming Okazaki fragments. Reassembling of 3′ overhang to appropriate length enables correct T-loop formation and adequate telomere function. If telomeric 3′ overhangs are not regenerated, structural and functional integrity of T-loop is altered, consequently leading to telomere attrition. Telomere shortening to a definite critical length (around 3–5 kb in human cells) leads cells in senescence and additional shortening induces apoptosis.[9] In shortened telomeres under critical length, the T-loop formation is altered, causing telomere fragility that can be identified in metaphase chromosomes by using quantitative fluorescence in situ hybridization (Q-FISH). Fragile telomeres are seen as multiple or highly extended telomeric signals at individual chromatid ends, split or lacking the corresponding signals.[10] Loss of TRF1 is main cause of telomere fragility because their replication is altered,[11] but telomere dysfunction can be caused by changes in other components of shelterin complex, structure of telomeric DNA, structure of TERRA, helicases, or nuclear lamina.[12] Telomere repeats are guanine-rich, which makes them very susceptible to oxidative damages; therefore telomere dysfunction may be initiated by oxidative stress shortening.[13] Oxidative lesions of guanine are highly unstable premutagenic lesions with adenine binding affinity causing transversions, mutations, thus enhancing telomere fragility.[14] Loss of telomere length maintenance leads to genome instability and is acknowledged as typical feature of carcinogenesis.[15] Most of the tumor cells upregulate telomerase activity to facilitate telomere elongation.

Initially, assessment of telomere length was done by Southern-blot-based techniques that gave estimate of average telomere length in population of cells. Development of Q-FISH enabled measurement of individual chromosome telomere length in single cells with the resolution of 200 base pairs, assessment of distribution of telomere length (p-arms vs. q-arms), ratio of shortest/longest telomeres, and telomere fragility. The method utilizes peptide nucleic acid PNA probes that are resistant to degradation by nucleases and peptidases and show highly specific hybridization with DNA.[16] Staining efficiency is almost 100%. Quantification of telomere length has been standardized using centromeric PNA probe for chromosome 2 and measurement tools for ISIS software (MetaSystems, Altlussheim, Germany) where centromere fluorescence is set to a value of 100%.[17] The telomere length is quantified from digital images of metaphase spreads and is expressed as telomere/centromere (T/C) ratio. The rate of fragile telomeres is expressed as the ratio between the number of fragile telomeres and total number of analyzed metaphases.

Chromosome orientation FISH method (CO-FISH) is used to identify the direction of a synthetized DNA chains, or particular DNA sequence. Method is used for identification of chromosome inversions associated with isochromosome formation, pericentric inversions, rate of sister chromatid exchanges at telomeres (T-SCE),[18] or centromeres (C-SCE).[19] Both variants of CO-FISH contribute to better understanding molecular mechanisms leading to genomic instability.

SAMPLES/TISSUES

Telomere length can be determined in human and other vertebrate cells and should be done in metaphases. The method is particularly convenient for species that contain interstitial telomeric sites in their genome, as well as for species that have ultra-long and heterogeneous telomeres, such as mice. A slide-based method can be applied to all cells that can be cultivated (primary cells as lymphocytes, fibroblast, and cell lines).

DESCRIPTION Q-FISH METHOD

Principle of the procedures are as follows: The target DNA (fixed on the slide surface) and Cy3-conjugated PNA probes (telomeric + centromeric) are denatured together at 80°C for 5 minutes under a coverslip. Hybridization takes place from 30 minutes up to 2 hours at room temperature (RT) in a moist chamber. After post-hybridization washes and dehydration in ethanol series slides are stained with DAPI (= 4′,6-diamidino-2-phenylindole)-antifade and covered by glass coverslips. Analysis is performed using a high-power oil objective of fluorescent microscope equipped with image capture and software for telomere measurement. The following filters (excitation/emission) are recommended: 360/340 for the DAPI counterstain and 550/640 for the Cy3 signals. Positive results are seen as red spots at the end of each chromosomal arm, and centromere of chromosome 2.

A standard cell biological and molecular equipment (heating plate, incubator, microcentrifuge, water bath for washing steps (with adjustable temperature), micropipettes, tips, glass Hellendahl jars (100 ml), glass slides, glass coverslips for fluorescence microscopy, fluorescent microscope equipped with filters for Cy3 and DAPI, digital imaging system and software for telomere length measurement (ISIS software (MetaSystems, Altlussheim, Germany)) are necessary to perform Q-FISH.

REAGENTS

- 1×TBS (Tris-Buffered Saline, pH 7.5)—500 ml per experiment
- Pre-treatment stock solution: Dilute 2,000× concentrated proteinase K in 1×PBS
- Pre-treatment solution (working solution): mix 40 µl pre-treatment stock solution with 80 ml TBS buffer at RT. Prepare fresh for each experiment.
- Telomere and centromere PNA probes ready to use probe (Cy3 conjugated PNA probes in hybridization solution = 70% formamide) available on request from manufacturer (DAKO/Agilent)
- Rinse solution ×1—Dilute 50× stock from DAKO/Agilent in pure water and store at RT
- Wash solution ×1—Dilute 50× stock from DAKO/Agilent in pure water and warm to 65°C ~2 h before experiment; solution might be milky, but it does not influence the results.
- Pure water (deionized or double-distilled water)
- 37% formaldehyde (e.g. Merck)
- Formaldehyde in TBS (working solution): add 8 ml of 37% formaldehyde to 72 ml 1×TBS buffer. Do not use more than four weeks.
- Cold ethanol series; 70%, 85%, 96% (e.g. 4°C in refrigerator)
- Mounting solution: Antifade containing DAPI as a counterstain. To preserve signal intensity, the best is to use Vectashield (VectorLabs) supplemented with 0.1 µg/ml of DAPI.

PROCEDURE

1. Prepare metaphase spreads on a glass microscope slides employing standard cytogenetic procedure. Slide quality is the most important factor for good hybridization results. Cells should be well fixed and well spread.
2. All following incubations are at RT if not otherwise specified.
3. Heat working solution of Wash solution to 65°C in a water bath and pre-worm heating incubator (e.g. Plate Shaker Thermostat PST-100 HL, Biosan, Riga, Latvia, LV-1067) on 80°C.
4. Prepare six glass Hellendahl jars (100 ml), and fill in ~80 ml each of (i) TBS, (ii) Formaldehyde in TBS (working solution), (iii) TBS, (iv) Pre-treatment solution (working solution), (v) TBS, (vi) TBS
5. Immerse slides in TBS (i) for 2 min
6. Transfer and incubate slides in jar (ii) for 2 min exactly
7. Wash slides in TBS (i) and then in TBS (iii) 5 min, each
8. Incubate slides in jar (iv) 10 min

9. Wash slides in TBS (v) and then in TBS (vi) 5 min, each
10. Immerse slides in cold ethanol series (70%, 85% and 96%) 3×2 min
11. Airdry slides in a vertical position
12. Add 10 µl of PNA probes/Cy3 to carefully chosen area on the slide
13. Cover the area with a 18×18 mm coverslip
14. Place the slides in the pre-heated incubator to 80°C for 5 min
15. Place the slides in moist chamber (in the dark) at RT for 2 hours
16. Briefly immerse slides in Rinse solution ×1 to remove coverslip
17. Immerse slides in the pre-heated Wash solution ×1 for 5 min at 65°C
18. Repeat steps 11 and 12
19. Apply 2×10 µl of mounting solution onto each slide and put 24×50 mm coverslip. For the counterstain color to develop, leave the slides 15 min in the dark. Slides are ready for microscopy. Since a variety of microscopes and image acquisitions systems are used in different laboratories, the manufacturer manual for use should be followed. Figure 17.2 presents captured metaphase for telomere length analysis.
20. For analysis of telomere length, using measurement tool for ISIS software (MetaSystems, Altlussheim, Germany) with centromere fluorescence of chromosome 2 set to value of 100% is described. Upon capturing, contrast and threshold of signals are adjusted; using option of inverse DAPI staining chromosomes are arranged in karyogram form. Afterwards, the karyogram is reverted to fluorescence mode. The telomere measurement software displays two horizontal lines overlaid to each chromosome that define the centromere signal on chromosome 2 (red lines) and telomere signals on p and q arms (green lines) (Figure 17.3).

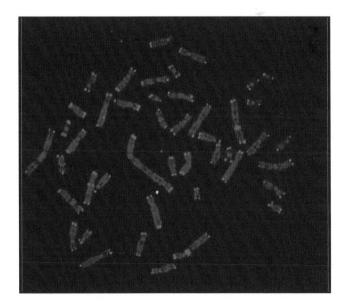

FIGURE 17.2   Captured metaphase hybridized with Cy3-labeled PNA probe for measurement of telomere length.

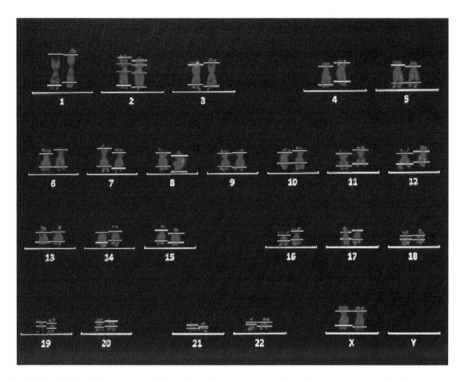

FIGURE 17.3   Fluorescence intensity of centromere signal is set at value of 100%. Each chromosome arm is bordered with green software line.

DESCRIPTION CO-FISH METHOD

CO-FISH: CO-FISH involve labeling of S-phase synthetized DNA with BrdU and BrdC during replication. After preparation, and slide making, newly synthetized DNA chain is removed using Hoechst/UV nicking and enzymatic digestion of newly synthetized DNA by exonuclease III. After this treatment, only the parental DNA strands will be detectable. Depending of the cell type, and type of experiment, BrdU and BrdC should be present one or two rounds of replication. After one round of labeling, two telomeric signals at each chromosome ends will be visible, whereas labeling with for two rounds of replication will display only one telomeric signal per chromosome ends. Accordingly, the FISH-procedure is similar to Q-FISH—however, to get CO-FISH results, a special sample preparation and handling is necessary and described next, as outlined first in the following: Label the S-phase synthetized DNA with BrdU and BrdC during replication. Split cells in 3:1 ratio either onto a T25 (= 25 ml culture flask) or T75 (= 75 ml culture flask) flasks. Wait for 45 minutes until cells are settled in or attached onto the surface. Make BrdU/BrdC mixture in a ratio 3:1 to achieve a final concentration of 10 µM of Brdu/BrdC. Add colcemid (10 µl/ml) or Colchicine (25 mg/ml) 6–7 hours prior to the 24 hours' deadline. Cells that pass two rounds of replication will display one telomeric signal per chromosome ends. Activity of homologous replication attempting to extend telomeres will be seen as sister-chromatid exchanges; i.e. both chromatid ends will display the telomeric signal.

REAGENTS

- BrdU/BrdC: Take 15 μl of BrdU and 5 μl of BrdC from stock (50mM conc.) to make a 3:1 ratio, mix well and add 1 μl of the mixture into 5 ml of T25 flask. N.B.: for a T75 add 3 μl of the mixture to make a final concentration of 10 μM of Brdu/BrdC.
- Exonuclease III (e.g Thermofisher): Take 1.5 μl of stock (200 U/μl) and mix it with 100 μl of buffer to make a final concentration of (3 U/μl)
- 70% formamide: 70ml formamide + 10 ml 20×SSC + 20 ml of ddH$_2$O to make a final volume of 100 ml
- Hoechst 33258: Take 50 μl of stock (0.5 μg/ml) and mix it with 50 ml of ddH$_2$O in a glass Hellendahl jar.
- Phosphate buffered saline: 1×PBS (e.g. Merck)
- 2×SSC: Dilute 20×SSC (commercially available, e.g. Merck) with ddH2O

PROCEDURE

1. Apply metaphase spread protocol and make slides and "age" them for few days under room temperature.
2. Put the slides in glass Hellendahl jar comprising working solution of Hoechst 33258 for 15 minutes at RT.
3. Remove and mount the slide with 2×SSC, and put the glass coverslip over the each slide. The thin layer of 2×SSC between slide and glass coverslip should be present during exposure to 365 nm UV light. Expose slides for 30 minutes.
4. Briefly wash the slides with 1×PBS by dipping to remove coverslips.
5. Add 20 μl of Exonuclease III (3 U/μl) onto each slide, and cover them with parafilm for 10 minutes.
6. Wash the slide in 1×PBS by dipping into a glass Hellendahl jar.
7. Dehydrate the slide by adding 1 ml of 100% ethanol onto slide surface for 1 minute.
8. Airdry the slide.
9. Add 20 μl/slide of PNA probe (Cy3 or FITC- labelled)—and leave it to hybridize for two minutes at 70°C in the pre-heated incubator. Keep in wet and dark containment for two hours.
10. Wash in 70% formamaide for 15 min twice.
11. Wash in 1×PBS for 10 minutes.
12. Dehydrate by serial dehydration (70%, 90% and 100%) with ethanol for 5 minutes.
13. Add 20 μl of DAPI Vectashield onto each slide, and put on 24x50mm coverslip. For the counterstain color to develop, leave the slides 15 min in the dark. Slides are ready for microscopy.

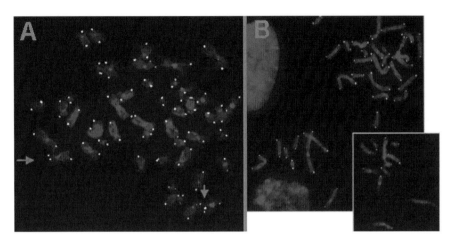

FIGURE 17.4   A) T-SCE staining using CO-FISH. Red arrows indicate for sister chromatid exchanges at telomeres. B) Metaphase stained using CO-FISH without T-SCE.

14. Since a variety of microscopes and image acquisitions systems are used in different laboratories, the manufacturer manual for use should be followed. Upon capturing, contrast and threshold of signals are adjusted. Cells that pass two rounds of replication will display one telomeric signal per chromosome ends. Telomeric sisterchromatid exchanges (T-SCEs) are seen as two signals on each chromatid (Figure 17.4).

YIELDS

Q-FISH for Telomere Length

Employing measurement tools, intensity of each telomere signals results are displayed graphically and numerically. Results of measurement are expressed as T/C ratio, i.e. percentage vs. centromere signal for each chromosome arm (Figure 17.5).

Q-FISH for Telomere Fragility

Fragile telomeres are identified as duplicated or extended signals on chromatid arm, missing the signals or fusions (Figure 17.6). Rate of fragile telomeres (fragility index) is calculated as ratio between the numbers of fragile telomeres/number of metaphases.

Co-FISH for Detection of Telomeric Sisterchromatid Exchanges (T-SCEs)

The incidence of T-SCE is expressed as number of chromosome displaying two signal at each chromosome ends per cell.

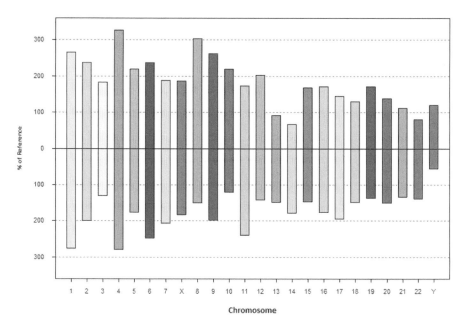

FIGURE 17.5   Telomere length for each chromosome arm expressed as T/C ratio.

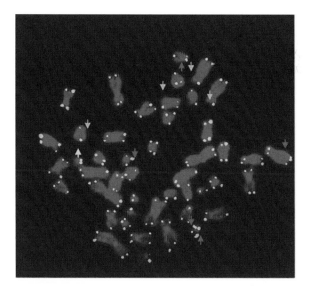

FIGURE 17.6   Telomere fragility: split signals—red arrows; lack of signals—purple arrows; telomere chromatid fusions—yellow arrow.

CONCLUSIONS

Telomeres are complex nucleoprotein structures whose integrity determines the stability of the genome. Many DNA damage repair proteins are located on telomeres. They are activated upon early DNA damage or directly participate in its repair. Telomeres shortening during each cell division, having the function of biological "clock", leads to elimination of cells with critically shortened telomeres. Rich in guanine, telomeres are susceptible to guanine oxidative damage. While oxidized bases are supplemented, break-induced repair can make disruption and elongation of telomeres inducing genomic instability, which is known as a hallmark of tumor cells. Molecular biology methods in combination with molecular cytogenetic methods will certainly in the near future clarify the mechanisms by which telomeres maintain genomic stability. This will certainly be important for the prevention and treatment of many diseases.

REFERENCES

1. Blackburn, E.H.; Greider, C.W.; Jack, W.; Szostak, J.W. The nobel prize in physiology or medicine 2009. *Nobel Found.* **2009**, 1, 10–105.
2. de Lange, T. How telomeres solve the end-protection problem. *Science.* **2009**, 326, 948–952.
3. de Lange, T. Shelterin-mediated telomere protection. *Annu. Rev. Genet.* **2018**, 52, 223–247.
4. Sfeir, A.; de Lange, T. Removal of shelterin reveals the telomere end-protection problem. *Science* **2012**, 336, 593–597.
5. Deng, Z.; Wang, Z.; Stong, N.; Plasschaert, R.; Moczan, A.; Chen, H.-S.; Hu, S.; Wikramasinghe, P.; Davuluri, R.V.; Bartolomei, M.S.; Riethman, H.; Lieberman, P.M. A role for CTCF and cohesin in subtelomere chromatin organization, TERRA transcription, and telomere end protection. *EMBO J.* **2012**; 31, 4165–4178.
6. Bettin, N.; Pegorar, C.O.; Cusanelli, E. The emerging roles of TERRA in telomere maintenance and genome stability. *Cells.* **2019**, 8, 246.
7. Muraki, K.; Nyhan, K.; Han, L.; Murnane, J. Mechanisms of telomere loss and their consequences for chromosome instability. *Front. Oncol.* **2012**, 2, 135.
8. Hayflick, L. Living forever and dying in attempt. *Exp. Gerontol.* **2003**, 38, 1231–1241.
9. Tomáška, Ľ.; Cesare, A.J.; AlTurki, T.M.; Griffith, J.D. Twenty years of t-loops: A case study for the importance of collaboration in molecular biology. *DNA Repair (Amst).* **2020**, 94, 102901.
10. Liu, H.; Xie, Y.; Zhang, Z.; Mao, P.; Liu, J.; Ma, W.; Zhao, Y. Telomerric recombination induced by DNA damage results in telomere extension and length heterogeneity. *Neoplasia.* **2018**, 20, 905–916.
11. Martínez, P.; Thanasoula, M.; Muñoz, P.; Liao, C.; Tejera, A.; McNees, C.; McNees, C.; Flores, J.M.; Fernández-Capetillo, O.; Tarsounas, M.; Blasco, M.A. Increased telomere fragility and fusions resulting from TRF1 deficiency lead to degenerative pathologies and increased cancer in mice. *Genes Dev.* **2009**, 23, 2060–2075.
12. Kychygina, A.; Dall'Osto, M.; Joshua, A.M.; Cadoret, J.C.; Piras, V.; Picket, H.A.; Crabbe, L. Progerin impairs 3D genome organization and induces fragile telomeres by limiting the dNTP pools. *Sci. Rep.* **2012**, 11, 13195.
13. von Zglinicki, T. Oxidative stress shortens telomeres. *Trends Biochem. Sci.* **2002**, 27, 339–344.

14. Barnes, R.P.; Fouquerel, E.; Opresko, P.L. The impact of oxidative DNA damage and stress on telomere homeostasis. *Mech. Ageing Dev.* **2019**, 177, 37–45.

15. Blackburn, E.H. Switching and signaling at the telomere. *Cell.* **2001**, 106, 661–673.

16. Nielsen, P.E.; Egholm, M.; Berg, R.H.; Buchardt, O. Sequence-selective recognition of DNA by strand displacement with a thymine-substituted polyamide. *Science.* **1991**, 254, 1497–1500.

17. Perner, S.; Brüderlein, S.; Hasel, C.; Waibel, I.; Holdenried, A.; Ciloglu, N.; Chopurian, H.; Nielsen, K.V.; Plesch, A.; Högel, J.; Möller, P. Quantifying telomere lengths of human individual chromosome arms by centromere-calibrated fluorescence in situ hybridization and digital imaging. *Am. J. Pathol.* **2003**, 163, 1751–1756.

18. Bailey, S.M.; Williams, E.S.; Cornforth, M.N.; Goodwin, E.H. Chromosome orientation fluorescence in situ hybridization or strand-specific FISH. *Methods Mol. Biol.* **2010**, 659, 173–183.

19. Giunta, S. Centromere chromosome orientation fluorescent in situ hybridization (Cen-CO-FISH) detects sister chromatid exchange at the centromere in human cells. *Bio. Protoc.* **2018**, 8, e2792.

18 FISH—in Three Dimensions—3D-FISH

Thomas Liehr

CONTENTS

INTRODUCTION

Molecular cytogenetics, particularly fluorescence in situ hybridization (FISH), can be applied as one of nowadays three major approaches to study the nuclear architecture (= 3D-FISH)[1]; the two other ways to access this problem are based on life-cell imaging,[2] and long read sequencing,[3] and are not further covered here. However, all studies applying such methods with the goal to unravel the peculiarities of interphases can be summarized as nucleomics.[1]

Besides high-resolution laser scanning microscopes, it is also possible to use standard fluorescence microscopes for 3D-FISH studies. Even though such studies can be performed in normally prepared, flattened nuclei,[4] there is a simple approach to retain original shape and size of interphases before, during and after FISH-procedure, as well as during microscopic evaluation, called suspension-FISH (S-FISH).[5]

As summarized elsewhere,[1] nucleomics can be traced back to 1885, when Carl Rabl suggested that interphase chromosomes are threadlike but separated into individual chromosomes.[6] Thus, he already suggested "chromosome territories", a wording shaped in 1909 by Theodor Boveri, and Eduard Strasburger at about the same time.[1] In the 1950s, the inactive X-chromosome in females has been characterized as a stainable Barr-body[7] within the nucleus. And since 1985, there was evidence that chromosome territories and/or subdomains of them are able to change shapes and structures during time in cell differentiation.[8] In the last decade finally, the next generation and long-read sequencing approaches could identify the topologically associating domains (TADs) and show that genes do interact in cis and trans on DNA-level.[4, 9, 10]

DOI: 10.1201/9781003223658-18

According to data obtained during the last decades, in the interphase nucleus the DNA is protein-covered, and there is never-ending transcription, de- and refolding, and repair going on.[11] Even though the nucleus is imagined nowadays as fluent, DNA-compartments of chromosomes, being subunits of the nucleus, are never really separated,[10–12] which surprisingly also includes a connection of maternal and paternal genomes via spindles and spindle-pole.[1, 13] Besides, chromosomes are arranged in a nucleus normally according to size and gene-density; an exception has yet been seen only in rod cells in retina of nocturnal mammals, where an "inverted" arrangement is needed to support night vision.[1, 14, 15]

Here the S-FISH approach is outlined.[5]

SAMPLES/TISSUES

Any kind of tissue can be used for 3D-FISH-anaylses, as long as intact interphase nuclei can be accessed in it. However, to keep interphase cells in their original spherical shape, to the best of our knowledge only the S-FISH approach has been reported yet. For this technique, only cells fixed in Carnoy's fixative from standard cytogenetic preparations are suited.[5]

DESCRIPTION OF METHODS

S-FISH was first reported in 2002 and was e.g. successfully applied for 3D-FISH studies of human sperm,[14] or in studied of great ape cells using human probes,[16] or others like positioning-characterization of small supernumerary marker chromosomes, (derivative) X-chromosomes, or chromosomes involved in chromosomal rearrangements in leukemia cases.[17]

How to Proceed

A) *Transfer prepared interphase cells from Carnoy's fixative into hybridization buffer:*
   1. Centrifuge ~50–100 µl of a concentrated cell suspension diluted in Carnoy's fixative (1,500 rpm/4°C/10 min).
   2. Wash in 500 µl methanol, incubate for 2 min and centrifuge as in step 1.
   3. Wash in 500 µl 0.9% NaCl, incubate for 2 min and centrifuge as in step 1.
   4. Incubate in 500 µl pepsin solution (950 µl of distilled water + 50 µl 0.2 N HCl + 5 µl pepsin stock solution) at 37°C for 5 min, and centrifuge as in step 1.
   5. Incubate in 500 µl 0.9% NaCl solution for 2 min at room temperature (RT), and repeat step 1. Leave 50 µl suspension in the tube.
   6. Denature at 95°C/5 min, pellet the cells by centrifugation for 10 min at 1,500 rpm, and repeat step 1. Leave 20 µl suspension in the tube.
   7. Prepare a 3× concentrated probe (compared to normal FISH-experiment) in 25 µl hybridization buffer; if necessary add 5–50 µg of COT1 DNA, denature at 95°C for 5 min, and (if needed) prehybridize at 37°C for 30–60 min.
   8. Now add this pre-hybridized probe (step 7) to suspension from step 6.

B) S-FISH procedure and evaluation:
1. Put vial with cells and hybridization buffer from step A/8 for 12–16 h (overnight) on 37°C.

C) S-FISH postwash
1. Add for washing 500 µl 0.4×SSC/68°C/2 min, and repeat step A/1.
2. Add for further washing 500 µl 4×SSC/RT/2 min, and repeat step A/1.
3. Add 150 µl DAPI solution (2 µl DAPI stock solution in 2 ml Vectashield Antifade) at RT/10 min. Then wash in 500 µl/0.9% NaCl (RT), and repeat step A/1.
4. Resuspend in 0.5% DAPI Vectashield gel (melted 250 mg agarose per 25 ml 0.9% NaCl at 600 W in a microwave = 1%; 2 ml of this to 2 ml of Vectashield Antifade); transfer at once to a 15 µl well-slide; add a coverslip.
5. A fluorescence microscope being able to acquire Z-stacks is necessary; Cell-P software (from Olympus) may be used for three-dimensional analysis of the results.

YIELDS

By including data from electron microscopy not much considered up to now, Joan-Ramon Daban could do the following statement recently:

> Experimental evidence indicates that the chromatin filament is self-organized into a multilayer planar structure that is densely stacked in metaphase and unstacked in interphase. This chromatin organization is unexpected, but it is shown that diverse supramolecular assemblies are multilayered. The mechanical strength of planar chromatin protects the genome integrity, even when double-strand breaks are produced. Here, it is hypothesized that the chromatin filament in the loops and topologically associating domains is folded within the thin layers of the multilaminar chromosomes. It is also proposed that multilayer chromatin has two states: inactive when layers are stacked and active when layers are unstacked. Importantly, the well-defined topology of planar chromatin may facilitate DNA replication without entanglements and DNA repair by homologous recombination.[18]

CONCLUSIONS

Based on data from nucleomics, some basic puzzle stones of chromosome and interphase structure seem to be solved. Still, this just opens manifold doors of new research directions, specifically in clinical genetics and tumor genetics. Yet there are just few such examples,[19–21] but more are expected to come.

REFERENCES

1. Liehr, T. Nuclear architecture. In: *Cytogenomics*; Liehr, T., Ed. Academic Press, New York, **2021**, pp. 297–305.
2. Potlapalli, B.P.; Schubert, V.; Metje-Sprink, J.; Liehr, T.; Houben, A. Application of Tris-HCl allows the specific labeling of regularly prepared chromosomes by CRISPR-FISH. Cytogenett. *Genome Res.* **2020**, 160, 156–165.

3. Eagen, K.P. Principles of chromosome architecture revealed by Hi-C. Trends *Biochem. Sci.* **2018**, 43, 469–478.

4. Maass, P.G.; Weise, A.; Rittscher, K.; Lichtenwald, J.; Barutcu, A.R.; Liehr, T.; Aydin, A.; Wefeld-Neuenfeld, Y.; Pölsler, L.; Tinschert, S.; Rinn, J.L.; Luft, F.C.; Bähring, S. Reorganization of inter-chromosomal interactions in the 2q37-deletion syndrome. *EMBO J.* **2018**, 37, e96257.

5. Steinhaeuser, U.; Starke, H.; Nietzel, A.; Lindenau, J.; Ullmann, P.; Claussen, U.; Liehr, T. Suspension (S)-FISH, a new technique for interphase nuclei. *J. Histochem. Cytochem.* **2002**, 50, 1697–1698.

6. Rabl, C. *Über Zelltheilung.* Morphologisches Jahrbuch. Gegenbaur, C., Ed. **1885**, 10, 214–330.

7. Barr, M.L.; Bertram, L.F.; Lindsay, H.A. The morphology of the nerve cell nucleus, according to sex. *Anat Rec.* **1950**, 107, 283–297.

8. Blobel, G. Gene gating: A hypothesis. *P.N.A.S. U.S.A.* **1985**, 82, 8527–8529.

9. Roy, S.S.; Mukherjee, A.K.; Chowdhury, S. Insights about genome function from spatial organization of the genome. *Hum. Genomics.* **2018**, 12, 8.

10. Melo, U.S.; Schöpflin, R.; Acuna-Hidalgo, R.; Mensah, M.A.; Fischer-Zirnsak, B.; Holtgrewe, M.; Klever, M.K.; Türkmen, S.; Heinrich, V.; Pluym, I.D.; Matoso, E.; de Sousa, S.B.; Louro, P.; Hülsemann, W.; Cohen, M.; Dufke, A.; Latos-Bieleńska, A.; Vingron, M.; Kalscheuer, V.; Quintero-Rivera, F.; Spielmann, M.; Mundlos, S. Hi-C identifies complex genomic rearrangements and TAD-shuffling in developmental diseases. *Am. J. Hum. Genet.* **2020**, 106, 872–884.

11. Cremer, T.; Cremer, M.; Cremer, C. Der Zellkern-eine Stadt in der Zelle: Teil 1: Chromosomenterritorien und Chromatindomänen. *Biol in uns. Zeit.* **2016**, 46, 290–299.

12. Lemke, J.; Claussen, J.; Michel, S.; Chudoba, I.; Mühlig, P.; Westermann, M.; Sperling, K.; Rubtsov, N.; Grummt, U. W.; Ullmann, P.; Kromeyer-Hauschild, K.; Liehr, T.; Claussen, U. The DNA-based structure of human chromosome 5 in interphase. *Am. J. Hum. Genet.* **2002**, 71, 1051–1059.

13. Weise, A.; Bhatt, S.; Piaszinski, K.; Kosyakova, N.; Fan, X.; Altendorf-Hofmann, A.; Tanomtong, A.; Chaveerach, A.; De Cioffi, M.B.; De Oliveira, E.; Walther, J.U.; Liehr, T.; Chaudhuri, J. P. Chromosomes in a genome-wise order: Evidence for metaphase architecture. Mol Cytogenet. **2016**, 9, 36.

14. Manvelyan, M.; Hunstig, F.; Bhatt, S.; Mrasek, K.; Pellestor, F.; Weise, A.; Simonyan, I.; Aroutiounian, R.; Liehr, T. Chromosome distribution in human sperm: A 3D multicolor banding-study. Mol Cytogenet. **2008**, 1, 25.

15. Solovei, I.; Kreysing, M.; Lanctôt, C.; Kösem, S.; Peichl, L.; Cremer, T.; Guck, J.; Joffe, B. Nuclear architecture of rod photoreceptor cells adapts to vision in mammalian evolution. Cell. **2009**, 137, 356–368.

16. Manvelyan, M.; Hunstig, F.; Mrasek, K.; Bhatt, S.; Pellestor, F.; Weise, A.; Liehr, T. Position of chromosomes 18, 19, 21 and 22 in 3D-preserved interphase nuclei of human and gorilla and white hand gibbon. *Mol. Cytogenet.* **2008**, 1, 9.

17. Liehr, T. Chromosome architecture studied by high-resolution FISH banding in three-dimensionally preserved human interphase nuclei. In: *Human Interphase Chromosomes, Biomedical Aspects*; Iourov, I.; Vorsanova, S.; Yurov, Y., Eds. Springer, Berlin, **2020**, pp. 147–155.

18. Daban, J.R. Supramolecular multilayer organization of chromosomes: Possible functional roles of planar chromatin in gene expression and DNA replication and repair. *FEBS Let.* **2020**, 594, 395–411.

19. Manvelyan, M.; Kempf, P.; Weise, A.; Mrasek, K.; Heller, A.; Lier, A.; Höffken, K.; Fricke, H.J.; Sayer, H.G.; Liehr, T.; Mkrtcyhan, H. Preferred co-localization of chromosome 8 and 21 in myeloid bone marrow cells detected by three dimensional molecular cytogenetics. *Int. J. Mol. Med.* **2009**, 24, 335–341.
20. Klonisch, T.; Wark, L.; Hombach-Klonisch, S.; Mai, S. Nuclear imaging in three dimensions: A unique tool in cancer research. *Ann. Anat.* **2010**, 192, 292–301.
21. Timme, S.; Schmitt, E.; Stein, S.; Schwarz-Finsterle, J.; Wagner, J.; Walch, A.; Werner, M.; Hausmann, M.; Wiech, T. Nuclear position and shape deformation of chromosome 8 territories in pancreatic ductal adenocarcinoma. *Analyt. Cell. Pathol.* **2011**, 34, 21–33.

19 FISH—On Fibers

Thomas Liehr

CONTENTS

INTRODUCTION

Fluorescence in situ hybridization (FISH) on stretched or highly extended chromosome fibers is known as Fiber-FISH,[1] or molecular combing.[2] In Figure 19.1, it is visualized that FISH has a principal resolution from a few dozen to hundred base pairs (in molecular combing), a few ten- to a few hundred-thousand base pairs (in Fiber-FISH), a few hundred-thousand base pairs to mega base pairs in interphase-FISH and mega base pair to chromosome size when accessing metaphases in FISH. Both high-resolution approaches of FISH—Fiber-FISH,[1] and molecular combing[2]—were described in the 1990s, and later on used in some specialized labs.[3–5] According to a PubMed search,[6] molecular combing was applied in research at about the same rates between 2001 and 2021 (Figure 19.2); on the other hand Fiber-FISH applications in human samples had a kind of hype between 1994 and 2000 and afterwards declined (Figure 19.2). In non-human-oriented research, Fiber-FISH has been used in about the same rates since 1994 (Figure 19.2). While molecular combing has recently been rediscovered and even commercialized,[7] for Fiber-FISH such a renaissance has not happened yet.

As Fiber-FISH, also sometimes called nuclear chromatin release, neither needs sophisticated equipment nor is complicated to perform, here the protocol for this approach shall be brought back to molecular cytogeneticists' minds.

SAMPLES/TISSUES

Each cell suspension prepared in Carnoy's fixative (methanol/acetic acid = 3:1) is suited as starting material to produce stretched DNA-fibers suited for Fiber-FISH.

DOI: 10.1201/9781003223658-19

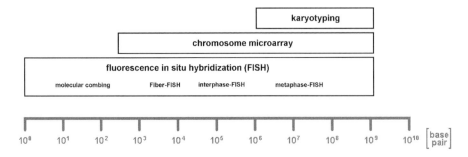

FIGURE 19.1 Schematic depiction of resolution of FISH approaches compared to karyotyping and chromosome micro array. The approximate resolutions accessible by metaphase-FISH, interphase-FISH, Fiber-FISH and molecular combing are shown.

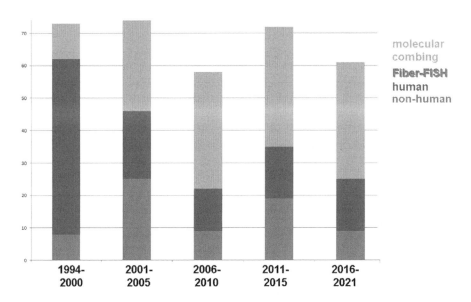

FIGURE 19.2 A Pubmed search[6] revealed the number of research papers published between 1994 and 2021 using molecular combing or Fiber-FISH (broken down in human and non-human oriented studies).

DESCRIPTION OF METHODS

The Fiber-FISH approach is a modification from Fidlerová et al.,[3] and was previously published elsewhere as a modified version.[8] Cytogenetically worked up material in Carnoy's fixative (methanol/acetic acid = 3:1), previously being stored at −20°C for up to several years, may be used.

1. Centrifuge the tube with Carnoy's fixative (methanol/acetic acid = 3:1), and cell pellet at 1,000 rpm for 5 min at room temperature (RT), and replace the supernatant by fresh Carnoy's fixative and mix the suspension by inversion of the closed cup.

2. Adjust the number of cells to ~1×10⁶ cells per ml by using a Fuchs-Rosenthal hemocytometer under a phase contrast microscope.

3. Scratch a ~1.5 cm circle in a clean and fat-free slide using a diamond pencil; and immerse slide in distilled water at RT.

4. Take the prepared slide out of water and clean below and outside the circle by a tissue.

5. Place ~20 µl of suspension from step 2 within the circle, and let dry for only 20 sec.

6. Transfer slide immediately into 1×PBS at RT for 30 sec to 1 min, according to the suspension.

7. Remove slide from 1×PBS, clean quickly below and outside the circle by a tissue, and add within the circle 100 µl of 0.5 M NaOH/30% ethanol for 30 sec; this step shall disrupt the nuclear shape and integrity. Do never allow fluid to leave the region of the marked circle.

8. Add now 100 ml fresh methanol and let air dry at RT—optimally under a hood.

9. Check dry slide(s) under phase contrast microscope—at this step nuclei should have a 'disrupted' shape and a kind of compact short tail with DNA-fibers should be visible.

10. Incubate slides at 60°C overnight and store until use in FISH at −20°C.

11. Do standard FISH, but avoid any kind of protein-degrading pretreatment.

YIELDS

The author of this paper used Fiber-FISH to clearly map the CMT1A-REP elements being ~1.4 Mb apart from each other and the in-between localized *PMP22* gene.[5, 8] Alterations of regions being in range of 100,000 to 1,000,000 base pairs can be optimally accessed by Fiber-FISH using locus-specific probes.

CONCLUSIONS

In case of necessity to study chromosomal changes at a resolution not accessible any more by interphase-FISH and being too large to be visualized in molecular combing, Fiber-FISH is the optimal alternative to correspondingly prepare cytogenetic material.

REFERENCES

1. Heng, H.H.; Squire, J.; Tsui, L.C. High-resolution mapping of mammalian genes by in situ hybridization to free chromatin. *Proc. Natl. Acad. Sci. U. S. A.* **1992**, 89, 9509–9513.

2. Bensimon, A.; Simon, A.; Chiffaudel, A.; Croquette, V.; Heslot, F.; Bensimon, D. Alignment and sensitive detection of DNA by a moving interface. *Science.* **1994**, 265, 2096–2098.

3. Fidlerová, H.; Senger, G.; Kost, M.; Sanseau, P.; Sheer, D. Two simple procedures for releasing chromatin from routinely fixed cells for fluorescence in situ hybridization. Cytogenet. *Cell Genet.* **1994**, 65, 203–205.

4. Heiskanen, M.; Peltonen, L.; Palotif, A. Visual mapping by high resolution FISH. *Trends Genet.* **1996**, 12, 379–382.

5. Rautenstrauss, B.; Fuchs, C.; Liehr, T.; Grehl, H.; Murakami, T.; Lupski, J.R. Visualization of the CMT1A duplication and HNPP deletion by FISH on stretched chromosome fibers. *J. Peripher. Nerv. Syst.* **1997**, 2, 319–322.

6. Pubmed. https://pubmed.ncbi.nlm.nih.gov [accessed on 03/2022].

7. Bisht, P.; Avarello, M.D.M. Molecular combing solutions to characterize replication kinetics and genome rearrangements. In: *Cytogenomics*; Liehr, T., Ed. Academic Press, New York, **2021**, pp. 47–72.

8. Fuchs, C.; Liehr, T.; Rautenstrauss, B. High-resolution FISH of stretched chromosome fibers. *Trends Genet.* **1997**, 13, 287.

20 FISH—and Single-Cell Gel Electrophoresis Assay (Comet Assay)

Galina Hovhannisyan, Tigran Harutyunyan and Rouben Aroutiounian

CONTENTS

DOI: 10.1201/9781003223658-20

INTRODUCTION

The Comet Assay (= CA, also called single cell gel electrophoresis) was developed in the 1980s as a relatively simple and fast way of detecting DNA damage and repair at the level of individual cells.[1, 2] The main steps of the CA include fixation of cell suspension with agarose onto glass microscope slides, lysis of cells to disrupt membranes and remove histones, and electrophoresis. Negatively charged, fragmented DNA migrates out of the nucleus in the electric field more rapidly than intact DNA, forming a structure resembling the tail of a comet, whereas undamaged DNA forms the head of the comet.[3] Images of comets stained with fluorescent dyes are analyzed under a microscope using image analysis software. The percentage of DNA in the tail reflects the level of DNA damage in individual cells.[4] The range of detection is between a few hundred DNA breaks per cell and a few thousand, encompassing levels of damage that can be repaired and tolerated by human cells.[5]

CA can be conducted under neutral or alkaline electrophoresis conditions. The neutral version of CA allows the detection of double-strand breaks. Currently, the most commonly used is the alkaline CA version, which detects a wider range of damage, including single- and double-strand breaks and alkali-labile sites.[3] The alkaline CA also identifies DNA lesions that are converted into strand breaks with various lesion-specific enzymes.[6] Up to now, twelve enzymes have been used; however, only the bacterial formamidopyrimidine DNA glycosylase (Fpg) and endonuclease III (EndoIII), which induce breaks at sites of oxidized purines and pyrimidines, respectively, are used extensively.[6, 7]

The ability to detect DNA damage, induced by genotoxic agents at subtoxic, physiologically relevant exposures,[8] has contributed to the widespread use of the CA in environmental biomonitoring and in vitro and in vivo genotoxicity testing in animals,[9–11] plants,[12] and humans.[5, 13–15]

COMET-FISH

The alkaline CA was combined with fluorescence in situ hybridization (FISH) to detect DNA damage and repair of specific DNA sequences. According to Glei et al.,[16]

> the nature of the measured Comet-FISH endpoint precludes us from stating basically that damage and repair are occurring within the specific gene, it is at least possible to evaluate whether the damage and repair are occurring within the vicinity of the gene of interest.

The FISH protocol has been adapted to the experimental limitations of CA and was first applied to comet preparations to determine the spatial distribution of telomeres, centromeres, and segments of the O6-methylguanine–DNA methyltransferase (*MGMT*) gene in human lymphocytes.[17] To avoid damaging of agarose layer with cells, the thermal denaturation of DNA in standard FISH was replaced with a chemical denaturation in Comet-FISH.[17] Protocols of the Comet-FISH technique and their modifications have been developed for different DNA probes application to a wide spectrum of biological material.[18–27] Several reviews provide technical and theoretical aspects of Comet-FISH, limitations as well as advantages of the assay and examples of its applications.[8, 28–32]

MAIN ACHIEVEMENTS OF COMET-FISH APPLICATION

Comet-FISH studies were carried out using DNA probes for genes, whole chromosomes, centromeric and telomeric regions, and other genomic loci, mainly to solve questions in in vitro, and also in exceptionally for in vivo case studies in human, animals, and plants. There are also limited examples for application padlock and strand-specific FISH probes on comets.

Comet-FISH in Studies for Gene Damage and Repair In Vitro

Among the various genes studied by Comet-FISH, the *TP53* gene is the most often accessed one, because of its important role in the etiology of many tumors and its association with genome stability.[33] It was shown that the repair rates of *TP53* gene region in gamma-irradiated bladder cancer cell lines RT4 and RT112,[34] and mitomycin C-treated RT4 cells,[35] were higher compared to overall genomic DNA. In addition, *TP53* gene region is more rapidly repaired than the *hTERT* gene region in gamma-irradiated RT4 and RT112 cells and in normal fibroblasts.[36]

Differences in the baseline levels of genetic instability of *TP53* and *HER-2/NEU1* gene loci in the breast cancer cell lines MCF-7, MDA-MB-468, and CRL-2336 relative to the normal cell lines and preferential repair of *TP53* in gamma-irradiated normal and malignant cells were found.[37] Application of FISH probes for 5' and 3' regions of dihydrofolate reductase (*DHFR*) and *MGMT* genes in CHO cells, and the human *TP53* gene confirmed faster *TP53* gene repair compared with total DNA.[38] Damage of *TP53* and *c-Myc* loci was revealed in human lymphocytes treated with pesticides terbuthylazine and carbofuran[39]; however, no significant effects were observed in lymphocytes, exposed to insecticides chlorpyrifos, imidacloprid, and α-cypermethrin.[40] The repair of the *TP53* and telomerase reverse transcriptase (*hTERT*) gene regions was more efficient in TK1+ compared to TK1- clones of the γ-irradiated Raji lymphoblastoid cell line.[41] It was shown that products of oxidative stress, such as hydrogen peroxide (H_2O_2), trans-2-hexenal, and 4-hydroxy-2-nonenal (HNE), cause DNA damage in genes *APC*, *KRAS*, and *TP53*, relevant for human colon cancer.[42]

There was no significant difference between primary mucosa cells of patients with carcinoma and controls in susceptibility of the epidermal growth factor receptor (*EGFR*) gene to benzo[a]pyrene-diolepoxide (BPDE), a major representative of tobacco-associated carcinogens despite the fact that this gene involved in several epithelial malignancies,[43] and no increase in BPDE-induced damage to the *cyclin D1* gene, which is of particular importance in in head and neck carcinogenesis.[44] Inhibitory effect of grapefruit juice on the genotoxicity induced by hydrogen peroxide was found in human lymphocytes, based on concentration/time-dependent return of intact *TP53* signals to the head of comets.[45]

Comet-FISH in Studies of Chromosomes and Specific Chromosomal Regions In Vitro

Comet-FISH also appears to be a useful approach for detecting and comparing damage to specific chromosomal regions of significance in leukemogenesis. DNA breaks at both 5q31 and 11q23 chromosomal regions were identified in TK6 lymphoblastoid cells exposed to melphalan, etoposide, or the benzene metabolite, and hydroquinone, at various concentrations.[46] Significantly higher damage levels in chromosomes 3, 5, and 8 compared with chromosome 1 was found in primary mucosa cells harvested during surgery from patients with and without carcinoma, treated with BPDE.[47] Results of Comet-FISH showed no difference between the sensitivity of chromosome 16 and 13 to mitomycin C and radiation in lymphocytes of Fanconi anemia (FA) patients, although the FA-A gene is located on chromosome 16.[48]

Telomere fragility was compared in normal human leukocytes and tumor cell lines treated with cytostatic agents. It was shown that telomeres in CHO and CCRF-CEM cells were about 2–3 times more sensitive towards BLM than global DNA, while in HT1080 telomeres were less fragile than total DNA.[49–51]

Comet-FISH In Vitro with Padlock and Strand-Specific Probes

In addition to conventionally labelled probes to particular DNA sequences, Comet-FISH was also carried out using padlock probes.[21] According to Shaposhnikov et al.,[52, 53] ends of padlock probes are designed complementary to adjacent sequences in the target DNA, so that on hybridization the two ends are juxtaposed and can readily be ligated to form a circle that is topologically linked to the target DNA. By amplifying circularized padlock probes through rolling-circle amplification in situ, specifically reacted probes are copied and then detected by hybridization with fluorescently labelled complementary DNA molecules. In comet preparations with padlock probes *Alu* sequences in HeLa cells,[52, 53] and gene-specific repair of three genes, *OGG1* (8-oxoguanine-DNA glycosylase-1), *XPD* (xeroderma pigmentosum group D), and *HPRT* (hypoxanthine-guanine phosphoribosyltransferase) were analyzed in human lymphocytes.[54]

CA was also combined with strand-specific FISH probes, which enables distinguishing of DNA damage and repair in the transcribed and the non-transcribed strand.[26] Applying this approach, it has been demonstrated that 8-oxoguanine is preferentially repaired in the transcribed strand of the ataxia-telangiesctasia mutated (ATM) gene in UV-irradiated human fibroblasts.[55]

Comet-FISH for In Vivo Studies

Comet-FISH has also been successfully applied to assess genetic instability in vivo. Analysis of *Ret*, *Abl1* (*cAbl*), and *Trp53* gene fragmentations on comet slides in peripheral blood cells from irradiated C57BL/6 and CBA/J mice made it possible to distinguish the effects of all doses as an indicator of DNA damage.[56] Prolonged exposure to possible carcinogens (phenylhydrazine, ethylene oxide, dichloromethane, and 1,2-dichloroethane) in pharmaceutical industry workers delays DNA repair and affects structural integrity of *TP53*.[57]

The possibility of using Comet-FISH in Pacific oyster *Crassostrea gigas* was confirmed using genotoxicant hydrogen peroxide model. The DNA damage was detected in nucleolar organizer regions (NORs) in oyster's hemocytes. As bivalves are widely-used for monitoring environmental pollution in aquatic ecosystems, the application of Comet-FISH to these organisms can provide additional information on the nature of the genomic regions affected by genotoxicants.[58]

In plant research, Comet-FISH with DNA probes, detecting *FokI* element-containing heterochromatin, NORs, or telomeres was first applied to *Vicia faba*. It was found that after treatment with specific endonucleases, the distribution of genomic loci in comet heads and tails reflects the expected distribution of the endonuclease-induced breaks.[59] Subsequently, DNA damage induced by maleic hydrazide in *Crepis capillaris* was localized in the genome using FISH with 5S and 25S rDNA probes.[60] FISH with 5S and 25S rDNA and telomeric/centromeric probes was applied to comets induced by maleic hydrazide, N-nitroso-N-methylurea (MNU), and γ-rays in barley.[61] The DNA damage involving the 25S rDNA sequences was analyzed in maleic hydrazide-treated *Crepis capillaris* with B chromosomes.[62] Comet-FISH with the 45S rDNA and telomeric probes was performed in *Vitis vinifera L.* treated with Cu^{2+}.[63] The data obtained suggests that Comet-FISH in plants may be a useful tool for environmental monitoring assessment.

SAMPLES/TISSUES

A major advantage of CA is that cells from various tissues of a wide variety of eukaryotic organisms can be studied.[3, 10, 11, 26, 63, 64] CA slides are typically prepared from fresh tissue; however, frozen tissue samples can also be applied[65]; in pre-frozen cells, the background DNA level does not increase and the outcome of the assay remains unchanged.[66, 67]

In principle, FISH can be applied to any comet slides. To date, Comet-FISH studies have been done on primary cells, including human,[17, 38–40, 45, 48–50, 54, 57] and mouse blood,[56] human oropharyngeal mucosa cells,[43, 44, 47] human colon cells from surgical tissue,[42] and skin fibroblasts from *Xeroderma pigmentosum* patients.[55]

Furthermore, Comet-FISH studies have been carried out on cell lines, including RT4 and RT112 (bladder carcinoma),[34–36] CHO (Chinese hamster ovary),[38, 51] MCF-7, MDA-MB468, and CRL2336 (breast cancer), GM1310B (lymphoblastoma) and AG11134 (normal mammary epithelial),[37] HT1080 (human fibrosarcoma) and CCRF-CEM (human T lymphoblastoma),[51] TK6 (human lymphoblastoma),[46] GM38 (normal fibroblast) and CSA and CSB (Cockayne syndrome fibroblast),[36] Raji (human B lymphoblastoma),[41] and HeLa (human breast cancer cells).[52, 53] Moreover, Comet-FISH was applied to the hemocytes of the Pacific oyster *Crassostrea gigas*,[58] and plant tissues, including seeds,[61] roots,[59] and leaves.[60, 62, 63]

The following protocol provides a detailed description of a standard Comet-FISH experiment suitable for human blood cells and cell lines. Blood samples can be collected and directly used or cultivated according to the experimental design. Cell lines should be cultivated and harvested according to standard procedures.

DESCRIPTION OF METHODS

MATERIALS

Equipment and Supplies (CA and Comet-FISH)

- Glass slides.
- Glass cover slips.
- Incubator (37°C).
- Water bath
- Staining jars.
- Moist chamber.
- Electrophoresis tank (horizontal) and power supply (500 mA, 25 V).
- Fluorescence microscope.
- Image analysis system (e.g., Comet Assay IV, Perceptive Instruments, Suffolk, UK).

Reagents and Solutions (CA)

- Normal melting point agarose (NMP agarose): 1% in H_2O.
- Low melting point agarose (LMP agarose): 0.9% in PBS
- Phosphate buffered saline (PBS).
- Lysis solution: 10 mM Tris-(hydroxymethyl)-aminomethane, pH 7.5, 100 mM Na_2EDTA, 2.5 M NaCl, 1% Triton X-100, pH 10, stored at 4°C.
- Electrophoresis solution: 1 mM Na_2EDTA, 300 mM NaOH, pH 13.1, stored at 4°C.
- Neutralization buffer: 0.4 M Tris-HCl, pH 7.5, 0.08 M Tris base, pH 7.2.
- SYBR-Green solution: (1:10,000) (30 µl per slide).

Reagents and Solutions (FISH)

- Telomere PNA FISH Kit/Cy3, or other DNA probes.
- Ethanol 75, 80, 95, and 100%.
- Rinse solution (included in the Telomere PNA FISH kit/Cy3).
- Hybridization buffer (included in the Telomere PNA FISH kit/Cy3). For other DNA probes, follow the manufacturer's instructions.
- Post-hybridization washing solution (included in the PNA FISH Kit/Cy3). For other DNA probes, follow the manufacturer's instructions.
- Phosphate-buffered detergent (PBD): 94 mM Na_2HPO_4(2 H_2O), 6 mM Na_2HPO_4/1 H_2O, 0.06% Triton X-100.
- SYBR Green.
- Antifade.
- Counterstaining solution: 1 µl Sybr Green stock solution, 500 µl water, and 500 µl antifade; store in the dark at—20°C in 500 µl aliquots.

METHOD

Cell Samples Preparation

1. Collect human blood samples, use heparin as an anticoagulant. Use any cell line, depending on the design of the experiment.

2. Cultivate the blood cells or cell lines according to standard procedure.
3. Treat cells with test compounds according to experiment design.
4. Get a suspension with concentration of about $1–2×10^6$ cells/ml.

Comet Assay

1. Cover the glass with a layer of 1% normal melting point agarose by dipping it in a vertical staining jar with melted agarose solution.
2. Dry slides at 37°C about 24 h in cell culture incubator to solidify agarose.
3. Drop 100 µl of cell/low–melting point agarose suspension (containing 10 µl of whole blood or cell line culture with 90 µl 0.9% low melting point agarose in 1× phosphate buffered saline = PBS) onto microscope slide.
4. Cover slides with cover slips to allow a homogeneous distribution of mixture of cells with low melting point agaroses on microscope slide.
5. Cool slides for 10 min at 4°C to solidify agarose.
6. Remove cover slips and immerse slides in cold lysis solution for 60 min at 4°C.
7. Place slides into an electrophoresis chamber containing cold (4°C) electrophoresis solution for 20 min.
8. Connect the electrophoresis tank to the power supply, and perform electrophoresis at 1.25 V/cm and 300 mA for 20 min at 4°C.
9. Remove slides from the electrophoresis tank, and wash them once for 20 min in neutralization buffer at room temperature (RT).

Comet-FISH

1. Dehydrate slides in absolute ethanol for at least three days.
2. Rehydrate slides in double-distilled H_2O for 15 min.
3. Denature DNA in 0.5 M NaOH for 25 min at RT.
4. Dehydrate slides in an ethanol series (70, 80, and 95%, each for 5 min).
5. Dry slides at RT.
6. Denature telomere PNA probes by preheating to 80°C in water bath for 3 min.
7. Pipette 10 µl denatured probe to an area of approximately 20×20 mm, cover with coverslip of appropriate size, and seal with rubber cement.
8. Incubate overnight at 37°C in a humid box.
9. Wash slides in prewarmed post-hybridization washing solution (from the PNA FISH Kit/Cy3) at 65°C in water bath, without agitation for 2.5 min, and cool the slides immediately in cold 1× phosphate-buffered detergent (PBD).
10. Follow the supplier's instructions for denaturation, hybridization, and post-hybridization washing when using other DNA probes.
11. Counterstain the slides with SYBR Green including 50% antifade (30 µl per slide), cover with a coverslip 24×60mm, and evaluate the results under fluorescence microscope.

Evaluation

1. Apply comet analysis software (e.g. Comet Assay IV analysis system, Perceptive Instruments, Suffolk, UK) for comet images.

2. Analyze visually the number of FISH signals per comet and the distribution of FISH signals between head and tail for Comet-FISH images.

YIELDS

Comet-FISH allows the analysis of damage and repair at the level of the overall DNA, and the localization of damage in specific loci of the genome on the same preparations (Figure 20.1). Methodological differences between Comet-FISH and FISH are aimed at preserving the gel layer, and practice shows that this goal is quite achievable. The following outputs are expected to be achieved:

1. **Assessment of overall DNA damage.** This requires an assessment of %DNA in comet tail.
2. **Assessment of loci-specific DNA damage.** The DNA locus can be considered intact if FISH signal is located in the head of a comet or damaged if it is located in the tail of a comet. Quantification of damage in specific domains can be expressed as the percent of spots in heads and tails.
3. **Assessment of overall DNA repair.** This requires an assessment of decrease of %DNA in tail over time.
4. **Assessment of loci-specific DNA repair.** Reverse shift of fluorescent signals from the comet tail to the head over time corresponds to the repair of a given DNA region. Quantification of repair in specific domains can be expressed as the percent of spots in heads and tails.
5. **Comparison of damage and repair of specific loci and whole DNA.** Comet-FISH allows to compare the rates of damage and repair of specific loci and whole DNA since both of these events can be tracked simultaneously in the same cell; thus, it is possible that individual loci of genome or chromosomes can be damaged more or less frequent than the entire DNA.

FIGURE 20.1 The illustration shows the green-colored nucleus of a bleomycin-treated human leukocyte in the form of a comet, which is obtained after the migration of damaged DNA to the anode during electrophoresis. Red dots on the comet's "head" and "tail" represent fluorescently colored telomeres.

IMPORTANT NOTES

- The FISH signal is able to migrate to the comet's tail if the break occurs not only within a specific loci of the genome, but also in the vicinity of the area of interest.[16]
- Certain genes or parts of genes or other loci of the genome remain in the head of comet even when there is a DNA break nearby, and it is likely that this localization within the head reflects the presence nearby of scaffold- or matrix-associated areas.[16]
- Cells immersed in gel retain their minimally deformed three-dimensional structure. However, the location of cells at different depths in the gel suggests special attention when analyzing images,[28] and even does not exclude the possibility that some signals immersed deep in the gel may be invisible.

CONCLUSIONS

FISH allows detecting genomic regions of interest immediately on comet preparations. Application of FISH on comets is based on replacing the thermal denaturation of DNA with an alkaline one to preserve the structure of the gel in which the cells are embedded. In that way, global genomic and sequence-specific DNA damage and repair of cells from almost any tissue can be comparatively investigated.

Comet-FISH studies were carried out using DNA probes for whole chromosomes, specific chromosome regions, centromeres, telomeres, genes, and any other genome loci. Various examples of the application of the method on human, animal, and plant cells confirmed sensitivity and the usefulness of this approach in genotoxicity testing and biomonitoring.

ACKNOWLEDGMENTS

This work was supported by the RA MESCS and BMBF (project #AG-01/20), and RA MESCS (project #21AG-1F068).

REFERENCES

1. Ostling, O.; Johanson, K.J. Microelectrophoretic study of radiation-induced DNA damages in individual mammalian cells. *Biochem Biophys Res Commun.* **1984**, 123(1), 291–298.
2. Singh, N.P.; McCoy, M.T.; Tice, R.R.; Schneider, E.L. A simple technique for quantitation of low levels of DNA damage in individual cells. *Exp Cell Res.* **1988**, 175(1), 184–191.
3. Cordelli, E.; Bignami, M.; Pacchierotti, F. Comet assay: A versatile but complex tool in genotoxicity testing. *Toxicol Res (Camb).* **2021**, 10(1), 68–78.
4. Karbaschi, M.; Ji, Y.; Abdulwahed, A.M.S.; Alohaly, A.; Bedoya, J.F.; Burke, S.L.; Boulos, T.M.; Tempest, H.G.; Cooke, M.S. Evaluation of the major steps in the conventional protocol for the alkaline comet assay. *Int J Mol Sci.* **2019**, 20(23), 6072.
5. Azqueta, A.; Muruzabal, D.; Boutet-Robinet, E.; Milic, M.; Dusinska, M.; Brunborg, G.; Møller, P.; Collins, A.R. Technical recommendations to perform the alkaline standard

and enzyme-modified comet assay in human biomonitoring studies. *Mutat Res Genet Toxicol Environ Mutagen.* **2019**, 43, 24–32.

6. Muruzabal, D.; Collins, A.; Azqueta, A. The enzyme-modified comet assay: Past, present and future. *Food Chem Toxicol.* **2021**, 147, 111865.

7. Muruzabal, D.; Sanz-Serrano, J.; Sauvaigo, S.; Treillard, B.; Olsen, A.K.; López de Cerain, A.; Vettorazzi, A.; Azqueta, A. Validation of the in vitro comet assay for DNA cross-links and altered bases detection. *Arch Toxicol.* **2021**, 95(8), 2825–2838.

8. Spivak, G.; Cox, R.A.; Hanawalt, P.C. New applications of the Comet assay: Comet-FISH and transcription-coupled DNA repair. *Mutat Res.* **2009**, 681(1), 44–50.

9. de Lapuente, J.; Lourenço, J.; Mendo, S.A.; Borràs, M.; Martins, M.G.; Costa, P.M.; Pacheco, M. The Comet assay and its applications in the field of ecotoxicology: A mature tool that continues to expand its perspectives. *Front Genet.* **2015**, 4(6), 180.

10. Gajski, G.; Žegura, B.; Ladeira, C.; Pourrut, B.; Del Bo', C.; Novak, M.; Sramkova, M.; Milić, M.; Gutzkow, K.B.; Costa, S.; Dusinska, M.; Brunborg, G.; Collins, A. The comet assay in animal models: From bugs to whales: (Part 1 Invertebrates). *Mutat Res.* **2019**, 779, 82–113.

11. Gajski, G.; Žegura, B.; Ladeira, C.; Novak, M.; Sramkova, M.; Pourrut, B.; Del Bo', C.; Milić, M.; Gutzkow, K.B.; Costa, S.; Dusinska, M.; Brunborg, G.; Collins, A. The comet assay in animal models: From bugs to whales: (Part 2 Vertebrates). *Mutat Res Rev Mutat Res.* **2019**, 781, 130–164.

12. Pietrini, F.; Iannilli, V.; Passatore, L.; Carloni, S.; Sciacca, G.; Cerasa, M.; Zacchini, M. Ecotoxicological and genotoxic effects of dimethyl phthalate (DMP) on Lemna minor L. and Spirodela polyrhiza (L.) Schleid. plants under a short-term laboratory assay. *Sci Total Environ.* **2022**, 1;806(Pt 4), 150972.

13. Azqueta, A.; Ladeira, C.; Giovannelli, L.; Boutet-Robinet, E.; Bonassi, S.; Neri, M.; Gajski, G.; Duthie, S.; Del Bo', C.; Riso, P.; Koppen, G.; Basaran, N.; Collins, A.; Møller, P. Application of the comet assay in human biomonitoring: An hCOMET perspective. *Mutat Res Rev Mutat Res.* **2020**, 783, 108288.

14. Bonassi, S.; Ceppi, M.; Møller, P.; Azqueta, A.; Milić, M.; Neri, M.; Brunborg, G.; Godschalk, R.; Koppen, G.; Langie, S.A.S.; Teixeira, J.P.; Bruzzone, M.; Da Silva, J.; Benedetti, D.; Cavallo, D.; Ursini, C.L.; Giovannelli, L.; Moretti, S.; Riso, P.; Del Bo', C.; Russo, P.; Dobrzyńska, M.; Goroshinskaya, I.A.; Surikova, E.I.; Staruchova, M.; Barančokova, M.; Volkovova, K.; Kažimirova, A.; Smolkova, B.; Laffon, B.; Valdiglesias, V.; Pastor, S.; Marcos, R.; Hernández, A.; Gajski, G.; Spremo-Potparević, B.; Živković, L.; Boutet-Robinet, E.; Perdry, H.; Lebailly, P.; Perez, C.L.; Basaran, N.; Nemeth, Z.; Safar, A.; Dusinska, M.; Collins, A. hCOMET project. DNA damage in circulating leukocytes measured with the comet assay may predict the risk of death. *Sci Rep.* **2021**, 11(1), 16793.

15. Milić, M.; Ceppi, M.; Bruzzone, M.; Azqueta, A.; Brunborg, G.; Godschalk, R.; Koppen, G.; Langie, S.; Møller, P.; Teixeira, J.P.; Alija, A.; Anderson, D.; Andrade, V.; Andreoli, C.; Asllani, F.; Bangkoglu, E.E.; Barančoková, M.; Basaran, N.; Boutet-Robinet, E.; Buschini, A.; Cavallo, D.; Costa Pereira, C.; Costa, C.; Costa, S.; Da Silva, J.; Del Bo', C.; Dimitrijević Srećković, V.; Djelić, N.; Dobrzyńska, M.; Duračková, Z.; Dvořáková, M.; Gajski, G.; Galati, S.; García Lima, O.; Giovannelli, L.; Goroshinskaya, I.A.; Grindel, A.; Gutzkow, K.B.; Hernández, A.; Hernández, C.; Holven, K.B.; Ibero-Baraibar, I.; Ottestad, I.; Kadioglu, E.; Kažimirová, A.; Kuznetsova, E, Ladeira, C, Laffon, B, Lamonaca, P, Lebailly, P, Louro, H, Mandina Cardoso, T, Marcon, F.; Marcos, R.; Moretti, M.; Moretti, S.; Najafzadeh, M.; Nemeth, Z.; Neri, M.; Novotna, B.; Orlow, I.; Paduchova, Z.; Pastor, S.; Perdry, H.; Spremo-Potparević, B.; Ramadhani, D.; Riso, P.; Rohr, P.; Rojas, E.; Rossner, P.; Safar, A.; Sardas, S.; Silva, M.J.; Sirota, N.; Smolkova,

B.; Staruchova, M.; Stetina, R.; Stopper, H.; Surikova, E.I.; Ulven, S.M.; Ursini, C.L.; Valdiglesias, V.; Valverde, M.; Vodicka, P.; Volkovova, K.; Wagner, K.H.; Živković, L.; Dušinská, M.; Collins, A.R.; Bonassi, S. The hCOMET project: International database comparison of results with the comet assay in human biomonitoring. Baseline frequency of DNA damage and effect of main confounders. *Mutat Res Rev Mutat Res.* **2021**, 787, 108371.

16. Glei, M.; Hovhannisyan, G.; Pool-Zobel, B.L. Use of Comet-FISH in the study of DNA damage and repair: Review. *Mutat Res.* **2009**, 681(1), 33–43.

17. Santos, S.J.; Singh, N.P.; Natarajan, A.T. Fluorescence in situ hybridization with comets. *Exp Cell Res.* **1997**, 232(2), 407–411.

18. Rapp, A.; Hausmann, M.; Greulich, K.O. The comet-FISH technique: A tool for detection of specific DNA damage and repair. Methods Mol Biol. **2005**, 291, 107–119.

19. Spivak, G. The Comet-FISH assay for the analysis of DNA damage and repair. *Methods Mol Biol.* **2010**, 659, 129–145.

20. Shaposhnikov, S.; Thomsen, P.D.; Collins, A.R. Combining fluorescent in situ hybridization with the comet assay for targeted examination of DNA damage and repair. *Methods Mol Biol.* **2011**, 682, 115–132.

21. Henriksson, S.; Nilsson, M. Padlock probes and rolling circle amplification for detection of repeats and single-copy genes in the single-cell comet assay. *Methods Mol Biol.* **2012**, 853, 95–103.

22. Schlörmann, W.; Glei, M. Detection of DNA damage by comet fluorescence in situ hybridization. *Methods Mol Biol.* **2012**, 920, 91–100.

23. Laubenthal, J.; Anderson, D. Fluorescence in situ hybridization on electrophoresed cells to detect sequence specific DNA damage. *Methods Mol Biol.* **2013**, 1054, 219–235.

24. Glei, M.; Schlörmann, W. Analysis of DNA damage and repair by comet fluorescence in situ hybridization (Comet-FISH). *Methods Mol Biol.* **2014**, 1094, 39–48.

25. Shaposhnikov, S.; El Yamani, N.; Collins, A.R. Fluorescent in situ hybridization on comets: FISH comet. *Methods Mol Biol.* **2015**, 1288, 363–373.

26. Mondal, M.; Guo, J. Comet-FISH for ultrasensitive strand-specific detection of DNA damage in single cells. *Methods Enzymol.* **2017**, 591, 83–95.

27. Hovhannisyan, G.; Aroutiounian, R. Comet-FISH In: *Fluorescence In Situ Hybridization (FISH), Application Guide*; Liehr, T., Ed. 2nd Edition. Springer, Berlin, **2017**, pp. 373–378.

28. Baugh, E.H.; Ke, H.; Levine, A.J.; Bonneau, R.A.; Chan, C.S. Why are there hotspot mutations in the TP53 gene in human cancers? *Cell Death Differ.* **2018**, 25(1), 154–160.

29. McKenna, D.J.; Rajab, N.F.; McKeown, S.R.; McKerr, G.; McKelvey-Martin, V.J. Use of the comet-FISH assay to demonstrate repair of the TP53 gene region in two human bladder carcinoma cell lines. *Radiat Res.* **2003**, 159(1), 49–56.

30. McKenna, D.J.; Gallus, M.; McKeown, S.R.; Downes, C.S.; McKelvey-Martin, V.J. Modification of the alkaline Comet assay to allow simultaneous evaluation of mitomycin C-induced DNA cross-link damage and repair of specific DNA sequences in RT4 cells. *DNA Repair (Amst).* **2003**, 2(8), 879–890.

31. McKenna, D.J.; Doherty, B.A.; Downes, C.S.; McKeown, S.R.; McKelvey-Martin, V.J. Use of the comet-FISH assay to compare DNA damage and repair in p53 and hTERT genes following ionizing radiation. *PLoS One.* **2012**, 7(11), e49364.

32. Shaposhnikov, S.; Frengen, E.; Collins, A.R. Increasing the resolution of the comet assay using fluorescent in situ hybridization: A review. Mutagenesis. **2009**, 24(5), 383–389.

33. Glei, M.; Hovhannisyan, G.; Pool-Zobel, B.L. Use of Comet-FISH in the study of DNA damage and repair: Review. *Mutat Res.* **2009**, 681(1), 33–43.

34. Hovhannisyan, G.G. Fluorescence in situ hybridization in combination with the comet assay and micronucleus test in genetic toxicology. *Mol Cytogenet.* **2010**, 3, 17.
35. Azqueta, A.; Collins, A.R. The essential comet assay: A comprehensive guide to measuring DNA damage and repair. *Arch Toxicol.* **2013**, 87(6), 949–968.
36. Spivak, G. New developments in comet-FISH. *Mutagenesis.* **2015**, 30(1), 5–9.
37. Kumaravel, T.S.; Bristow, R.G. Detection of genetic instability at HER-2/neu and p53 loci in breast cancer cells sing Comet-FISH. Breast Cancer Res Treat. **2005**, 91(1), 89–93.
38. Horváthová, E.; Dusinská, M.; Shaposhnikov, S.; Collins, A.R. DNA damage and repair measured in different genomic regions using the comet assay with fluorescent in situ hybridization. *Mutagenesis.* **2004**, 19(4), 269–276.
39. Mladinic, M.; Zeljezic, D.; Shaposhnikov, S.A.; Collins, A.R. The use of FISH-comet to detect c-Myc and TP 53 damage in extended-term lymphocyte cultures treated with terbuthylazine and carbofuran. *Toxicol Lett.* **2012**, 211(1), 62–69.
40. Zeljezic, D.; Vinkovic, B.; Kasuba, V.; Kopjar, N.; Milic, M.; Mladinic, M. The effect of insecticides chlorpyrifos, α-cypermethrin and imidacloprid on primary DNA damage, TP 53 and c-Myc structural integrity by comet-FISH assay. *Chemosphere.* **2017**, 182, 332–338.
41. McAllister, K.A.; Yasseen, A.A.; McKerr, G.; Downes, C.S.; McKelvey-Martin, V.J. FISH comets show that the salvage enzyme TK1 contributes to gene-specific DNA repair. *Front Genet.* **2014**, 5, 233.
42. Glei, M.; Schaeferhenrich, A.; Claussen, U.; Kuechler, A.; Liehr, T.; Weise, A.; Marian, B.; Sendt, W.; Pool-Zobel, B.L. Comet fluorescence in situ hybridization analysis for oxidative stress-induced DNA damage in colon cancer relevant genes. *Toxicol Sci.* **2007**, 96(2), 279–284.
43. Reiter, M.; Welz, C.; Baumeister, P.; Schwenk-Zieger, S.; Harréus, U. Mutagen sensitivity and DNA repair of the EGFR gene in oropharyngeal cancer. *Oral Oncol.* **2010**, 46(7), 519–524.
44. Reiter, M.; Baumeister, P.; Hartmann, M.; Schwenk-Zieger, S.; Harréus, U. Chemoprevention by celecoxib and mutagen sensitivity of cyclin d1 in patients with oropharyngeal carcinoma. *In Vivo.* **2014**, 28(1), 49–53.
45. Razo-Aguilera, G.; Baez-Reyes, R.; Alvarez-González, I.; Paniagua-Pérez, R.; Madrigal-Bujaidar, E. Inhibitory effect of grapefruit juice on the genotoxicity induced by hydrogen peroxide in human lymphocytes. *Food Chem Toxicol.* **2011**, 49(11), 2947–2953.
46. Escobar, P.A.; Smith, M.T.; Vasishta, A.; Hubbard, A.E.; Zhang, L. Leukaemia-specific chromosome damage detected by comet with fluorescence in situ hybridization (comet-FISH). *Mutagenesis.* **2007**, 22(5), 321–327.
47. Harréus, U.A.; Kleinsasser, N.H.; Zieger, S.; Wallner, B.; Reiter, M.; Schuller, P.; Berghaus, A. Sensitivity to DNA-damage induction and chromosomal alterations in mucosa cells from patients with and without cancer of the oropharynx detected by a combination of Comet assay and fluorescence in situ hybridization. *Mutat Res.* **2004**, 563(2), 131–138.
48. Mohseni Meybodi, A.; Mozdarani, H. DNA damage in leukocytes from Fanconi anemia (FA) patients and heterozygotes induced by mitomycin C and ionizing radiation as assessed by the comet and comet-FISH assay. *Iran Biomed J.* **2009**, 13(1), 1–8.
49. Arutyunyan, R.; Gebhart, E.; Hovhannisyan, G.; Greulich, K.O.; Rapp, A. Comet-FISH using peptide nucleic acid probes detects telomeric repeats in DNA damaged by bleomycin and mitomycin C proportional to general DNA damage. *Mutagenesis.* **2004**, 19(5), 403–408.
50. Arutyunyan, R.; Rapp, A.; Greulich, K.O.; Hovhannisyan, G.; Haroutiunian, S.; Gebhart, E. Fragility of telomeres after bleomycin and cisplatin combined treatment measured in human leukocytes with the Comet-FISH technique. *Exp Oncol.* **2005**, 27(1), 38–42.

51. Hovhannisyan, G.; Rapp, A.; Arutyunyan, R.; Greulich, K.O.; Gebhart, E. Comet-assay in combination with PNA-FISH detects mutagen-induced DNA damage and specific repeat sequences in the damaged DNA of transformed cells. *Int J Mol Med.* **2005**, 15(3), 437–442.

52. Shaposhnikov, S.; Azqueta, A.; Henriksson, S.; Meier, S.; Gaivão, I.; Huskisson, N.H.; Smart, A.; Brunborg, G.; Nilsson, M.; Collins, A.R. Twelve-gel slide format optimised for comet assay and fluorescent in situ hybridisation. *Toxicol Lett.* **2010**, 195(1), 31–34.

53. Shaposhnikov, S.; Larsson, C.; Henriksson, S.; Collins, A.; Nilsson, M. Detection of Alu sequences and mtDNA in comets using padlock probes. *Mutagenesis.* **2006**, 21(4), 243–247.

54. Henriksson, S.; Shaposhnikov, S.; Nilsson, M.; Collins, A. Study of gene-specific DNA repair in the comet assay with padlock probes and rolling circle amplification. *Toxicol Lett.* **2011**, 202(2), 142–147.

55. Guo, J.; Hanawalt, P.C.; Spivak, G. Comet-FISH with strand-specific probes reveals transcription-coupled repair of 8-oxoGuanine in human cells. *Nucleic Acids Res.* **2013**, 41(16), 7700–7712.

56. Amendola, R.; Basso, E.; Pacifici, P.G.; Piras, E.; Giovanetti, A.; Volpato, C.; Romeo, G. Ret, Abl1 (cAbl) and Trp53 gene fragmentations in comet-FISH assay act as in vivo biomarkers of radiation exposure in C57BL/6 and CBA/J mice. *Radiat Res.* **2006**, 165(5), 553–5561.

57. Zeljezic, D.; Mladinic, M.; Kopjar, N.; Radulovic, A.H. Evaluation of genome damage in subjects occupationally exposed to possible carcinogens. *Toxicol Ind Health.* **2016**, 32(9), 1570–1580.

58. Pérez-García, C.; Rouxel, J.; Akcha, F. Development of a comet-FISH assay for the detection of DNA damage in hemocytes of Crassostrea gigas. *Aquat Toxicol.* **2015**, 161, 189–195.

59. Menke, M.; Angelis, K.J.; Schubert, I. Detection of specific DNA lesions by a combination of comet assay and FISH in plants. *Environ Mol Mutagen.* **2000**, 35(2), 132–138.

60. Kwasniewska, J.; Grabowska, M.; Kwasniewski, M.; Kolano, B. Comet-FISH with rDNA probes for the analysis of mutagen-induced DNA damage in plant cells. *Environ Mol Mutagen.* **2012**, 53(5), 369–375.

61. Kwasniewska, J.; Kwasniewski, M. Comet-FISH for the evaluation of plant DNA damage after mutagenic treatments. *J Appl Genet.* **2013**, 54(4), 407–415.

62. Kwasniewska, J.; Mikolajczyk, A. Influence of the presence of B chromosomes on DNA damage in Crepis capillaris. *PLoS One.* **2014**, 9(1), e87337.

63. Castro, C.; Carvalho, A.; Gaivão, I.; Lima-Brito, J. Evaluation of copper-induced DNA damage in Vitis vinifera L. using Comet-FISH. *Environ Sci Pollut Res Int.* **2021**, 28(6), 6600–6610.

64. Russo, C.; Acito, M.; Fatigoni, C.; Villarini, M.; Moretti, M. B-Comet Assay (Comet Assay on buccal cells) for the evaluation of primary DNA damage in human biomonitoring studies. *Int J Environ Res Public Health.* **2020**, 17(24), 9234.

65. Hobbs, C.A.; Recio, L.; Winters, J.; Witt, K.L. Use of frozen tissue in the Comet Assay for the evaluation of DNA damage. *J Vis Exp.* **2020**, 157.

66. Pant, K.; Springer, S.; Bruce, S.; Lawlor, T.; Hewitt, N.; Aardema, M.J. Vehicle and positive control values from the in vivo rodent comet assay and biomonitoring studies using human lymphocytes: Historical database and influence of technical aspects. *Environ Mol Mutagen.* **2014**, 55(8), 633–642.

67. Ladeira, C.; Koppen, G.; Scavone, F.; Giovannelli, L. The comet assay for human biomonitoring: Effect of cryopreservation on DNA damage in different blood cell preparations. *Mutat Res Genet Toxicol Environ Mutagen.* **2019**, 843, 11–17.

21 Molecular Karyotyping

Ilda Patrícia Ribeiro, Luís Miguel Pires,
Susana Isabel Ferreira, Mariana Val,
Joana Barbosa Melo and Isabel Marques Carreira

CONTENTS

INTRODUCTION

Nowadays, several diseases are linked to copy-number variants (CNVs), and various genome-wide methods can be used for its detection. However, it is important to highlight that the presence of CNVs by itself does not mean an abnormal or pathogenic phenotype. The clinical relevance of CNVs depends on its functional impact, having a high probability of a phenotypic effect if the region of imbalance maps critical gene/s or an important regulatory region.

The most commonly used method for CNVs detection has been array-based comparative genomic hybridization (aCGH). This technique was first described in 1997 by Solinas Toldo,[1] and its first applications were published by the groups

DOI: 10.1201/9781003223658-21

of Pinkel, Snidjers and Buckley in the following years.[2–4] Initially, this technique was introduced as matrix-CGH, and later coined as array-CGH (aCGH).[3] In 2005, Vermeesch and colleagues proposed to call this technology as molecular karyo-typing.[5] Currently, the use of aCGH technology is massified worldwide either at diagnostic or investigation point of view. Many reports were published concerning chromosome imbalances associated with cancer as well as in constitutional cytoge-netics, with postnatal and prenatal applications. The aCGH technology use whole genome probes immobilized on glass slides (a microarray) that allow, due to the deep resolution, the diagnosis of smaller genomic imbalances, being the method of choice to small genomic imbalances identification with a wide range of clinical applications. It was estimated that the introduction of aCGH technique in the clinical practice allowed increasing the detection of apparently pathogenic genomic imbal-ances in as much as 20% of cases that have had normal karyotyping tests.[6]

THE aCGH TECHNOLOGY AND DATA INTERPRETATION

Initially, the CGH technique was developed using metaphase chromosomes for anal-ysis of chromosomal imbalances. This technique is based on the co-hybridization of two samples of genomic DNA (test and reference), fluorescently labeled and mixed in a 1:1 ratio, to normal metaphase chromosome spreads on a glass slide.[7] The ratios of fluorescence intensities for each chromosome or chromosome segment show its rela-tive copy number in the test genome compared with the control/reference genome.[8] So, the result of hybridization is detected by the analysis of the intensities of the two different fluorochromes, being the regions of gain or loss of DNA sequences, such as deletions, duplications, or amplifications, detected due to the fluctuations in the inten-sities ratio of the two fluorochromes along the target chromosomes.[7] Besides, CGH technique can be used to scan an entire genome for imbalances; its resolution (like cytogenetic methods) is limited to alterations of approximately 5–10 Mb.[9] Between 1992 and 2000, this chromosome-based CGH approach was employed to human genetics, more specially in the solid tumors field.[10] Presently, CGH technique is still used for comparison of closely related species in molecular cytogenetic studies and to identify cryptic sex chromosomes within a species.[11, 12] Nevertheless, metaphase spreads were replaced by arrays of DNA fragments such as BAC (bacterial artificial chromosomes) or PAC (P1 artificial chromosome) clones and more recently by oli-gonucleotides and SNP (single-nucleotide polymorphism) probes, being this modi-fied version of the CGH test, named aCGH. In the aCGH technique, thousands to millions of different 'spots' comprising several identical, immobilized single strand DNA (ssDNA) probes complementary to a part of the genome are printed on a glass slide, allowing much higher resolution and consequently the identification of sub-microscopic aberrations, in comparison with CGH analysis and conventional cyto-genetics.[13] In the aCGH technique, equal amounts of test and reference genomic DNAs are differentially labeled with fluorochromes (usually Cyanine 5—red and Cyanine 3—green, respectively), followed by co-hybridization onto a microarray, a glass slide containing the DNA targets for whole genome. Since the DNAs have been denatured, the single strands when applied to the glass slide hybridize with the arrayed single-strand probes, by Watson-Crick base pairing.[14] Unbound labeled

DNA is washed off the slide. The reference sample should be gender-matched with the sample in study. After the hybridization, the slides are scanned into TIFF image files by a microarray scanner, which uses laser light to measure the relative intensity of emission, usually at green (Cy3—for control) and red (Cy5—for biological sample in study) wavelengths being all targets' intensities measured, and the image files are quantified in a specific feature extraction software. The green to red fluorescence $\log_2$ ratio is calculated for each spot. The copy number alterations are calculated by differences in fluorescence intensities along any given spot(s), enabling the detection of gains and losses throughout the genome with high resolution. Therefore, equal amount of green and red fluorescence for a specific spot signifies that this region of the genome to which that spot is mapped harbors an equal copy-number in the tested and control samples. In the other hand, a ratio favoring red fluorescence implies an excess of that genomic segment in the test genome (i.e., a copy-number gain) in comparison with the control, whereas a green signal shows a copy number loss (Figure 21.1).

The principle of aCGH technique can be demonstrated with the following simple example (Figure 21.2A). DNA from a male with 9p trisomy and a partial 9q (9q13-q31.2) trisomy, labeled in red, is co-hybridized with DNA from a normal male (46,XY) labeled in green against an array slide. At each target sequence of the array slide without copy number variants, a balanced, yellow color is observed. Since at chromosome 9 (specifically 9p and 9q13-q31.2) there is more DNA labeled with red due to trisomy 9, this means that in the competitive hybridization higher quantity of 'DNA red' will be linked in comparison with DNA labelled with green (from control) for this chromosome, and consequently a gain of chromosome 9 material will be visualized after this experiment (represented in a positive scale, by a right deviation of the blue color in the Figure 21.2A).

A genotype–phenotype correlation is crucial to interpret the CNVs detected. Not all CNVs are associated with unfavorable clinical phenotypes, being in some cases classified as benign—without impact in the phenotype. In other cases, CNVs are linked to a spectrum of clinical phenotypes that may range from benign to pathological, including a very challenging class of CNVs named VUS (variant of unknown/ uncertain clinical significance). Therefore, CNVs could be phenotypically benign, because of phenotypic differences created by evolution, or could be responsible for disease or for disease susceptibility. Additionally, the CNVs could be categorized as germ line (constitutional), where every cell of the body is altered, or somatic, where only a subgroup of cells in the organism harbors the alteration, such as in cancer or in mosaics. The interpretation of the results and consequently the assessment of the pathogenicity of a CNV could be challenging and needs to consider the size and location of the copy number alteration, breakpoint information, whether it has occurred *de novo*, and gene content, namely whether the region contains genes of known importance. Determining whether a specific CNV is present in affected (or unaffected) relatives can also be important as well as verify in databases if other unrelated but affected individuals with similar clinical phenotypes are described, namely the use of CNV databases such as DECIPHER (www.dechipergenomics.org/) and if those CNVs are also reported among healthy individuals, using CNV databases such as the Database of Genomic Variants (http://dgv.tcag.ca/dgv/app/home).

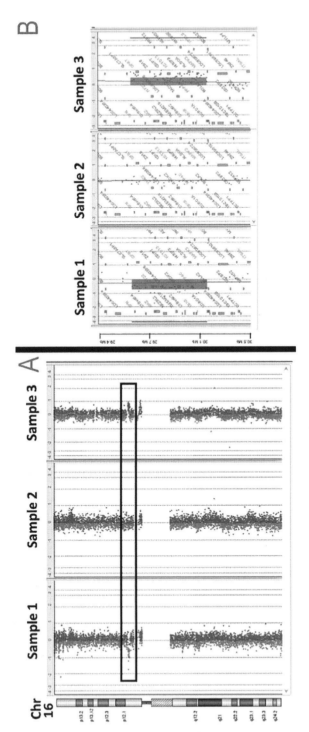

FIGURE 21.1   aCGH results of chromosome 16 from three different samples, using CytoGenomics software for analysis.

A) Chromosome view.

B) Gene view, presenting the gene content in this chromosome region. **Sample 1** presents a deletion of 541 kb at 16p11.2, involving 30 genes arr[GRCh37] 16p11.2(29656684_30197341)x1; **Sample 2** without CNVs at 16p11.2; and **Sample 3** presents a duplication of 544 kb at 16p11.2, involving 34 genes arr[GRCh37] 16p11.2(29652999_30197341)x3.

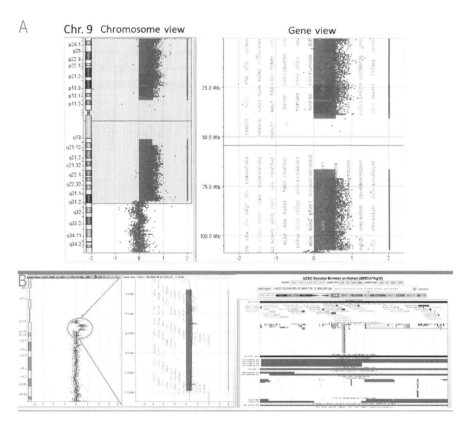

FIGURE 21.2 A) Chromosome view and respective gene content of a 9p trisomy and a partial 9q trisomy. Results obtained using aCGH technique and analyzed with Cytogenomics software. In orange are represented the OMIM Morbid Map genes, in blue the OMIM genes and in grey the RefSeq genes. Trisomy is represented as a positive scale. B) Plot of chromosome 22 displaying a gain at 22q11.21, detected by aCGH technique and analyzed with Cytogenomics software. The duplicated region is shown in the UCSC Genome Browser (https://genome.ucsc.edu/) with some selected tracks to assist in the genotype–phenotype correlation and consequently in the clinical relevance determination. Example of these tracks are chromosome band, RefSeq genes, OMIM genes, DECIPHER CNVs, ClinGen CNVs and DGV.

There are also several public databases and sources such as PubMed (https://pubmed.ncbi.nlm.nih.gov/), OMIM (www.omim.org/) and human genome browsers (UCSC—https://genome.ucsc.edu/, Ensembl—www.ensembl.org/index.html) that provide useful tracks of genome-oriented data (Figure 21.2B).[15]

DIFFERENT PLATFORMS: PAST AND AT PRESENT

The aCGH technology is based in the deposit and immobilization of small amounts of single-strand DNA (ssDNA), known as probes, on a solid support, such as a glass slide, in an ordered way. The first-generation aCGH approach included probes derived from bacterial artificial chromosomes (BACs) or complementary DNA (cDNA)–cloned genomic segments. Nevertheless, the BAC-based arrays present some constraints, namely the difficulty in the calculation of CNV sizes and boundaries as well as the fact that its production is prone to errors.[16] The progress of aCGH platforms and the reference human genome sequencing allowed to design oligonucleotide probes with higher resolution. The size of the probes is very variable; for example, BAC-based arrays use large-insert genomic clones of 100–150 kb, while oligonucleotide-based arrays use probes of ~60-mer.[17] The resolution of aCGH varies according to the size of the genomic fragments and their density—the genomic distance between DNA probes.[5] For example, a microarray design with probes for whole genome that are 1 Mb apart will be unable to identify copy number variations of small sequence.[14] So, the resolution of the array increases directly with the decrease of the size of the nucleic acid probes and the reduction of the distance (more contiguous) between the targets on the microarray. The higher resolution allows more accurate identification of chromosomal breakpoints and gene alterations. Nowadays, the great majority of the commercial and academic laboratories are using oligonucleotide- or single-nucleotide polymorphism (SNP)–based arrays.[17] The advantages of oligonucleotide-based arrays consist of more flexibility in terms of probe selection, which enable higher probe density and customization of array content. Arrays designed specifically for copy-number analysis use longer oligonucleotide probes that offer a better signal-to-noise ratios in comparison to SNP-platforms, having however platforms with SNP-detecting arrays that include a mixture of SNP and copy-number probes to address this issue.[17] It is important to stress that single oligonucleotide probes do not provide accurate determination of copy number, there being required multiple consecutive probes with the same copy-number change to correctly establish a gain or a loss.[17] Usually, at least three consecutive probes with abnormal $\log_2$ ratios are needed to identify a copy number alteration. However, the number of consecutive probes required could vary between arrays of oligonucleotide probes and SNP probes. There are different platforms of oligonucleotide-based microarray, which means that depending on the selected platform, the number of oligonucleotide probes used for a whole-genome single-sample analysis can vary, and consequently the density of probes and the resolution are also different (Figure 21.3). The more recent array-based method on the market is SNP array genotyping, also recognized as SNP chips. The SNP platforms perform allelic discrimination to interrogate polymorphisms at specific positions in the genome, being its great advantage the capability of also identify long stretches of homozygosity, which might represent uniparental disomy (UPD) or consanguinity not suspected in the clinical history.[17] The SNP array genotype data can be downstream evaluated based on the B-allele frequency (BAF), which can reveal, when the BAF deviates from the trimodal expected distribution (AA = 0, AB = 0.5, and BB = 1), allelic imbalances, namely CNVs, regions of absence of heterozygosity (AOH) or uniparental disomy (UPD), all detected within the same assay.[16, 18] It is also important to stress that SNP-based aCGH can only detect quite long stretches of isodisomy, and this approach misses the UPD based on heterodisomy.[19]

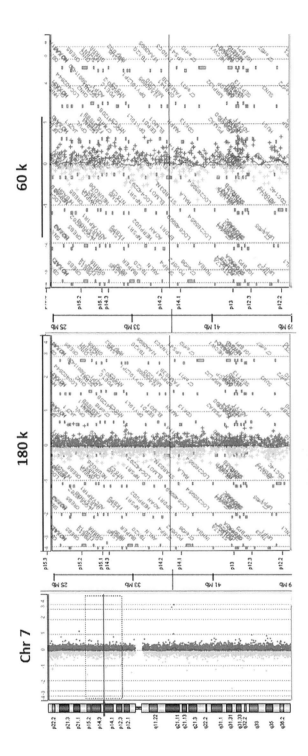

FIGURE 21.3 Representation of the number of oligonucleotide probes present at chromosome 7 in two different platforms, 180K and 60K of Agilent oligonucleotide based-microarray, using Agilent Genomic Workbench v6.5 software.

ADVANTAGES AND LIMITATIONS

In a single assay, aCGH allows to simultaneously identify genome-wide copy number alterations with high resolution. This technique presents several benefits and constraints. Among the advantages of aCGH technique, we can highlight:

- high-resolution whole genomic view for all euchromatic regions of genomes,
- ability to simultaneously detect aneuploidies, deletions, duplications, and/or amplifications of any locus represented on an array,
- detection of submicroscopic chromosomal abnormalities,
- only needing DNA (or RNA for gene expression microarrays), so not needing any living or dividing cells, which enable the use of both fresh and some archival DNA, including uncultured specimens and DNA obtained from postmortem samples, such as some formalin fixed paraffin embedded (FFPE) tissues,
- short turnaround time usually not exceeding few days and more cost-effective than multiple fluoresence in situ hybridization (FISH) assays,

Considering the limitations of aCGH, we can highlight:

- only detecting unbalanced chromosomal aberrations,
- alterations in polyploidy is not depictable (except using SNP array), since this technique relies on normalization of the intensity ratios,
- mosaic with low expression rate is usually, in regular routine settings, not detected. Cells with a chromosomal imbalance need to be present in at least 20% of the studied tissue, otherwise the aberrant cell clone may be missed,[20]
- small indels (insertion or deletion of bases) are not detected, in areas that are not so well covered by probes,
- without information available about centromeric-regions and heterochromatic regions, which corresponding to about 10% of the human genome,[21]
- being unable to define the direction of inserted sequences within the genomic context (i.e., direct vs inverted orientation).[16]

APPLICATIONS

The application of aCGH into the clinical practice has introduced a paradigm shift in the diagnostic workup and accelerated the identification of the molecular basis of several genetic diseases.[22] Initially, CGH was developed as a research tool for the investigation of CNVs in cancer. However, nowadays, aCGH technique is mainly used as a routine diagnostic tool, being the first-tier clinical diagnostic test for patients with intellectual disability, congenital anomalies and autistic spectrum disorder, and more recently has also been applied for the detection of genomic imbalances in prenatal genetic diagnosis. Since CNVs are common in the genome, the clinical significance and the genotype-phenotype correlation of aCGH results can be challenging, parental analysis and the use of several internet-based databases being pivotal. The application of aCGH technique in three main areas will be discussed:

MOLECULAR KARYOTYPING APPROACH IN HUMAN GENETIC DIAGNOSTICS

Array CGH has been applied to high-resolution analysis of constitutional abnormalities. The most important benefit of molecular karyotyping in human genetics is its capability to detect submicroscopic CNVs correlated to clinical phenotype in children and adults with normal GTG-banding based karyotype. This high-resolution technique allows the identification of novel disease-causing CNVs that can affect individual genes, exons, and regulatory regions. Therefore, besides the detection of CNVs for well-known genomic disorders, aCGH has also revealed new genomic disorders and disease-causing genes.

MOLECULAR KARYOTYPING APPROACH IN PRENATAL DIAGNOSTICS

Several studies have also showed the feasibility of performing aCGH for prenatal diagnosis using a wide range of sample types. In 2006, Rickman et al. analyzed 30 native prenatal samples by aCGH and demonstrated its potential for aneuploidy detection in DNA isolated from 1 ml of native amniotic fluid.[23] In this study, 29/30 samples were accurately diagnosed, except for a case of triploidy. Since aCGH technique allows the chromosomal aneuploidy and submicroscopic imbalances detection in whole genome with higher resolution, it is recommended as a first-tier approach in prenatal diagnosis for detection of CNVs in fetuses with structural anomalies observed by ultrasound.[24, 25] So, aCGH has been progressively used in prenatal diagnosis with high clinical impact in the clarification of the karyotype findings significance and also in the diagnosis of not identifiable imbalances using only karyotyping.[26]

MOLECULAR KARYOTYPING APPROACH IN CANCER RESEARCH

The aCGH technology was first developed as a research tool for the investigation of genomic alterations in cancer and has proven valuable in the identification of DNA copy number signatures and profiles for different types of tumors. This technique has been applied to a large number of cancer studies with reproducible results, some examples being head and neck cancer,[27, 28] basal cell carcinomas,[29] cholangiocarcinoma,[30] multiple myeloma,[31] bladder cancer,[32, 33] acute myeloid leukemia,[34, 35] among others.

CLINICAL CASES

Selected clinical cases will be presented as example of aCGH applications in postnatal human genetic diagnosis, in prenatal diagnosis and in cancer characterization. All cases were analyzed by aCGH technique using Agilent SurePrint G3 Human Genome microarray 180 K (Agilent Technologies, Santa Clara, CA, USA).

For all clinical post and prenatal cases, three questions are to be answered:

   i) How to interpret this result? How to classify this CNV? What is the most probable mechanism of origin of the alteration?
  ii) What is the origin of this CNV?
 iii) What can we offer to this couple? What is the future strategy for the family?

Clinical cases 1 and 2 are two examples of aCGH application in postnatal human genetic diagnosis (using DNA extracted from peripheral blood), with different outcomes.

CLINICAL CASE 1

A 11-year-old boy with renal insufficiency, retinopathy and ataxia was referred for aCGH that identified a 105 kb homozygous deletion at chromosome 2, arr[GRCh37] 2q13(110862477_110967912)x0 (Figure 21.4A).

   i) How to interpret this result? How to classify this CNV? What is the most probable mechanism of origin of the alteration?

There are several CNVs described in this region at DECIPHER database. Two genes, *MALL* and *NPHP1*, are mapped in this region. *NPHP1* gene is present in OMIM Morbid Map and is associated with nephronophthisis 1, juvenile, an autosomal recessive cystic kidney disease that leads to renal failure. Almost 15% of nephronophthisis patients present extrarenal symptoms affecting other organs, such as eyes, liver, bones and central nervous system.[36] Around two thirds of the patients with familial juvenile nephronophthisis have a homozygous deletion at the 2q13.[37]

    This homozygous deletion seems to explain the clinical phenotype of the boy. In order to establish the most probable mechanism of origin of the alteration, the aCGH study of the progenitors were recommended.

   ii) What is the origin of this CNV?

Both progenitors, not relatives, present 105kb heterozygous deletion, arr[GRCh37] 2q13(110862477_110967912)x1 (Figure 21.4B). The parents do not exhibit any clinical phenotype. Thus, deletion of one copy is not harmful for the parents. Although inherited, the 2q13 deletion is clearly pathogenic for the child, because it is in homozygosity.

   iii) What can we offer to this couple? What is the future strategy for the family?

This couple received genetic counselling, where the implication of the molecular findings was explained, and prenatal diagnosis was offered for future pregnancies.

    Later on, this couple had a new pregnancy and requested an invasive prenatal diagnosis. DNA from amniotic fluid was analyzed by aCGH technique, and it was verified that the fetus has inherited the deletion from one of the parents (Figure 21.4C). Thus, it is expected that the fetus did not present any phenotype like the carrier progenitors.

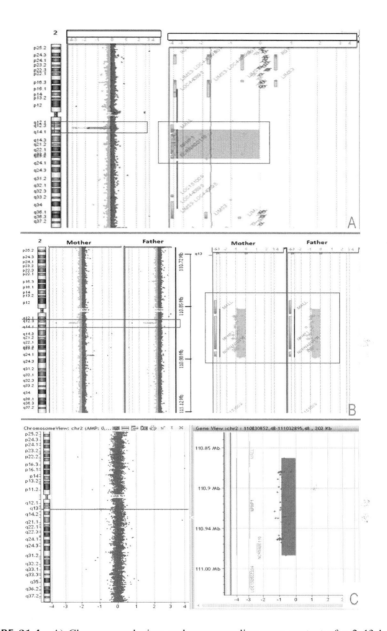

FIGURE 21.4 A) Chromosomal view and corresponding gene content of a 2q13 homo-zygous deletion identified in the 11-year-old boy. Results obtained using aCGH technique and analyzed using Agilent Genomic Workbench v6.5 software. The results are according to Human Genome build 19 and include imbalances with at least three consecutive probes with abnormal log2 ratios. Deletion is represented in a negative scale. B) Chromosomal view and respective gene content of the 2q13 deletion identified in the mother and the father of the 11-year-old boy. Results obtained using aCGH technique and analyzed using Agilent Genomic Workbench v6.5 software. The results are according to Human Genome build 19 and include imbalances with at least three consecutive probes with abnormal log2 ratios. Deletion is represented in a negative scale. C) Chromosomal view and respective gene content of the 2q13 deletion identified in a sample of amniotic fluid during pregnancy of this couple that have a heterozygous deletion in arr[GRCh37] 2q13(110862477_110967912)x1 and a son that inherited from both parents this deletion. Results obtained using aCGH technique and analyzed with Cytogenomics software. Deletion is represented in a negative scale.

Clinical Case 2

A newborn with several congenital anomalies, such as axial hypotonia, retrognathism, fetal growth restriction, high palate, and hypogonadism was referred for aCGH that identified two genomic anomalies (Figure 21.5): arr[GRCh37] 10p12.1(26182512_26815179)x3,15q11.2q13.1(22765628_28940098)x4. The mother has a brother with global psychomotor developmental delay and a son with hyperactivity and global psychomotor developmental delay.

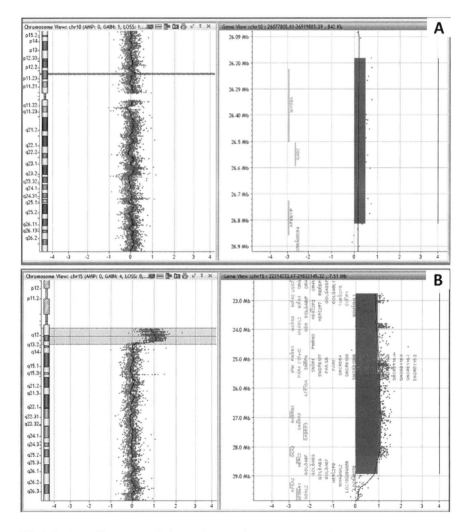

FIGURE 21.5 Chromosomal view and respective gene content of

A) 10p12.1 duplication and

B) 15q11.2q13.1 triplication.

Results obtained using aCGH technique and analyzed with CytoGenomics software. Gain of genetic material is represented in a positive scale.

i)   *How to interpret this result? How to classify this CNV? What is the most probable mechanism of origin of the alteration?*
   •   duplication of 632 Kb at 10p12.1 includes three genes described at OMIM database (OMIM ID: 606808-*MYO3A*, *138275-GAD2*, 609036-*APB-B1IP*). The *MYO3A* gene is described at OMIM Morbid Map and related to autosomal recessive deafness. CNVs in duplication are not reported for this region in healthy individuals at Database of Genomic Variants. In the DECIPHER database, there is one patient with cognitive deficit that has a similar duplication (ID: 300716) with unknown origin, which was classified as likely pathogenic. However, this reported patient also presented another genomic alteration, a 657Kb deletion at 1p36.23 of unknown origin, also classified as likely pathogenic. At the ClinGen database, there are bigger CNVs in duplication than those of our case, which were classified either as benign or pathogenic or even likely pathogenic.
   •   triplication of 6.1Mb at 15q11.2q13.1 includes several genes, 24 reported at OMIM database and 10 at OMIM Morbid Map database (OMIM ID: 608145-*NIPA1*, 603856-*MKRN3*, 605283-*MAGEL2*, 602117-*NDN*, 182279-*SNRPN*, 601623-*UBE3A*, 137192-*GABRB3*, 137142-*GABRA5*, 611409-*OCA2*, 605837-*HERC2*). In the literature, triplication of 15q11.2q13.1 is related with hypotonia, motor delay, intellectual disability and autism.[38] These disabilities fit with the clinical phenotype of our patient.
ii) *What is the origin of this CNV?*

Both parents were also analyzed by aCGH. It was possible to conclude that the 10p12.1 duplication of our patient was inherited from the mother (phenotypically normal) and was classified, considering the present knowledge as a CNV of unknown clinical significance. The tetrasomy of 15q11.2q13.1 is *de novo* and classified as pathogenic. arr[GRCh37]10p12.1(26182512_26815179)×3mat,15q11.2q13.1(22765628_28940098)×4 dn.

iii) *What can we offer to this couple? What is the future strategy for the family?*

This couple was referred to genetic counselling. Monitoring and testing future pregnancies were recommended.

CONCLUSION FOR CLINICAL CASES 1 AND 2

These two clinical cases show the challenges in the interpretation and classification of CNVs, where the detailed clinical characteristics of the patients, the analysis of specific databases and the progenitors are pivotal to perform the genotype-phenotype correlations. Sometimes the current knowledge is yet scarce to infer without doubt about the clinical significance of some CNVs, which is even more challenging in prenatal cases. It is important, whenever possible to revisit some of these VUS, specifically when a new pregnancy occurs in that family.

CLINICAL CASES 3 AND 4

Clinical cases 3 and 4 are two examples of aCGH application in prenatal diagnosis, with different outcomes.

CLINICAL CASE 3

A 29 weeks and 5 days' pregnancy, showed by ultrasound a fetus with cleft lip, cleft palate and defect in the ventricular septum. Amniotic fluid was collected and conventional cytogenetic and aCGH techniques requested. Both techniques identified two genomic anomalies in this female fetus: 46,XX,rec(4)dup(4p)inv(4)(p15.33q31.3) arr[GRCh37]4p16.3p1533(45882_11887210)×3,4q313q35.2(155125113_ 190469337)×1.

  i) *How to interpret this result? How to classify this CNV? What is the most probable mechanism of origin of the alteration?*

The fetus 1 presented (Figure 21.6A):

- a 11.8 Mb 4pter duplication that harbors several genes, including 71 reported at OMIM database and 21 reported at OMIM Morbid map database;
- a 35.3 Mb 4qter deletion that harbors several genes, including 73 reported at OMIM database and 19 reported at OMIM Morbid map database.

The imbalances were also observed in conventional cytogenetics analysis of fetus 1 (Figure 21.6C), allowing a correlation with the ultrasound abnormalities seen on the fetus. They are also suggestive of a recombinant that could have originated from a balanced structural rearrangement in one of the progenitors. Therefore, karyotyping study of both parents was requested.

  ii) *What is the origin of this CNV?*

Regarding the most probable biological mechanism, it can be concluded that the male progenitor is a carrier of a pericentric inversion of chromosome 4 (Figure 21.6C and 21.6D). Thus, from the conventional cytogenetic analysis of the fetus and the progenitors, it was demonstrated that the fetus inherited a recombinant chromosome 4 resulting from the crossing over in the pericentric inversion loop of the paternal inverted chromosome 4 during spermatogenesis.

  Fetus 1–46,XX,rec(4)dup(4p)inv(4)(p15.33q31.3)pat. arr[GRCh37]4p16.3p
    1533(45882_11887210)×3,4q313q35.2(155125113_190469337)×1
  Father: 46,XY,inv(4)(p15.33q31.3)

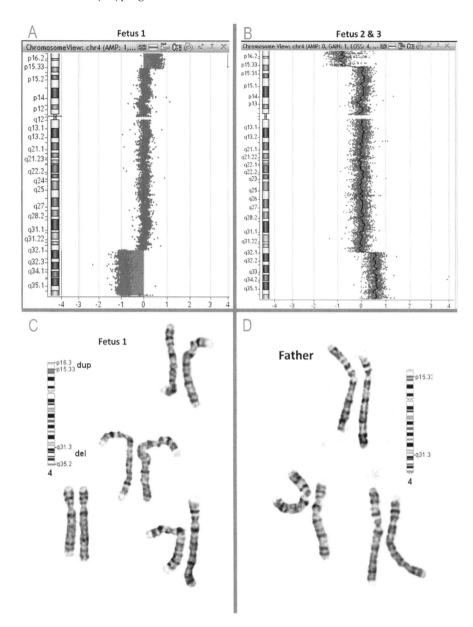

FIGURE 21.6 A) Chromosome view of the 4pter duplication and 4qter deletion identified in a sample of amniotic fluid (Fetus 1) and B) of the 4pter deletion and 4qter duplication identified in fetus 2 and 3. Results obtained using aCGH technique and analyzed with Cytogenomics software. Duplication is represented in a positive scale and deletion in a negative scale. C) Pair of chromosomes 4 obtained from representative G-banded metaphases of the amniotic cells from the fetus D) and of the peripheral blood from the father.

These two genomic imbalances identified in the fetus seem to be in the origin of the ultrasound malformations detected and consequently are correlated to the clinical phenotype.

iii) *What can we offer to this couple? What is the future strategy for the family?*

This couple was offered genetic counselling where it was well explained the consequence of this pericentric inversion in this phenotypically normal father for future pregnancies and the need of family studies (parents, brothers, sisters and other family members at risk). To all carriers, invasive prenatal diagnosis should be offered.

This couple had two more pregnancies. Trophoblastic cells from fetuses 2 and 3 were collected and aCGH revealed two genomic imbalances in mirror compared with fetus 1: 4p16.3p15.33 deletion and 4q31.3q35.2 duplication.

Fetuses 2 and 3 had similar alterations but different from fetus 1 (Figure 21.6B) that resulted from the meiotic recombination in the spermatogenesis of the male progenitor.

Fetus 2 and 3: 46,XX,rec(4)dup(4q)inv(4)(p15.33q31.3)pat—arr[GRCh37] 4p16.3p15.33(72447_1,887210)x1,4q31.3q35.2(155125113_190976439)x3.

Although in fetus 2 and 3 the genomic imbalances in chromosome 4 are in opposite way to that observed in fetus 1 (mirror, i.e., fetus 1 presented 4pter duplication and fetus 2 and 3 deletion and fetus 1 presented 4qter duplication and fetus 2 and 3 a deletion), the mechanism of origin is the same: inheritance of a recombinant chromosome 4 as a result of a paternal pericentric inversion in chromosome 4.

Deletions in the 4p16.3 region observed in the fetus are the main cause of Wolf-Hirschhorn syndrome, which presents as clinical manifestations growth delay, seizures, intellectual disability and distinct craniofacial features. In prenatal, 4pter deletion/4qter duplication were already reported with clinical manifestations like our patient.[39]

This case shows the clinical implications for the offspring of a balanced structural rearrangement in one of the progenitors and the need to offer to this couple prenatal diagnosis in future gestations and genetic counselling to explain the clinical implications of these genetic findings, the risk of recurrence and the need to offer family studies to other relatives.

CLINICAL CASE 4

Cells cultured from amniotic fluid of a fetus were analyzed by conventional cytogenetics, being the clinical indication for the invasive prenatal diagnosis a positive biochemical screening. The G-banded metaphases of the amniotic cells revealed a male fetus with two structural alterations: a paracentric inversion in chromosome 3 and a reciprocal translocation between chromosomes 2 and 5 (Figure 21.7A). 46,XY,t(2;5)(q33;q22)dn,inv(3)(p26.2p23).

i) *How to interpret this result? How to classify this CNV? What is the most probable mechanism of origin of the alteration?*

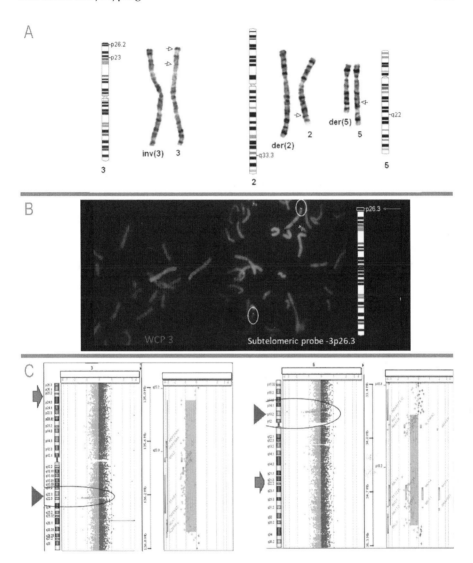

FIGURE 21.7 A) Representation of the pair of chromosomes 3, 2 and 5 obtained from representative G-banded metaphases of the amniotic cells from the fetus. The arrows show the structural imbalances observed: a paracentric inversion in chromosome 3 and a reciprocal translocation between the long arms of chromosomes 2 and 5. B) Results of FISH technique using a whole chromosomic painting (WCP) probe for chromosome 3 (WCP3, Cambio) labelled with SpectrumOrange and a subtelomeric probe for 3p26.3 (D3S4559, TotelVysion, Vysis) labelled with SpectrumGreen. C) Chromosome view and respective gene content of the 3q22.2q22.3 and 5p13.2 deletion identified in the DNA from amniotic cells of the fetus (red arrows). Results obtained using aCGH technique and analyzed using Agilent Genomic Workbench v6.5 software. The blue arrows represent the chromosome bands involved in the structural rearrangements identified by conventional cytogenetics and without genomic imbalances in the region observed by aCGH. The results are according to Human Genome build 19 and include imbalances with at least three consecutive probes with abnormal log2 ratios. Deletion is represented in a negative scale.

In order to ascertain the origin of these structural alterations, both parents were also analyzed by conventional cytogenetics, and it was possible to infer that these alterations of the fetus were *de novo*.

In order to further characterize the inversion, molecular cytogenetics (FISH technique) (Figure 21.7B) was also performed in the amniotic cells from the fetus using a whole chromosome painting (WCP) probe for chromosome 3 (WCP3, Cambio) labelled with SpectrumOrange that showed a continuum pattern in the chromosome 3; e.g. the inversion in the chromosome 3 did not change the pattern observed in this or any chromosome with this probe. A subtelomeric probe for 3p26.3 (D3S4559, TotelVysion, Vysis) labelled with SpectrumGreen was needed to confirm that this terminal region in the short arm of the chromosome 3 (3p26.3) was not involved in the inversion.

With these two techniques we can concluded that the fetus carries two structural chromosomal alterations that are *de novo* and seem to be balanced. Being *de novo*, the risk for the fetus of intellectual disability would be 6–10%.

In order to ascertain whether there were micro imbalances aCGH technique was performed and revealed losses of genetic material in two of the chromosomes involved (chromosomes 3 and 5) in the structural rearrangements observed by conventional cytogenetics arr[GRCh37]3q22.2q22.3(135247918_136673227)×1 ,5p13.2(33825339_35133937)×1 (Figure 21.7C):

- Deletion of 1.5 Mb at 3q22.2q22.3, where are mapped seven genes, being six described at OMIM database (OMIM ID: 604944-*PPP2R3A*, 614802-*MSL2*, 232050-*PCCB*, 604358-*STAG1*, 600508-*NCK1*, 605621-*IL20RB*) and 1 at OMIM Morbid Map (OMIM ID: 232050-*PCCB*).
- Deletion of 1.3 Mb at 5p13.2, where are mapped thirteen genes, being nine described at OMIM database (OMIM ID: 606184-*ADAMTS12*, 609445-*RXFP3*, 606202-*SLC45A2*, 604489-*AMACR*, 612045-*C1ATNF3*, 606586-*RAI14*, 603153-*RAD1*, 612471-*AGXT2*, 176716-*PRLR*) and 2 at OMIM Morbid Map (OMIM ID: 606202-*SLC45A2*, 604489-*AMACR*).

ii) *What is the origin of this CNV?*

DNA from the peripheral blood of the progenitors were also analyzed by aCGH and did not reveal these imbalances (data not shown), which were therefore classified as *de novo* in the fetus.

iii) *What can we offer to this couple? What is the future strategy for the family?*

This clinical case is a clear example of the importance of complementary techniques like conventional cytogenetics and aCGH to achieve a correct risk of recurrence for the offspring as well as to identify the mechanisms of origin of the chromosome imbalances and its accurate breakpoints. The conventional cytogenetics in this case reveals a double structural chromosome rearrangement *de novo* and apparently balanced with the inherent risks for the fetus. With complementarity of the aCGH technique, we can conclude that besides these structural rearrangements detected by conventional cytogenetics, two additional imbalances in two of these chromosomes, but at different locations were present, which increase the risk of intellectual disability and developmental disorders in the carrier/fetus. All these results were explained to the couple in order to decide about the pregnancy in course. Monitoring and testing future pregnancies were recommended.

CLINICAL CASES 5 AND 6

Clinical cases 5 and 6 are two examples of aCGH application in cancer genetic characterization in order to detect the genetic changes associated with cancer development and progression. In cancer research, this technology due to the study of whole genome allows to identify new chromosomal regions and genes related to different cancer types as well as to perform the correlation of the identified genomic profiles with tumor stage, risk factors, disease evolution, response to treatment and patients' survival.

Clinical Case 5

DNA from tumor tissue of a 47 years-old woman diagnosed with floor of the mouth tumor, stage I, human papilloma virus (HPV) negative and with tobacco consumption, was analyzed by aCGH. This whole genome approach revealed imbalances in several chromosomes, such as, chromosomes 1, 2, 3, 7 and 9, where there are mapped several genes with key roles for the carcinogenesis process (Figure 21.8). We observed that tumor cells of this patient displayed numerous rearrangements on several chromosomes simultaneously, while showing a small number of chromosomes without or with few aberrations. The interpretation of the large-scale genomic profiling of cancer samples has been challenging due to the intra- and inter-tumor heterogeneity displayed by solid tumors. The aCGH technique has helped in the identification and translation to clinical practice, of specific diagnostic and prognostic cancer biomarkers, which are of utmost importance in order to improve patient's treatment selection, prophylactic screening strategies and ultimately the patients quality of life.

Clinical Case 6

A primary cell culture established from a surgical tongue tumor resected sample was characterized by aCGH and banding cytogenetics. This primary culture presented a huge karyotypic heterogeneity with different numerical and structural chromosomal abnormalities identified in the analyzed metaphases (Figure 21.9A). Results from aCGH helped to establish the specific chromosomal bands involved in copy number gains and losses, assisting in the delimitation of the size of the imbalances and the establishment of specific breakpoints as well as to identify additional small copy number gains and losses not detected by conventional cytogenetics due to their reduced size. With the integration of these two techniques, we can observe aneuploidies and also to infer and characterize some structural rearrangements that are in the origin of the CNVs and aneuploidy (Figure 21.9 B–C).

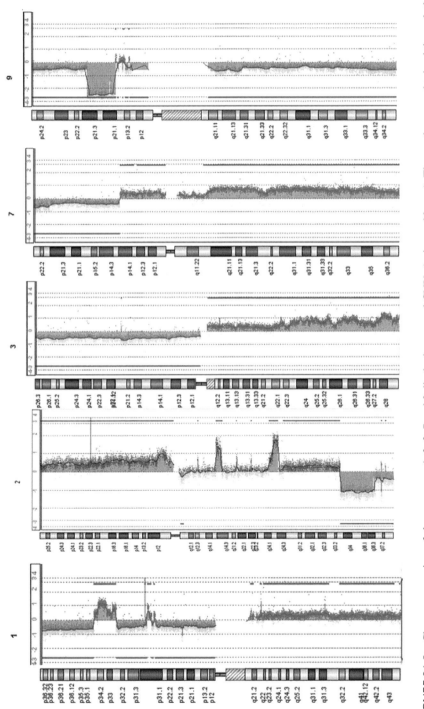

FIGURE 21.8 Chromosome view of chromosomes 1, 2, 3, 7 and 9 displaying several CNVs (gains and losses). The results were obtained through the analysis of DNA from a floor of the mouth tumor using aCGH technique and Agilent Genomic Workbench v6.5 software. The results are according to Human Genome build 19 (GRCh37) and include imbalances with at least three consecutive probes with abnormal log2 ratios. Deletion is represented in a negative scale and amplification in a positive scale.

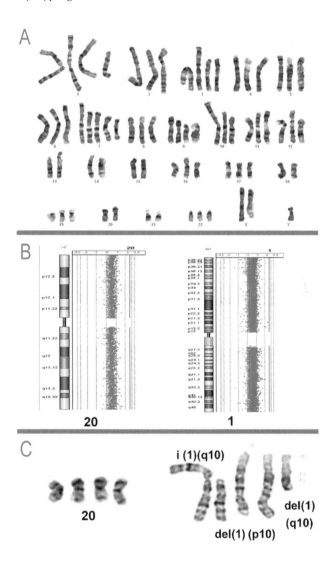

FIGURE 21.9 A) Karyogram of a representative G-banded metaphase for a primary cell culture established from a tongue tumor. B) Chromosome view of chromosomes 20 and 1 displaying whole chromosome gain detected by aCGH technique and analyzed through Agilent Genomic Workbench v6.5 software. Gains are represented in a positive scale. C) Representation of the chromosomes 20 and 1 obtained from a representative G-banded metaphase of the primary cell culture, showing different structural rearrangements that culminated in copy number gain of these chromosomes.

THE IMPACT OF MOLECULAR KARYOTYPING IN CLINICAL PRACTICE

Molecular karyotyping is a strong tool for the detection of submicroscopic chromosomal anomalies in clinical practice. The implementation of aCGH technique in postnatal and prenatal studies seems to increase the detection rate of chromosomal imbalances in comparison of conventional karyotyping. Several large-scale studies reveal that aCGH has a 10%–20% detection rate of chromosomal abnormalities in children with mental retardation/developmental delay with or without congenital anomalies, only 3%–5% of these genomic alterations being detectable by other techniques, such as conventional cytogenetics.[14] In 2013, Wapner and colleagues showed in a large multicenter NICHD sponsored study the clinical utility of aCGH in prenatal diagnosis.[40] In this study, aCGH presents, in pregnancies with fetal structural anomalies and a normal karyotype, an incremental diagnostic yield of about 6% in comparison to karyotype detection rate.[40] In miscarriages, the use of aCGH is limited to few studies; nevertheless, there seems to be an increased detection rate of chromosomal abnormalities in about 10% of the cases with normal karyotype and nearly 50% in samples with culture failure when karyotyping was impossible.[41] aCGH technique has some important advantages for diagnosing human diseases, namely the higher resolution and the shorter turnaround time due to not need cell culture, being a useful tool to identify new disease-causing genes, which has allowed to expand the knowledge of the etiology of several genetic disorders. Additionally, SNP-based microarrays are also able to diagnosis uniparental disomy, improving the diagnostic capabilities in the pre- and postnatal settings.

CONCLUSIONS

Molecular karyotyping is a powerful tool to identify with high resolution chromosomal and gene alterations for diagnostic and prognostic purposes. It is important to stress that aCGH cannot detect all genetic anomalies and that a normal aCGH result does not reduce the risk for other genetic conditions or even, in prenatal context, for birth defects, related to balanced rearrangements or single gene mutations, which are the detection limitations of this technique. Therefore, the results of molecular karyotyping are frequently complementary to those obtained with conventional karyotyping and Next Generation Sequencing (NGS) mutation studies.

The implementation of aCGH technique in the post and prenatal field has been a significant impact on the diagnosis of human constitutional diseases, since allowed to detect genomic copy number alterations undetectable by other cytogenetic and molecular technologies. This technique allows not only identifying numerous well-characterized genetic syndromes but also to find new genomic disorders and disease-causing genes, improving the knowledge of human genetic disorders and consequently the capability to diagnosis and treat the patients. The implementation of this technique in the clinical practice can help therapeutic selection and novel molecular targets identification, improving precision medicine in a wide range of human genetic disorders and cancer.

REFERENCES

1. Solinas-Toldo, S.; Lampel, S.; Stilgenbauer, S.; Nickolenko, J.; Benner, A.; Dohner, H.; Cremer, T.; Lichter, P.; Matrix-based comparative genomic hybridization: Biochips to screen for genomic imbalances. Genes Chromosomes *Cancer.* **1997**, 20, 399–407.

2. Snijders, A.M.; Nowak, N.; Segraves, R.; Blackwood, S.; Brown, N.; Conroy, J.; Hamilton, G.; Hindle, A.K.; Huey, B.; Kimura, K.; Law, S.; Myambo, K.; Palmer, J.; Ylstra, B.; Yue, J.P.; Gray, J.W.; Jain, A.N.; Pinkel, D.; Albertson, D.G. Assembly of microarrays for genome-wide measurement of DNA copy number. *Nat. Genet.* **2001**, 29, 263–264.

3. Pinkel, D.; Segraves, R.; Sudar, D.; Clark, S.; Poole, I.; Kowbel, D.; Collins, C.; Kuo, W.L.; Chen, C.; Zhai, Y.; Dairkee, S.H.; Ljung, B.M.; Gray, J.W.; Albertson, D.G. High resolution analysis of DNA copy number variation using comparative genomic hybridization to microarrays. *Nat. Genet.* **1998**, 20, 207–211.

4. Buckley, P.G.; Mantripragada, K.K.; Benetkiewicz, M.; Tapia-Paez, I.; Diaz De Stahl, T.; Rosenquist, M.; Ali, H.; Jarbo, C.; De Bustos, C.; Hirvela, C.; Sinder Wilen, B.; Fransson, I.; Thyr, C.; Johnsson, B.I.; Bruder, C.E.; Menzel, U.; Hergersberg, M.; Mandahl, N.; Blennow, E.; Wedell, A.; Beare, D.M.; Collins, J.E.; Dunham, I.; Albertson, D.; Pinkel, D.; Bastian, B.C.; Faruqi, A.F.; Lasken, R.S.; Ichimura, K.; Collins, V.P.; Dumanski, J.P. A full-coverage, high-resolution human chromosome 22 genomic microarray for clinical and research applications. *Hum. Mol. Genet.* **2002**, 11, 3221–3229.

5. Vermeesch, J.R.; Melotte, C.; Froyen, G.; Van Vooren, S.; Dutta, B.; Maas, N.; Vermeulen, S.; Menten, B.; Speleman, F.; De Moor, B.; Van Hummelen, P.; Marynen, P.; Fryns, J.P.; Devriendt, K. Molecular karyotyping: Array CGH quality criteria for constitutional genetic diagnosis. *J. Histochem. Cytochem.* **2005**, 53, 413–422.

6. Lee, C.; Hyland, C.; Lee, A.S.; Hislop, S.; Ihm, C. Copy number variation and human health. In: *Genomic and Personalized Medicineed*; Willard, P.D.H.F.; Ginsburg, G.S., Eds. Academic Press, New York, **2009**, pp. 108–119.

7. Kallioniemi, A.; Kallioniemi, O.P.; Sudar, D.; Rutovitz, D.; Gray, J.W.; Waldman, F.; Pinkel, D. Comparative genomic hybridization for molecular cytogenetic analysis of solid tumors. *Science.* **1992**, 258, 818–821.

8. du Manoir, S.; Speicher, M.R.; Joos, S.; Schröck, E.; Popp, S.; Döhner, H.; Kovacs, G.; Robert-Nicoud, M.; Lichter, P.; Cremer, T. Detection of complete and partial chromosome gains and losses by comparative genomic in situ hybridization. *Hum. Genet.* **1993**, 90, 590–610.

9. Lichter, P.; Joos, S.; Bentz, M.; Lampel, S. Comparative genomic hybridization: Uses and limitations. *Semin Hematol.* **2000**, 37, 348–357.

10. Gebhart, E.; Liehr, T. Patterns of genomic imbalances in human solid tumors (Review). *Int J Oncol* **2000**, 16, 383–399.

11. Spangenberg, V.; Kolomiets, O.; Stepanyan, I.; Galoyan, E.; de Bello Cioffi, M.; Martynova, E.; Martirosyan, I.; Grishaeva, T.; Danielyan, F.; Al-Rikabi, A.; Liehr, T.; Arakelyan, M. Evolution of the parthenogenetic rock lizard hybrid karyotype: Robertsonian translocation between two maternal chromosomes in Darevskia rostombekowi. *Chromosoma.* **2020**, 129, 275–283.

12. Deon, G.A.; Glugoski, L.; Vicari, M.R.; Nogaroto, V.; Sassi, F.M.C.; Cioffi, M.B.; Liehr, T.; Bertollo, L.A.C.; Moreira-Filho, O. Highly rearranged karyotypes and multiple sex chromosome systems in armored catfishes from the genus Harttia (Teleostei.; Siluriformes). *Genes.* **2020**, 11, 1366.

13. Szczaluba, K.; Demkow, U. Array comparative genomic hybridization and genomic sequencing in the diagnostics of the causes of congenital anomalies. *J. Appl. Genet.* **2017**, 58, 185–198.

14. Theisen, A. Microarray-based comparative genomic hybridization (aCGH). *Nature Educ.* **2008**, 1, 45.

15. de Leeuw, N.; Dijkhuizen, T.; Hehir-Kwa, J.Y.; Carter, N.P.; Feuk, L.; Firth, H.V.; Kuhn, R.M.; Ledbetter, D.H.; Martin, C.L.; van Ravenswaaij-Arts, C.M.A.; Scherer, S.W.; Shams, S.; Van Vooren, S.; Sijmons, R.; Swertz, M.; Hastings, R. Diagnostic interpretation of array data using public databases and internet sources. *Hum. Mutat.* **2012**, 33, 930–940.

16. Zepeda Mendoza, C.J.; Gonzaga-Jauregui, C. Genomic disorders in the genomics era. In: *Genomics of Rare Diseases*; Gonzaga-Jauregui, C.; Lupski, J.R., Eds. Academic Press, New York, **2021**, pp. 35–59.

17. Miller, D.T.; Adam, M.P.; Aradhya, S.; Biesecker, L.G.; Brothman, A.R.; Carter, N.P.; Church, D.M.; Crolla, J.A.; Eichler, E.E.; Epstein, C.J.; Faucett, W.A.; Feuk, L.; Friedman, J.M.; Hamosh, A.; Jackson, L.; Kaminsky, E.B.; Kok, K.; Krantz, I.D.; Kuhn, R.M.; Lee, C.; Ostell, J.M.; Rosenberg, C.; Scherer, S.W.; Spinner, N.B.; Stavropoulos, D.J.; Tepperberg, J.H.; Thorland, E.C.; Vermeesch, J.R.; Waggoner, D.J.; Watson, M.S.; Martin, C.L.; Ledbetter, D.H. Consensus statement: Chromosomal microarray is a first-tier clinical diagnostic test for individuals with developmental disabilities or congenital anomalies. *Am. J. Hum. Genet.* **2010**, 86, 749–764.

18. Brady, P.D.; Vermeesch, J.R. Genomic microarrays: A technology overview. *Prenat. Diagn.* **2012**, 32, 336–343.

19. Liehr, T. Cases with uniparental disomy; **2022**. http://cs-tl.de/DB/CA/UPD/0-Start.html.

20. Weise, A.; Liehr, T. Molecular karyotyping. In: *Cytogenomics*; Liehr, T., Ed. Academic Press, New York, **2021**, pp. 73–85.

21. Liehr, T. Repetitive elements, heteromorphisms, and copy number variants. In: *Cytogenomics*; Liehr, T., Ed. Academic Press, New York, **2021**, pp. 373–388.

22. Cheung, S.W.; Bi, W. Novel applications of array comparative genomic hybridization in molecular diagnostics. *Exp. Rev. Mol. Diagn.* **2018**, 18, 531–542.

23. Rickman, L.; Fiegler, H.; Shaw-Smith, C.; Nash, R.; Cirigliano, V.; Voglino, G.; Ng, B.L.; Scott, C.; Whittaker, J.; Adinolfi, M.; Carter, N.P.; Bobrow, M. Prenatal detection of unbalanced chromosomal rearrangements by array CGH. *J. Med. Genet.* **2006**, 43, 353–361.

24. Genetics ACoOaGCo. Committee opinion No. 581: The use of chromosomal microarray analysis in prenatal diagnosis. *Obstet. Gynecol.* **2013**, 122, 1374–1377.

25. Society for Maternal-Fetal Medicine Electronic address pso; Dugoff, L.; Norton, M.E.; Kuller, J.A. The use of chromosomal microarray for prenatal diagnosis. *Am. J. Obstet. Gynecol.* **2016**, 2015, B2–B9.

26. Cheung, S.W.; Pursley, A.N. Comparative genomic hybridization in the study of human disease; **2011**, Wiley. https://onlinelibrary.wiley.com/doi/epdf/10.1002/9780470015902.a0005955.pub2.

27. Ribeiro, I.P.; Caramelo, F.; Esteves, L.; Menoita, J.; Marques, F.; Barroso, L.; Migueis, J.; Melo, J.B.; Carreira, I.M. Genomic predictive model for recurrence and metastasis development in head and neck squamous cell carcinoma patients. *Scient. Rep.* **2017**, 7, 13897.

28. Ribeiro, I.P.; Esteves, L.; Santos, A.; Barroso, L.; Marques, F.; Caramelo, F.; Melo, J.B.; Carreira, I.M. A seven-gene signature to predict the prognosis of oral squamous cell carcinoma. *Oncogene.* **2021**, 40, 3859–3869.

29. Cardoso, J.C.; Ribeiro, I.P.; Caramelo, F.; Tellechea, O.; Barbosa de Melo, J.; Marques Carreira, I. Basal cell carcinomas of the scalp after radiotherapy for tinea capitis in childhood: A genetic and epigenetic study with comparison with basal cell carcinomas evolving in chronically sun-exposed areas. *Exp. Dermatol.* **2021**, 30, 1126–1134.

30. Tavares, I.; Martins, R.; Ribeiro, I.P.; Esteves, L.; Caramelo, F.; Abrantes, A.M.; Neves, R.; Caetano-Oliveira, R.; Botelho, M.F.; Barbosa de Melo, J.; Diogo, D.; Tralhao, J.G.; Carreira, I.M. Development of a genomic predictive model for cholangiocarcinoma using copy number alteration data. *J. Clin. Patho.* **2022**, 75, 274–278.

31. Couto Oliveira, A.; Ribeiro, I.P.; Pires, L.M.; Goncalves, A.C.; Paiva, A.; Geraldes, C.; Roque, A.; Sarmento-Ribeiro, A.B.; Barbosa de Melo, J.; Carreira, I.M. Genomic characterisation of multiple myeloma: Study of a Portuguese cohort. *J. Clin Pathol,* **2022**, 75, 422–425.

32. Pinto-Leite, R.; Carreira, I.; Melo, J.; Ferreira, S.I.; Ribeiro, I.; Ferreira, J.; Filipe, M.; Bernardo, C.; Arantes-Rodrigues, R.; Oliveira, P.; Santos, L. Genomic characterization of three urinary bladder cancer cell lines: Understanding genomic types of urinary bladder cancer. *Tumour Biol.* **2014**, 35, 4599–4617.

33. Spasova, V.; Mladenov, B.; Rangelov, S.; Hammoudeh, Z.; Nesheva, D.; Serbezov, D.; Staneva, R.; Hadjidekova, S.; Ganev, M.; Balabanski, L.; Vazharova, R.; Slavov, C.; Toncheva, D.; Antonova, O. Clinical impact of copy number variation changes in bladder cancer samples. *Exp. Ther. Med.* **2021**, 22, 901.

34. Capela de Matos, R.R.; Ney Garcia, D.R.; Othman, M.A.K.; Moura Ferreira, G.; Melo, J.B.; Carreira, I.M.; Meyer, C.; Marschalek, R.; Costa, E.S.; Land, M.G.P.; Liehr, T.; Ribeiro RC.; Silva, M.L.M. A new complex karyotype involving a KMT2A-r variant three-way translocation in a rare clinical presentation of a pediatric patient with acute myeloid leukemia. Cytogenet. *Genome Res.* **2019**, 157, 213–219.

35. Ronaghy, A.; Yang, R.K.; Khoury, J.D.; Kanagal-Shamanna, R. Clinical applications of chromosomal microarray testing in myeloid malignancies. *Curr. Hematol. Malign. Rep.* **2020**, 15, 194–202.

36. Wolf, M.T.F. Nephronophthisis and related syndromes. *Curr. Opin. Pediatr.* **2015**, 27, 201–211.

37. Ala-Mello, S.; Koskimies, O.; Rapola, J.; Kääriäinen, H. Nephronophthisis in Finland: Epidemiology and comparison of genetically classified subgroups. *Europ. J. Hum. Genet.* **1999**, 7, 205–211.

38. Lu, Y.; Liang, Y.; Ning, S.; Deng, G.; Xie, Y.; Song, J.; Zuo, N.; Feng, C.; Qin, Y. Rare partial trisomy and tetrasomy of 15q11-q13 associated with developmental delay and autism spectrum disorder. *Mol. Cytogenet.* **2020**, 13, 21.

39. Bi, W.; Cheung, S.-W.; Breman, A.M.; Bacino, C.A. 4p16.3 microdeletions and microduplications detected by chromosomal microarray analysis: New insights into mechanisms and critical regions. *Am. J. Med. Genet. A.* **2016**, 170, 2540–2550.

40. Wapner, R.J.; Martin, C.L.; Levy, B.; Ballif, B.C.; Eng, C.M.; Zachary, J.M.; Savage, M.; Platt, L.D.; Saltzman, D.; Grobman, W.A.; Klugman, S.; Scholl, T.; Simpson, J.L.; McCall, K.; Aggarwal, V.S.; Bunke, B.; Nahum, O.; Patel, A.; Lamb, A.N.; Thom, E.A.; Beaudet, A.L.; Ledbetter, D.H.; Shaffer, L.G.; Jackson, L. Chromosomal microarray versus karyotyping for prenatal diagnosis. *N. Engl. J. Med.* **2012**, 367, 2175–2184.

41. Gao, J.; Liu, C.; Yao, F.; Hao, N.; Zhou, J.; Zhou, Q.; Zhang, L.; Liu, X.; Bian, X.; Liu, J. Array-based comparative genomic hybridization is more informative than conventional karyotyping and fluorescence in situ hybridization in the analysis of first-trimester spontaneous abortion. *Mol. Cytogenet.* **2012**, 5, 33.

22 FISH—Mitochondrial DNA

Tigran Harutyunyan

CONTENTS

INTRODUCTION

Mitochondria are the powerhouses of cells, generating over 90% of energy in the form of adenosine triphosphate (ATP) required for the normal cell functioning.[1] Moreover, mitochondria are involved in cell cycle regulation and intrinsic apoptosis. In addition, each mitochondrion contains up to several thousand copies of its own DNA molecule. In human cells, the mitogenome consists of a circular molecule called mitochondrial chromosome, or better mitochondrial DNA (mtDNA), that harbors 16,569 bp and 37 genes, including 13 proteins (for the OXPHOS system), 22 tRNAs, and 2 rRNAs. In addition, at least eight mitochondrial-derived peptides have been identified which are encoded by 12S and 16S rRNA genes.[2]

Recent data indicates horizontal transfer of mitochondria as a mechanism of intercellular signaling in normal and pathological conditions.[3] In addition, numtogenesis,

DOI: 10.1201/9781003223658-22

a de novo insertion of mtDNA in the nuclear genome, was observed at high frequency in cancer cell, which can also result in genetic diseases.[4] It was shown that environmental mutagens can increase the frequency of de novo mtDNA insertions in human chromosomes after exposure to DNA double-strand break (DSB) inducing agents.[5] Also, Lutz-Bonengel et al.[6] and Wei et al.[7] demonstrated the presence of amplified NUMTs (nuclear mtDNA sequences) or mega-NUMTs in nuclear genome of healthy individuals, resembling paternal mtDNA. In addition, implication of mtDNA in the innate immunity and COVID-19 pathogenesis was recently shown.[8] Therefore, availability of tools for identification of mitochondrial exchange and integration of mtDNA in the nuclear genome has clinical, forensic, phylogenetic and research importance.

Sequencing of eukaryotic genomes revealed the presence of NUMTs in nuclear DNA (nDNA) which occurred by insertion of corresponding sequences during evolution, with variant frequencies within different species.[9] In the human genome, over 1,000 NUMTs have been identified, with the sizes ranging from 39 to 16,106 bp and encompassing ~400 kb of DNA stretches. Interestingly, the number of these insertions in chromosomes positively correlate with the relative chromosome length indicating random insertion of mtDNA. However, there is also evidence that insertion loci of NUMTs in (human) genome are non-random, since enrichment with repetitive elements in 2 kb of both flanking regions were identified.[9] In recent years, it was also shown that NUMTs might cause bias in the interpretation of nDNA sequencing data due to high homology with mtDNA sequences. In particular, biparental inheritance of mtDNA was suggested after sequencing and analysis for heteroplasmy in the offspring of three heteroplasmic fathers, which indicated for the presence of maternally and paternally inherited mtDNA haplotypes.[10] However, the authors did not completely rule out NUMT contamination of the sequencing data. Recently, the results of Lutz-Bonengel et al.[6] have challenged this hypothesis by demonstrating the presence of heteroplasmy in the offspring carrying paternal U4c1 mitogenome haplogroup along with maternal V haplogroup. Sequencing studies revealed the presence of maternal V mitotype only in cells devoid of nucleus (e.g. thrombocytes and hair shafts), while cells depleted from mtDNA had only U4c1 mitotype. Fluorescence in situ hybridization (FISH) analysis confirmed the insertion of mtDNA in 14q31 chromosome loci, proving the presence of mega-NUMT in nDNA.[6] Thus, NUMTs should be considered in phylogenetic, medical genetic, and forensic studies to avoid false-positive results.

While the biological role of inherited NUMTs is not determined, the occurrence of de novo NUMTs and mtDNA insertions in the nDNA, defined as numtogenesis, can be observed in pathological conditions, too.[4] The analysis of whole-genome sequencing data from over 2,600 cancers demonstrated elevation in frequencies of somatically acquired mtDNA insertions being highest in HER2+ breast cancer cells and squamous cell lung cancers.[11] Elevation of de novo NUMTs was also observed in colorectal adenocarcinoma genomes, which contained up to 4.2-fold higher NUMTs compared to blood-derived normal genomes.[12]

Furthermore, human diseases caused by de novo insertions of mtDNA in nDNA are described, although in a limited number of cases. Such insertions were identified in single patients with severe plasma factor VII deficiency,[13]

Usher syndrome type 1C,[14] Pallister-Hall syndrome,[15] mucolipidosis-IV,[16] lissencephaly,[17] and X-linked hyper-IgM syndrome combined with immunodeficiency syndrome.[18] Thus, identification of mtDNA insertions in nDNA is of diagnostic importance.

Recently detected was horizontal exchange of mitochondria between different cells defined as "momiome" (**mo**bile functions of **mi**tochondria and the mitochondrial gen**ome**).[19] Thus, randomly exchanged mitochondria can be defined as permutational mitochondria. Permutational mitochondria mainly occur via tunneling nanotubes (TNTs), extracellular vesicles, and Cx43 gap junctions which can be observed between normal and injured/transformed cells.[3] In the co-culture system of human vascular smooth muscle cells (VSMCs) and mesenchymal stem cells (MSCs), the shuttling of mitochondria from VSMCs significantly elevated the proliferation of MSCs,[20] while the enhanced mitochondrial transfer from MSCs to epithelial cells can rescue epithelial injury.[21]

Interestingly, the presence of a particular cell type in the tumor microenvironment is important for the horizontal transfer of mitochondria. In the co-culture of bone marrow stromal cells, endothelial cells, and MCF7 cells, mitochondria transfer was observed only from endothelial cells to MCF7 cells.[22] It was shown that horizontal mitochondrial transfer in the tumor microenvironment can restore the OXPHOS and support the survival and chemoresistance in cancer cells.[23] Thus, inhibition of permutational mitochondria occurrence in cancer cells may be a promising future approach for anticancer therapy.

The potential of the environmental mutagens to induce translocation of the mtDNA in the nuclear genome, defined as numtomutagenesis, has been demonstrated. First studies in yeast showed that numtomutagenesis can occur in *Saccharomyces cerevisiae* after induction of DSBs using I-SceI nuclease.[24] Ionizing radiation can induce numtomutagenesis in chicken (*Gallus gallus*) genome after X-ray-irradiation of eggs.[25] In our own studies, the numtomutagenic activity of DSB-inducing anticancer drug doxorubicin was analyzed in metaphase chromosomes of human peripheral blood lymphocytes. FISH analysis allowed to identify the number of mtDNA insertions in each chromosome, which positively correlated with the frequency of doxorubicin-induced micronuclei (marker of chromosome damage).[5] Sequencing of the mtDNA insertion junctions in cancer cells demonstrated microhomologies suggesting the involvement of the non-homologous end joining repair mechanism in the insertion of the mtDNA fragments in nDNA.[26] Thus, DSB-inducing environmental factors can also have numtomutagenic activity, which should be considered as a new marker in mutagenicity testing.

Mutations of mtDNA can result in mitochondrial dysfunction and mtDNA depletion syndromes with a broad spectrum of clinically heterogeneous symptoms due to abnormal energy production by OXPHOS.[27] In addition, mtDNA copy number variations were detected in cancer cells and in patients infected with Human Immunodeficiency Virus 1 (HIV-1).[28, 29]

Circulating cell-free mtDNA which occurs due to intracellular mitochondrial damage, is an early indicator of severe illness and mortality from COVID-19.[30] Thus, rapid detection of mtDNA quantitative changes, as well as integration of mtDNA in nDNA and permutational mitochondria, has clinical benefits.

FISH is a gold standard of the molecular cytogenetic diagnostics enabling quantitative and qualitative analysis with application of specific probes for the target sequences of variable size. Janes et al.[31] analyzed depletion of mtDNA in fibroblasts treated with 2',3'-dideoxycytidine, a standard medication for the treatment of patients with HIV-1, using FISH; and Lucic et al.[32] demonstrated the applicability of FISH for the detection of HIV-1 in cells infected with the virus in vitro.

In tumor cells of breast cancer patient the insertion of mtDNA in chromosome 10 was identified using fiber FISH technique.[26] Caro et al.[33] demonstrated the accumulation of mtDNA fragments in nDNA with age in rat tissues. FISH analysis confirmed the presence of mtDNA in centromeric regions in 20 out of 40 chromosomes, while Koo et al.[34] developed fiber FISH for the detection of NUMTs in MCF-12A/MDA-MB-231 normal epithelial cells.

In this chapter, FISH analysis of mtDNA insertions in metaphase chromosomes of whole blood lymphocytes is described. However, this protocol can be also useful for the analysis of permutational mitochondria and alterations of mtDNA copy numbers under pathological conditions.

SAMPLES/TISSUES/MATERIALS

FISH enables the detection of mtDNA on a single-cell level in any cells. Slides for FISH analysis of numtomutagenesis can be prepared from both heparinized whole blood or blood collected in EDTA containing vacutainers. Nevertheless, if metaphase chromosomes should be used in the analysis the heparinized blood is more suitable in contrast to EDTA.

HUMAN WHOLE BLOOD CULTURE

- RPMI-1640 medium with L-glutamine (Cat. No. 11875093, Gibco™)
- Fetal bovine serum (Cat. No. F2442, Sigma-Aldrich)
- Penicillin-Streptomycin solution (Cat. No. P4333, Sigma-Aldrich)
- Phytohemagglutinin (Cat. No. M5030-BC, MERCK)

SLIDES WITH METAPHASE CHROMOSOMES

- Colcemid solution (Cat. No. 10295892001, Sigma-Aldrich)
- Potassium chloride (Cat. No. P9541, Sigma-Aldrich)
- Hypotonic solution of KCl: 0.075 M freshly prepared KCl solution used at 37°C
- Methanol 100 % (Cat. No. 34860, Sigma-Aldrich)
- Glacial acetic acid (Cat. No. 1000631011, Sigma-Aldrich)
- Carnoy's fixative: methanol/glacial acetic acid freshly prepared mix in 3:1 (v/v) ratio and used at −20°C

FISH PROBE FOR MTDNA

- mtDNA
- Primers for mtDNA amplification and labeling

- EDTA 0.5 M (Cat. No. 324506, Sigma-Aldrich)
- Ethanol 100 % (Cat. No. 459836, Sigma-Aldrich)
- tRNA (Cat. No. 10109541001, Sigma-Aldrich)
- Sodium acetate buffer solution 3 M, pH 5.2 (Cat. No. S7899, Sigma-Aldrich)
- Dextran sulfate buffer: 2 g + 10 ml 50 % deionized formamide/2×SSC/50 mM phosphate buffer for 3 h, 70°C. Store at −20°C.

PRETREATMENT AND POSTFIXATION

- Ethanol solutions: 70 %, 90 %, and 100 %
- PBS (Cat. No. 806544, Sigma-Aldrich)
- Hydrochloric acid solution 0.2 M (Cat. No. 13–1720, Sigma-Aldrich)
- Pepsin (Cat. No. P7012, Sigma-Aldrich)
- Pepsin stock solution: 1 g pepsin + 50 ml double-distilled water. Aliquot in 0.5 ml, store at −20°C
- Pepsin pretreatment solution: 5 ml of 0.2 M HCl + 95 ml of distilled water at 37°C
- Postfix solution: 5 ml 2 % paraformaldehyde + 4.5 ml PBS + 0.5 ml $MgCl_2$ (1 M)

mtDNA-FISH AND WASHING

- mtDNA FISH probe
- COT DNA (Cat. No. 11581074001, Sigma-Aldrich)
- Fluoroshield with DAPI (Cat. No. F6057, Sigma-Aldrich)
- 20×SSC (Cat. No. S6639, Sigma-Aldrich)
- Formamide (Cat. No. F9037, Sigma-Aldrich)
- Tween 20 (Cat. No. P9416, Sigma-Aldrich)
- Denaturation buffer: 70 ml formamide + 10 ml 20×SSC + 20 ml distilled water
- 0.4×SSC solution: 10 ml 20×SSC + 490 ml distilled water
- 4×SSC/Tween solution: 100 ml 20×SSC + 400 ml distilled water + 0.25 ml Tween 20

DESCRIPTION OF METHODS

Although mtDNA-FISH is applicable for both interphase nuclei and metaphase chromosomes, the analysis of insertions of mtDNA in nDNA using metaphase chromosomes is recommended. This approach permits the identification of chromosomes with mtDNA insertion(s) and insertion loci, enabling the analysis of the distribution of mtDNA insertions in the genome. For the analysis of numtomutagenesis, human whole blood should be exposed to the mutagen and cultivated under standard conditions.

HUMAN WHOLE BLOOD CULTIVATION

1. Add 1 ml of human whole blood to 9 ml RPMI-1640 medium, containing 10% fetal bovine serum, 1% penicillin/streptomycin, and 10 μg/ml

phytohemagglutinin-L, and incubate at 37°C. The total cultivation period should be 72 h.

2. After 24 or 48 h growth add mutagen (e.g. doxorubicin) to the culture and incubate at 37°C.

PREPARATION OF METAPHASE SPREADS

1. 1.5 h before harvesting (at 70.5 h of cultivation), add 0.1 ml colcemid to the culture and incubate at 37°C to block the cell cycle in the metaphase.
2. Transfer the culture into a 15 ml centrifuge tube.
3. Centrifuge at room temperature for 7 min at 1500 rpm. Discard the supernatant except for about 0.5 ml remaining above the cell pellet and gently resuspend the pellet.
4. Add 10 ml of prewarmed (37°C) 0.075 M hypotonic solution of KCl, and incubate for 15 min at 37°C.
5. Repeat step 3.
6. Quickly add 5 ml of Carnoy's fixative (−20°C), and vortex immediately to avoid clumps. Add another 5 ml of Carnoy's fixative (−20°C), and incubate for 10 min at room temperature.
7. Repeat step 3.
8. Resuspend the pellet in 5 ml of Carnoy's fixative (−20°C), and repeat step 3.
9. Repeat step 8 twice.
10. Add 0.5 ml of Carnoy's fixative (−20°C), and store the suspension at −20°C until use.
11. Depending on the density of the suspension drop (20–50 µl) 2–3 drops onto the precooled (4°C) clean and humid slides, and let the slide dry on the warming plate at 51°C.
12. Keep the slides at room temperature for short-term storage, while slides for long-term storage should be kept at −20°C.

HOMEMADE mtDNA-FISH PROBES

1. The mtDNA for FISH can be amplified using specific pair of primers (forward: 5′-GAGCCGGAGCACCCTATGT-3′, reverse: 5′-GGGGAACG TGTGGGCTATTT-3′) in the following PCR reaction mixture of 50 µl:
   - 5 µl of Advantage 2 PCR Buffer (Takara, Kyoto, Japan),
   - 4 µl of dNTP Mix, 5 pM/µl of each primer (1 µl each),
   - 1 µl of Advantage 2 Titanium Taq Polymerase Mix (Takara),
   - 6 µl of mtDNA (~50 pg)
   - 30 µl of H_2O.
2. The following PCR conditions for amplification of the mtDNA insert can be set: denaturation at 95°C for 2 min, followed by 32 cycles of 95°C for 15 s, 61°C for 30 s, 72°C for 1 min 30 s, then 72°C for 10 min and termination at 4°C.

3. Labeling can be performed using nine pairs of primers specific for mtDNA amplification,[35] with the PCR conditions mentioned previously. The optional reaction mixture is as follows:
   - 2 μl of buffer,
   - 2 μl of Label-mix (Atto488 NT Labeling Kit, Jena Bioscience, Jena, Germany),
   - 2 μl of 25 mM $MgCl_2$,
   - 0.12 μl of AmpliTaq DNA Polymerase,
   - 2 μl of fluorochromes,
   - 12.08 μl of H_2O.
4. The specificity of the obtained probe can be confirmed using positive control with mtDNA insertion in chromosomes or using cells with undamaged mitochondria. In addition, a negative control (non-human chromosomes or mitochondria) is recommended to be used.
5. Mix the solution using the 20 μl pipette tip, and incubate at 15°C for 90 min.
6. Terminate the reaction with 1 μl of 0.5 M EDTA, mix vigorously, spin down and incubate at 65°C for 10 min.
7. Precipitate the labeled probe with 10 μl of tRNA + 5 μl sodium acetate + 100 μl 100 % ethanol at −20°C during the overnight.
8. Centrifuge the mixture at 15,000 rpm at 4°C for 15 min.
9. Discard the supernatant and dry the pellet using a speed vac.
10. Dilute the pellet in 20 μl of dextran sulfate buffer, and mix at 65°C for 5 min.
11. Keep the probe at −20°C until use.

SLIDE PRETREATMENT

1. Dehydrate slides in an ethanol series (70 %, 90 %, 100 %, 3 min each) and air-dry at room temperature.
2. Add 0.5 ml of pepsin to 100 ml of pepsin pretreatment solution, and put slides for 3–5 min in a Coplin jar at 37°C in the water bath.
3. Wash slides in the PBS for 5 min at room temperature.
4. Add 100 μl of postfix solution and cover with coverslips for 10 min at room temperature.
5. Remove coverslips; repeat step 3 and step 1.

mtDNA-FISH PROCEDURE

1. Mix 1 μl of the probe with 7 μl dextran sulfate in the Eppendorf tube containing COT DNA. The optimal concentration of the COT DNA should be determined experimentally.
2. Prewarm the mixture at 45°C for 20 min.

3. Add 100 µl of denaturation buffer to each slide, cover with coverslip, and incubate on a warming plate at 81°C for 3 min and in parallel begin the denaturation of the probe at 75°C in PCR machine.
4. Remove the coverslip and put slides in a 70 % ethanol at −20°C for 3 min.
5. Dehydrate slides in ethanol series of 90 % and 100 % at room temperature for 3 min each and air-dry.
6. Add 8 µl of the probe onto half of the slide, and cover with 24×24 coverslip and seal with rubber cement.
7. Place the slides in the humid chamber and incubate at 37°C overnight.

WASHING AND COUNTERSTAINING OF SLIDES

1. Remove the rubber cement and coverslips, and put the slides in the Coplin jar filled with prewarmed 0.4×SSC at 65°C for 5 min in the water bath.
2. Put the slides in the new Coplin jar with 4×SSC/Tween on the shaker at room temperature for 5 min.
3. Wash the slides in the PBS for 5 min, then dehydrate in ethanol series (70 %, 90 %, 100 %, 3 min each), and air-dry at room temperature.
4. Add 30 µl of fluoroshield containing DAPI for counterstaining, cover with 24×60 mm coverslip, and incubate at 4°C for at least 20 min.
5. Analyze the results under a fluorescent microscope.

YIELDS

ANALYSIS OF mtDNA INSERTIONS IN METAPHASE CHROMOSOMES

1. For the detection of mtDNA insertions in metaphase chromosomes by FISH, at least 1,800 metaphase spreads are recommended to be analyzed.
2. Capture mtDNA signals using fluorescent microscope quipped with 100× objective and the camera with high sensitivity.
3. Statistical analysis can be performed using t-test for the comparison of mtDNA insertion frequencies between untreated and treated variants.
4. Construction of the map with mtDNA insertions in each chromosome according to Harutyunyan et al.[5] is recommended for the description of the distribution of mtDNA insertions in the genome.
5. For evaluation of the quantitative changes in mitochondrial network, the application of free-to-use program ImageJ (developed at the NIH) is recommended.

CONCLUSIONS

Functional mitochondria are important players of intracellular and intercellular signaling. However, exchange of mitochondria between cells, mutations of mtDNA, or translocation of mtDNA in the nuclear genome can have deleterious effects for the organism. Moreover, DSB-inducing environmental agents are capable to increase numtomutagenic events in the cells. Thus, methods for rapid detection of aberrant

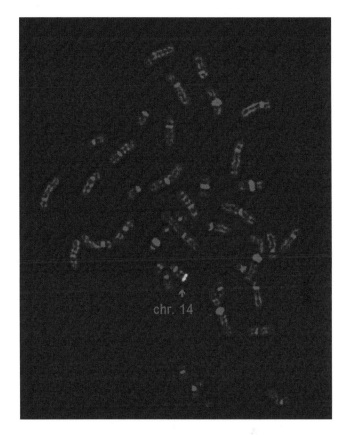

FIGURE 22.1 Insertion of mtDNA in chromosome 14 identified by mtDNA-FISH; the same patient as presented in [5] was studied here.

behavior of mitochondria or mtDNA will have diagnostic importance in many pathological conditions, including cancer, viral infections, and mtDNA disorders.

FISH analysis of mtDNA (Figure 22.1) enables quantitative and qualitative analysis of permutational mitochondria, numtogenesis, and numtomutagenesis in human cells. Further studies are required for identification of molecular pathways regulating mtDNA escape, transfer, and insertion in human chromosomes.

ACKNOWLEDGMENTS

This work was supported by the RA MESCS and BMBF (project #AG-01/20), and RA MESCS (project #21AG-1F068).

REFERENCES

1. Yan, C.; Duanmu, X.; Zeng, L.; Liu, B.; Song, Z. Mitochondrial DNA: Distribution, mutations, and elimination. *Cells*. **2019**, 8(4), 379.

2. Merry, T.L.; Chan, A.; Woodhead, J.; Reynolds, J.C.; Kumagai, H.; Kim, S.J.; Lee, C. Mitochondrial-derived peptides in energy metabolism. *Am J Physiol Endocrinol Metab.* **2020**, 319(4), E659–E666.

3. Valenti, D.; Vacca, R.A.; Moro, L.; Atlante, A. Mitochondria can cross cell boundaries: An overview of the biological relevance, pathophysiological implications and therapeutic perspectives of intercellular mitochondrial transfer. *Int J Mol Sci.* **2021**, 22(15), 8312.

4. Singh, K.K.; Choudhury, A.R.; Tiwari, H.K. Numtogenesis as a mechanism for development of cancer. *Semin Cancer Biol.* **2017**, 47, 101–109.

5. Harutyunyan, T.; Al-Rikabi, A.; Sargsyan, A.; Hovhannisyan, G.; Aroutiounian, R.; Liehr, T. Doxorubicin-induced translocation of mtDNA into the nuclear genome of human lymphocytes detected using a molecular-cytogenetic approach. *Int J Mol Sci.* **2020**, 21(20), 7690.

6. Lutz-Bonengel, S.; Niederstätter, H.; Naue, J.; Koziel, R.; Yang, F.; Sänger, T.; Huber, G.; Berger, C.; Pflugradt, R.; Strobl, C.; et al. Evidence for multi-copy Mega-NUMTs in the human genome. *Nucleic Acids Research.* **2021**, 49(3), 1517–1531.

7. Wei, W.; Pagnamenta, A.T.; Gleadall, N.; Sanchis-Juan, A.; Stephens, J.; Broxholme, J.; Tuna, S.; Odhams, C.A.; Genomics England Research Consortium; NIHR BioResource; et al. Nuclear-mitochondrial DNA segments resemble paternally inherited mitochondrial DNA in humans. *Nat Commun.* **2020**, 11(1), 1740.

8. Singh, K.K.; Chaubey, G.; Chen, J.Y.; Suravajhala, P. Decoding SARS-CoV-2 hijacking of host mitochondria in COVID-19 pathogenesis. *Am J Physiol Cell Physiol.* **2020**, 319(2), C258–C267.

9. Calabrese, F.M.; Balacco, D.L.; Preste, R.; Diroma, M.A.; Forino, R.; Ventura, M.; Attimonelli, M. NumtS colonization in mammalian genomes. *Sci Rep.* **2017**, 7(1), 16357.

10. Luo, S.; Valencia, C.A.; Zhang, J.; Lee, N.C.; Slone, J.; Gui, B.; Wang, X.; Li, Z.; Dell, S.; Brown, J.; et al. Biparental inheritance of mitochondrial DNA in humans. *Proc Natl Acad Sci U S A.* **2018**, 115(51), 13039–13044.

11. Yuan, Y.; Ju, Y.S.; Kim, Y.; Li, J.; Wang, Y.; Yoon, C.J.; Yang, Y.; Martincorena, I.; Creighton, C.J.; Weinstein, J.N.; et al. Comprehensive molecular characterization of mitochondrial genomes in human cancers. *Nat Genet.* **2020**, 52(3), 342–352.

12. Srinivasainagendra, V.; Sandel, M.W.; Singh, B.; Sundaresan, A.; Mooga, V.P.; Bajpai, P.; Tiwari, H. K.; Singh, K.K. Migration of mitochondrial DNA in the nuclear genome of colorectal adenocarcinoma. *Genome Med.* **2017**, 9(1), 31.

13. Borensztajn, K.; Chafa, O.; Alhenc-Gelas, M.; Salha, S.; Reghis, A.; Fischer, A.M.; Tapon-Bretaudière, J. Characterization of two novel splice site mutations in human factor VII gene causing severe plasma factor VII deficiency and bleeding diathesis. *Br J Haematol.* **2002**, 117(1), 168–171.

14. Ahmed, Z.M.; Smith, T.N.; Riazuddin, S.; Makishima, T.; Ghosh, M.; Bokhari, S.; Menon, P.S.; Deshmukh, D.; Griffith, A.J.; Riazuddin, S.; et al. Nonsyndromic recessive deafness DFNB18 and Usher syndrome type IC are allelic mutations of USHIC. *Hum Genet.* **2002**, 110(6), 527–531.

15. Turner, C.; Killoran, C.; Thomas, N.S.; Rosenberg, M.; Chuzhanova, N.A;, Johnston, J.; Kemel, Y.; Cooper, D.N.; Biesecker, L.G. Human genetic disease caused by de novo mitochondrial-nuclear DNA transfer. *Hum Genet.* **2003**, 112(3), 303–309.

16. Goldin, E.; Stahl, S.; Cooney, A.M.; Kaneski, C.R.; Gupta, S.; Brady, R.O.; Ellis, J.R.; Schiffmann, R. Transfer of a mitochondrial DNA fragment to MCOLN1 causes an inherited case of mucolipidosis IV. *Hum Mutat.* **2004**, 24(6), 460–465.

17. Millar, D.S.; Tysoe, C.; Lazarou, L.P.; Pilz, D.T.; Mohammed, S.; Anderson, K.; Chuzhanova, N.; Cooper, D.N.; Butler, R. An isolated case of lissencephaly caused by

the insertion of a mitochondrial genome-derived DNA sequence into the 5' untranslated region of the PAFAH1B1 (LIS1) gene. *Hum Genomics.* **2010**, 4(6), 384–393.

18. Li, X.; Xu, D.; Cheng, B.; Zhou, Y.; Chen, Z.; Wang, Y. Mitochondrial DNA insert into CD40 ligand gene-associated X-linked hyper-IgM syndrome. *Mol Genet Genomic Med.* **2021**, 9(5), e1646.

19. Singh, B.; Modica-Napolitano, J.S.; Singh, K. Defining the momiome: Promiscuous information transfer by mobile mitochondria and the mitochondrial genome. *Semin Cancer Biol.* **2017**, 47, 1–17.

20. Vallabhaneni, K.C.; Haller, H.; Dumler, I. Vascular smooth muscle cells initiate proliferation of mesenchymal stem cells by mitochondrial transfer via tunneling nanotubes. *Stem Cells Dev.* **2012**, 21(17), 3104–3113.

21. Ahmad, T.; Mukherjee, S.; Pattnaik, B.; Kumar, M.; Singh, S.; Kumar, M.; Rehman, R.; Tiwari, B.K.; Jha, K.A.; Barhanpurkar, A.P.; et al. Miro1 regulates intercellular mitochondrial transport & enhances mesenchymal stem cell rescue efficacy. *EMBO J.* **2014**, 33(9), 994–1010.

22. Pasquier, J.; Guerrouahen, B.S.; Al Thawadi, H.; Ghiabi, P.; Maleki, M.; Abu-Kaoud, N.; Jacob, A.; Mirshahi, M.; Galas, L.; Rafii, S.; et al. Preferential transfer of mitochondria from endothelial to cancer cells through tunneling nanotubes modulates chemoresistance. *J Transl Med.* **2013**, 11, 94.

23. Sahinbegovic, H.; Jelinek, T.; Hrdinka, M.; Bago, J. R.; Turi, M.; Sevcikova, T.; Kurtovic-Kozaric, A.; Hajek, R.; Simicek, M. Intercellular mitochondrial transfer in the tumor microenvironment. *Cancers (Basel).* **2020**, 12(7), 1787.

24. Ricchetti, M.; Fairhead, C.; Dujon, B. Mitochondrial DNA repairs double-strand breaks in yeast chromosomes. *Nature.* **1999**, 402(6757), 96–100.

25. Abdullaev, S.A.; Fomenko, L.A.; Kuznetsova, E.A.; Gaziev, A.I. Experimental detection of integration of mtDNA in the nuclear genome induced by ionizing radiation. *Radiats Biol Radioecol.* **2013**, 53(4), 380–388.

26. Ju, Y.S.; Tubio, J.M.; Mifsud, W.; Fu, B.; Davies, H.R.; Ramakrishna, M.; Li, Y.; Yates, L.; Gundem, G.; Tarpey, P.S.; et al. Frequent somatic transfer of mitochondrial DNA into the nuclear genome of human cancer cells. *Genome Res.* **2015**, 25(6), 814–824.

27. Basel, D. Mitochondrial DNA Depletion Syndromes. *Clin Perinatol.* **2020**, 47(1), 123–141.

28. Reznik, E.; Miller, M.L.; Şenbabaoğlu, Y.; Riaz, N.; Sarungbam, J.; Tickoo, S.K.; Al-Ahmadie, H.A.; Lee, W.; Seshan, V.E.; Hakimi, A.A.; et al. Mitochondrial DNA copy number variation across human cancers. *Elife.* **2016**, 5, e10769.

29. Schank, M.; Zhao, J.; Moorman, J.P.; Yao, Z. Q. The impact of HIV- and ART-induced mitochondrial dysfunction in cellular senescence and aging. *Cells.* **2021**, 10(1), 174.

30. Scozzi, D.; Cano, M.; Ma, L.; Zhou, D.; Zhu, J.H.; O'Halloran, J.A.; Goss, C.; Rauseo, A.M.; Liu, Z.; Sahu, S.K.; et al. Circulating mitochondrial DNA is an early indicator of severe illness and mortality from COVID-19. *JCI insight.* **2021**, 6(4), e143299.

31. Janes, M.S.; Hanson, B.J.; Hill, D.M.; Buller, G.M.; Agnew, J.Y.; Sherwood, S.W.; Cox, W.G.; Yamagata, K.; Capaldi, R.A. Rapid analysis of mitochondrial DNA depletion by fluorescence in situ hybridization and immunocytochemistry: Potential strategies for HIV therapeutic monitoring. *J Histochem Cytochem.* **2004**, 52(8), 1011–1018.

32. Lucic, B.; Wegner, J.; Stanic, M.; Jost, K.L.; Lusic, M. 3D immuno-DNA fluorescence in situ hybridization (FISH) for detection of HIV-1 and cellular genes in primary CD4+ T cells. *Methods Mol Biol.* **2021**, 2157, 239–249.

33. Caro, P.; Gómez, J.; Arduini, A.; González-Sánchez, M.; González-García, M.; Borrás, C.; Viña, J.; Puertas, M.J.; Sastre, J.; Barja, G. Mitochondrial DNA sequences are present

inside nuclear DNA in rat tissues and increase with age. *Mitochondrion.* **2010**, 10(5), 479–486.

34. Koo, D.H.; Singh, B.; Jiang, J.; Friebe, B.; Gill, B.S.; Chastain, P.D.; Manne, U.; Tiwari, H.K.; Singh, K.K. Single molecule mtDNA fiber FISH for analyzing numtogenesis. *Anal Biochem.* **2018**, 552, 45–49.

35. Ramos, A.; Santos, C.; Barbena, E.; Mateiu, L.; Alvarez, L.; Nogués, R.; Aluja, M.P. Validated primer set that prevents nuclear DNA sequences of mitochondrial origin co-amplification: A revision based on the New Human Genome Reference Sequence (GRCh37). *Electrophoresis.* **2011**, 32(6–7), 782–783.

23 FISH—in Birds

Rafael Kretschmer, Michelly da Silva dos Santos, Ivanete de Oliveira Furo, Edivaldo Herculano Correa de Oliveira and Marcelo de Bello Cioffi

CONTENTS

INTRODUCTION

Birds are the most diverse biological group among tetrapods, with more than 10,900 existing species.[1] Molecular and morphological studies support the division of living birds into three monophyletic groups: Palaeognathae, Galloanserae, and Neoaves.[2] Due to this wide variation, birds have been used as model species in numerous biological studies, focusing on ecology, behavior, genetics, and cytogenetics, and others.[3–6]

DOI: 10.1201/9781003223658-23

Cytogenetics is a useful tool in bird cytotaxonomy, especially for groups with a great chromosome variability. Most of the karyological data in birds are based on classical cytogenetic techniques, involving conventional staining and chromosomal banding methods, such as C-banding, (used to detect constitutive heterochromatin), G-banding (to identify regions rich in adenine and thymine (AT), and cytosine and guanine (CG)), and AgNOR silver staining method applied to reveal nucleolar organizer regions.[7–9] Despite its significant contribution to the characterization of karyotypes and provision of basic genomic information used in cytotaxonomy, banding resolution stands as a limitation in the identification of chromosome homology and comparative studies.

In this sense, the emergence of the fluorescence *in situ* hybridization (FISH) technique allowed an advance towards more accurate cytogenetic studies, enabling the identification and localization of DNA sequences or specific regions along the chromosomes.[10] Therefore, FISH helped to overcome limitations of comparative chromosomal studies based on banding methods, especially in birds, with typically high 2n and many microchromosomes (i.e. punctiform elements not easily distinguishable from each other).[6]

Chromosome painting data are available for 117 species, which represent less than 1% of all known bird species.[11] So far, whole chromosome painting probes (WCP) from only six different bird species have been used in avian comparative cytogenetics: chicken (*Gallus gallus*—GGA, Galliformes, 2n=78), stone-curlew (*Burhinus oedicnemus*—BOE, Charadriiformes, 2n=42), white hawk (*Leucopternis albicollis*—LAL, Falconiformes, 2n= 68), griffon vulture (*Gyps fulvus*—GFU, Falconiformes, 2n=66), eared dove (*Zenaida auriculata*—ZAU, Columbiformes, 2n=76), and monk parakeet (*Myiopsitta monachus*—MMO, Psittaciformes, 2n=48).[12–17] Among them, *G. gallus* probes stand out as the most used ones due to the economic importance of chickens, its well-known genome, and by having a karyotype very similar to the putative avian ancestral one.[6] Overall, comparative chromosome studies using WCP in birds have detected numerous inter- and intrachromosomal rearrangements, even in species with apparently conserved karyotypes.[6, 11, 18]

In addition to WCP, the development of probes using specific DNA sequences for FISH has allowed cytogenetic mapping of different repetitive sequences in distinct birds species, such as telomeric sequences, 18S/28SrDNA, 5S rDNA, and microsatellites, establishing chromosomal landmarks and providing information concerning different aspects such as karyotype organization and evolution and, besides, allowing insights on origin and differentiation of sex chromosomes.[19–26]

For all these reasons, FISH has become an indispensable tool for cytogenomic investigations in birds, and in this chapter, we present the current FISH protocol used in our laboratories. We also include details about the isolation and preparation of the most commonly used probes in avian cytogenetics. In addition, we highlight some steps related to the quality of avian chromosome preparations, which is a crucial condition for successful hybridizations.

MATERIAL

Besides the standard cell biological and molecular cytogenetic equipment, including standard reagents (e.g., colchicine, methanol, acid acetic, ethanol, among others), specialized items required are listed next.

FISH Probe Generation and Labeling

- AmpliTaq® DNA Polymerase, Stoffel Fragment (Cat. No.: N8080038, Applied Biosystems, USA)
- Biotin 14-dATP (Cat. No.: 19524016, Invitrogen, USA)
- Biotin16 NT Labeling Kit (Cat. No.: PP-310S-BIO16, Jena Bioscience, Germany)
- Cy5-dUTP (Cat. No.: 25005446, GE Healthcare Life Science, USA)
- Deionized formamide (Cat. No.: 1 09684 2500, Merck, Germany; aliquot and store at +4°C (See Note 1)
- Deoxinucleotides 100mM (Cat. No.: DNTP100–1KT, Sigma, USA)
- Dextran sodium sulfate (Cat. No.: D8906–10G, Sigma, USA)
- Digoxigenin NT Labeling Kit (Cat. No.: PP-310S-DIGX, Jena Bioscience, Germany)
- Digoxigenin-11-dUTP (Cat. No.: 11093088910, Roche Diagnostics, Switzerland)
- dNTP label mix: 2mM dATP/dCTP/dGTP, 1mM dTTP)
- DOP primer (5′- CCG ACT CGA GNN NNN NAT GTG G -3′)
- Double-distilled water (ddH$_2$0) = Aqua ad iniectabilia (Cat. No.: 2351544, Braun, Germany; aliquot and store at −20°C)
- EDTA 0.5 M (e.g. Merck, Germany; store at −20°C).
- Hybridization buffer: Dissolve 2g dextran sodium sulfate in 10 ml 50% deionized formamide/2×SSC/50 mM phosphate buffer for 3 h at 70°C; adjusted pH to 7 with phosphate buffer; hydrochloric acid destabilizes buffer solution. Aliquot and store at −20°C.
- Nick translation mix (Cat. No.: 11 745 808 910, Roche Diagnostics, Switzerland)
- Ribonucleic acid transfer—tRNA (Cat. No.: R8508, Sigma, USA)
- Sodium acetate solution (3 M, pH 5.2; e.g. Merck, Germany; store at −20°C)
- Spectrum-Orange dUTP (Cat. No.: 6J9415, Abbott, USA)
- Spectrum-Green dUTP (Cat. No.: 6J9410, Abbott, USA)
- TexasRed-dUTP (Cat. No.: C-7631, Molecular Probes, USA)
- TexasRed-dUTP/Spectrum-Orange-dUTP/Spectrum-Green-dUTP/ Streptavidin-Cy3/Streptavidin-Cy5 and Avidin-FITC working solutions: Reconstitute 1 mg with 1.0 ml of double-distilled water (ddH$_2$0), dispense into suitable aliquots and freeze

Slide Pretreatment

- 1×PBS (phosphate-buffered saline—Cat. No.: L1825, Biochrom, Germany; store at room temperature)
- Pepsin stock solution (Cat. No.: P-7012, Sigma, USA)
- Pepsin solution (0.005%): 99 µl H$_2$O, 10µl HCl and 2.5µl Pepsin (20 mg/ml)
- RNase A (Cat. No.: R4642, Sigma, USA)
- RNase solution (10µg/ml): 1.5 µl RNase A (10 mg/ml) and 1.5 ml 2×SSC.

FISH Procedure

- Avidin-FITC (Cat. No.: A2901, Sigma, USA)
- Anti-digoxigenin fluorescein (Cat. No.: 11207741910, Roche Diagnostics, Switzerland)
- Anti-digoxigenin rhodamin (Cat. No.: 11207750910, Roche Diagnostics, Switzerland)
- Denaturation buffer: 70% (v/v) deionized formamide (See Note 1), 20% (v/v) filtered double distilled water, 10% (v/v) 20×SSC; make fresh as required
- Detection solution 1: anti-biotin-labeled probes using a layer of Cy3-a or Cy5-avidin (1:1000)
- Detection solution 2: anti-digoxigenin-labeled probes using monoclonal anti-digoxin antibody produced in mice (1:500)
- Detection solution 3: FITC-labeled probes were detected with a layer of rabbit anti-FITC (1:200) followed by a layer of goatanti-rabbit-FITC antibodies (1:100)
- Formamide solution: 2×SSC/50% deionized formamide, pH 7.0 (See Note 1)
- Hybridization buffer: Dissolve 2g of dextran sodium sulfate in 10 ml of 50% formamide/2×SSC/50 mM phosphate buffer for 1 hour at 70°C. Aliquot and stock at −20°C
- 1×PBS(phosphate-buffered saline—Cat. No.: L1825, Biochrom, Germany; store at room temperature).
- Phosphate buffer: prepare 0.5 M Na_2HPO_4 and 0.5 M NaH_2PO_4, mix these two solutions (1:1) to get pH 7.0, and then aliquot and store at −20°C
- 20×SSC = saline sodium citrate (Cat. No.: 15557–036; Invitrogen, USA; store at room temperature); set up 1× and 2× SSC before use
- Streptavidin-Cy5 (Cat. No.: 25800881, GE Healthcare Life Sciences, USA)
- Streptavidin-Cy3 (Cat. No.: S6402, Sigma, USA)
- Tween 20 = polyoxyethylene-sorbitan monolaurate (Cat. No.: 10670–1000, Sigma, Germany, store at room temperature)
- Vectashield Mounting Medium with DAPI FR/10 ml (Cat. No.: H-1200, Vector, USA)
- Washing buffer (4×SCCT): 4×SSC, 0.05% Tween 20; make fresh as required.

METHODS

Slide Preparation: Dropping the Cell Suspension

To obtain chromosome preparations from birds demands biological samples; hence, it is important to notice that such procedures usually require previous approval by an Animal Experimentation Ethics Committee and permission to sample the animals, either for *in situ* or *ex situ* studies in birds. Avian mitotic chromosomes can be directly obtained from samples of different tissues, including kidney, liver, bulb feather, and bone marrow,[27] as well as from embryos/eggs.[28] However, chromosomes can also be obtained from lymphocyte cultures,[29] and the most used protocol is culture of fibroblast cells.[24] The general protocol for mitotic chromosome preparation consists of treatment with colchicine, hypotonic solution, and fixation in methanol and acid acetic (3:1).

The quality of chromosomal preparations, including i) high amount of metaphase plates, ii) correct spreading of the chromosomes, and iii) good chromosomal morphologies, is important for successful FISH experiments. If not accurately done, cytoplasm and other cellular materials may remain on the metaphase plates, blocking the proper probe hybridization and evaluable results. Therefore, the chromosomes obtained from fibroblast culture usually provides the best preparations.

To obtain high-quality metaphase spreads, some conditions such as humidity and temperature should be considered while dropping the mitotic cells on the slides:

1. Wash and thoroughly clean a glass slide, and prepare a fresh fixative solution (3:1 methanol: acetic acid).
2. Place a coupling jar with distillate water in a microwave to heat. Hold the slide over boiling water to create a film of condensation. Alternatively, place the slide under a rack in a water bath at 55°C.
3. Pipette about 20 µl of the cell suspension on the slide.
4. Pipette immediately 20 µl fixative on the slide.
5. Dry the preparation directly on the air.

FISH PROBES

Commercially Available Probes

We have used two kinds of commercially available probes for FISH in bird species:

1. Microsatellite probes: here labeled oligonucleotides containing mono-, di-, and tri- microsatellites have been used. These sequences can be ordered as directly labeled with Cy3 or Cy5 at 5′ terminal during synthesis (Sigma, St. Louis, MO, USA). Among the most used microsatellites are $(CA)_{15}$, $(CAC)_{10}$, $(CAG)_{10}$, $(CGG)_{10}$, $(GA)_{15}$, $(GAA)_{10}$, and $(GAG)_{10}$.
2. Telomeric probes: Telomeric (TTAGGG)n repeats can be detected by FISH using a Telomere PNA FISH Kit/FITC (Cat. No.: K5325, DAKO) or Telomere PNA FISH Kit/Cy3 (Cat. No.: K5326, DAKO). Telomeric probes can also be generated by PCR (please see next).

Homemade Probes

Several methods are available to obtain probes for FISH experiments. Usually, the probes for FISH experiments in birds can be obtained through PCR amplification (rDNA and telomeric) or isolated by flow cytometry followed by PCR amplification (whole chromosome probes). Next, we list the methods for obtaining the most used probes in our laboratories.

- rDNAs: The distribution of the 45S gene has been widely investigated in different groups of birds, being a useful marker for evolutionary studies.[26] These genes are highly conserved and organized in multigene families consisting of many copies, in which the 45S rDNA encodes for 18S, 5.8S, and 28S rRNAs.[30] The 5S ribosomal gene, which encodes the 5S rRNA,[30] has been investigated only in a few species.

  1. For sequence amplification, mix the reagents listed next in a 0.6 ml tube (DNAse free).
  2. Perform amplification in a thermocycler according to the program described next.
  3. Check 2 μl of amplification product on a 1% agarose gel with appropriate size markers. The product should yield a visible smear between ~1,400bp.
  4. Store the amplified DNA at −20°C.

- 18S-rDNA directed PCR amplification can be obtained using 18SF (5'CCGAGGACCTCACTAAACCA-3') and 18SR (5'-CCGCTTTGG TGACTCTTGAT-3') as primers to generate a fragment of approximately 1,400 base pairs. The primers were developed from the fish *Hoplias malabaricus*,[31] and due to the high sequence conservation, it has been applied successfully in several birds species.[26] Here, we amplified the 18S rDNA gene using the Emu (*Dromaius novaehollandiae*) DNA as a template. It is known that the Emu has this multigene family in a pair of microchromosomes,[32] and we confirmed this result (Figure 23.1 A). Reagents and details for the reaction are shown in Table 23.1.

- 5S-rDNA directed PCR: The probes were obtained from the DNA of the fish *Leporinus obtusidens* by PCR using the primers 5SF (5'-TACGCCCGATCTCGTCCGATC-3') and 5SR (5'-CAGGCTGGTAT GGCCGTAAGC-3'), according to.[33] These fragments of approximately 200 bp were obtained and successfully applied in FISH experiments in

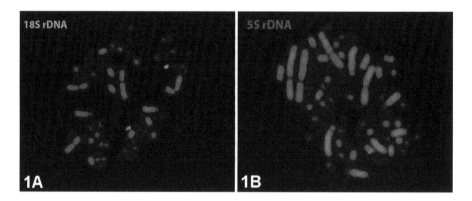

FIGURE 23.1 18S (A, green) and 5S (B, red) rDNA distribution in the Emu (*Dromaius novaehollandiae*).

TABLE 23.1

a) How to Prepare 25 μl Reaction Master Mix (per Sample) for 18S-rDNA

Reagent	Amount	Final Concentration
Genomic DNA	25 ng	-
Buffer 10×	2.5 μl	1×
MgCl₂ (50 mM)	0.75 μl	1.5 mM
Primer reverse (10 mM)	0.5 μl	0.2 mM
Primer forward (10 mM)	0.5 μl	0.2 mM
dNTP Mix (10mM each)	3.1 μl	2.5mM each (dATP, dCTP, dGTP, dTTP)
Adjust with ddH₂O	16.45 μl	-
Taq polymerase 5U/μl	0.2 μl	1.0 U

b) Thermocycler Program to Produce 18S-rDNA Probes

Number of Cycles	Reaction	Temperature, Time
1	Initial denaturation	94°C—5′
	Denaturation	95°C—1′00″
35 cycles	Annealing	60°C—1′00″
	Extension	72°C—1′30″
1	Final extension	72°C—7′00″
-	Hold	4° ∞

birds.[34] Reagents and details for the reaction are shown in Tables 23.1 and 23.2. As an example, here we amplified the 5S rDNA gene using the Emu DNA as a template. This gene was found in a small macrochromosome, probably the ninth pair in the Emu (Figure 23.1 B).

1. For the sequence amplification, mix the reagents listed next in a 0.6 ml tube (DNAse free).
2. Perform amplification in a thermocycler according to the program described next.
3. Check 2 μl of amplification product on a 1% agarose gel with appropriate size markers. The product should yield a visible smear between ~200bp.
4. Store the amplified DNA at −20°C.

- Telomere directed PCR amplification can be obtained using Primer: F (5′ TTAGGG-3′)5 e R (5′ CCCTAA-3′)5.[35] In this case, the amplification reaction can be placed with (Table 23.3) or without DNA as a template (Table 23.4).

1. For the sequence amplification, mix the reagents listed next in a 0.6 ml tube (DNAse free).
2. Perform amplification in a thermocycler according to the program described next (Table 23.2).
3. Check 5 μl of amplification product on a 1% agarose gel with appropriate size markers. The product size should be up to 25,000bp or 25Kb.
4. Store the amplified DNA at −20°C.

TABLE 23.2
Thermocycler Program to Produce 5S-rDNA Probes Are Shown

Number of cycles	Reaction	Temperature, Time
1	Initial denaturation	94°C—5'
	Denaturation	94°C—45"
35 cycles	Annealing	59°C—45"
	Extension	72°C—1'00"
1	Final extension	72°C—10'
-	Hold	4°∞

For reagents for this reaction, see Table 23.1.

TABLE 23.3
a) Reagents for the Telomere Directed PCR Amplification Reaction with DNA (50 µl Reaction Master Mix per Sample)

Reagent	Amount	Final Concentration
Genomic DNA	Variable	1–100 ng DNA
Buffer 10×	5.0 µl	1×
$MgCl_2$ (50 mM)	2.0 µl	2.0 mM
Primer 18SR (10 mM)	1.5 µl	0.3 mM
Primer 18SF (10 mM)	1.5 µl	0.3 mM
dNTP Mix	3.7 µl	2.5 mM each (dATP, dCTP, dGTP) 1.25 mM (dTTP)
Adjust with ddH2O	33.9 µl	-
Dye	1.0 µl	-
Taq polymerase 5U/µl	0.4 µl	1.0 U

b) Thermocycler Program to Produce Telomeric DNA Probes

Number of cycles	Reaction	Temperature, Time
1	Initial denaturation	95°C—10'
	Denaturation	94°C—45"
34 cycles	Annealing	50°C—1'00"
	Extension	68°C—1'00"
1	Final extension	68°C—7'
-	Hold	12°∞

PROBES OBTAINED BY FLOW SORTING

Whole chromosome painting probes from birds have been generated by flow-sorting. For this, chromosome suspensions can be prepared according to standard protocol,[36] sorted on a dual laser cell sorter (FAC-Star Plus, Becton Dickinson) as described in Yang et al. (1999),[37] and amplified by degenerate oligonucleotide-primes PCR (= DOP-PCR).[36, 38] An example is highlighted in Figure 23.2.

TABLE 23.4

a) Reagents for the Telomere Directed PCR Amplification Reaction without DNA (50 μl Reaction Master Mix per Sample)

Reagent	Amount	Final Concentration
Buffer 10×	5.0 μl	1×
MgCl$_2$ (50 mM)	2.5μl	2.5 mM
Primer 18SR (10 mM)	0.5 μl	0.1 mM
Primer 18SF (10 mM)	0.5 μl	0.1 mM
dNTP Mix	3.7 μl	2.5mM each (dATP, dCTP, dGTP) 1.25 mM (dTTP)
Adjust with ddH$_2$O	36.8 μl	-
Dye	0.6 μl	-
Taq polymerase 5U/μl	0.4 μl	1.0 U

b) Thermocycler Program to Produce Telomeric DNA Probes

Number of Cycles	Reaction	Temperature, Time
10 cycles	Denaturation	94°C—1'00"
	Annealing	55°C—0'30"
	Extension	72°C—1'00"
30 cycles	Denaturation	94°C—1'00"
	Annealing	60°C—0'30"
	Extension	72°C—1'00"
-	Hold	4°∞

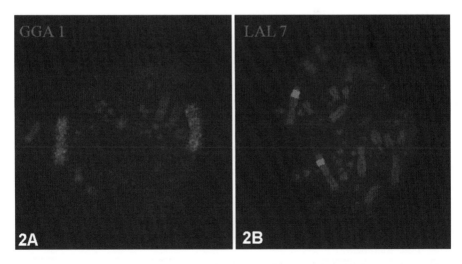

FIGURE 23.2 Examples of FISH experiments with flow sorted whole chromosome probes derived from *Gallus gallus* (GGA) (**A**) and *Leucopternis albicollis* (LAL) (**B**) onto white-tipped dove *Leptotila verreauxi* (LVE). The probes were labeled with biotin and detected with Cy3-streptavidin. Both signals are on LVE 1.

Probe Labeling

Probe labeling is an essential step for a successful FISH experiment and usually, labeling methods such as Nick translation and DOP-PCR are effective. The Nick translation procedure follows the manufacturer's instructions and is appropriate when the DNA sequence of interest was already isolated by flow sorting. In addition, DOP-PCR labeling is also applied after the flow sorting procedure to label sequences of whole chromosomes.[36, 38] By these methods, probes can be directly or indirectly labeled. In the first case, nucleotides carrying a specific fluorochrome are incorporated into the probe-DNA. In the second case, nucleotides carrying haptens (like biotin and digoxigenin) are incorporated, requiring a further detection step with appropriate fluorochrome-conjugated antibodies to enable the visualization of the hybridization sites.

Nick Translation

1. Add 1 µg of probe DNA to 4 µl biotin- or digoxigenin nick translation mix and fill up to 20 µl with ddH$_2$O; continue with step 3.
2. OR: add 1 µg of probe DNA to 4 µl of Nick translation mix and 4 µl of 5× concentrated fluorophore labeling mixture* and fill up to 20 µl with ddH$_2$O.
   - *Fluorophore labeling mixture:
   - 5 µl 2.5 mM dATP
   - 5 µl 2.5 mM dCTP
   - 5 µl 2.5 mM dGTP
   - 3.4 µl 2.5 mM dTTP
   - 4 µl of either 1mM of Cy5 dUTP/Spectrum-Green dUTP/Spectrum-Orange dUTP/Texas Red-dUTP
   - 27.6 µl ddH$_2$O
3. Mix carefully using the tip of a 20 µl pipette, and incubate at 15°C for 90 min.
4. Stop the reaction by adding 2 µl of EDTA 0.5M (pH 8.0) and incubating at 65°C for 10 min.

DOP-PCR Labeling

DOP-PCR is particularly used for the amplification of chromosome painting probes, which are usually generated by flow-sorted chromosomes. The amplification can be obtained using the degenerate 6 MW primer sequence: CCGACTCGAG NNNNNNATGTGG.

1. For a standard reaction for secondary DOP-PCR amplification, mix the reagents listed in Table 23.5 in a 0.6 ml tube (DNAse free).
2. Perform the amplification in a thermocycler according to the program described next.
3. Check 2 µl of amplification product on a 1% agarose gel with appropriate size markers. The product should yield a visible smear between ~200bp or 1.5 kb.
4. Store the amplified DNA at −20°C (up to several years).

TABLE 23.5

a) Reagents for the DOP-PCR Amplification Reaction (25 µl Reaction Master Mix per Sample)

Reagent	Amount	Final Concentration
Genomic DNA	Variable	1–100 ng DNA
Buffer 10×	2.5 µl	1×
MgCl$_2$ (50 mM)	1.25 µl	2.5 mM
6MW primer (20 mM)	2 µl	2 mM
dNTP Mix	2.5 µl	2.5mM each (dATP, dCTP, dGTP) 1.25 mM (dTTP)
Adjust with ddH2O	12.5 µl	-
Dye	2,5 µl	-
Taq polymerase 5U/µl	0.25 µl	1.25 U

b) Thermocycler Program to Do DOP-PCR

Number of Cycles	Reaction	Temperature, Time
1	Initial denaturation	94°C—5'
28 cycles	Denaturation	94°C—1'30"
	Annealing	62°C—1'30"
	Extension	72°C—3'00"
1	Final extension	72°C—8'00"
-	Hold	4°∞

PRECIPITATION OF DNA PROBE

The procedure is the same to precipitate the probes labeled by nick translation, DOP-PCR, or general PCR amplification. First, per each µg of the probe, it is necessary to add 2.5 vol of ethanol (100%) and 0.1 vol of sodium acetate (3 M, pH 5.2) to the final volume of labeled probes.

1. Mix carefully and incubate at −80°C for 30 min or at −20°C for at least 2 h.
2. Centrifugate at 13,000 rpm for 20 min at −4°C, and remove the supernatant with a micropipette without touching the pellet.
3. Dry the pellet in a vacuum centrifuge or let it at room temperature (RT) until it dries.
4. Add 1 µl of ddH$_2$0, spin, and let at 37°C for 5 min to help dissolve.
5. Add 20 µl of the hybridization buffer, vortex it, and let it shake at 65°C for 5 min. In the end, a final concentration of 50 ng/µl of labeled probe is obtained.

FLUORESCENCE IN SITU HYBRIDIZATION (FISH)

Chromosome painting follows three principal steps: DNA Precipitation, Slide pretreatment, and FISH. In the case of using rDNA or telomeric sequences as probes, it is necessary to add an additional step with RNase before the slides pretreatment with pepsin, to remove nuclear RNA and prevent nonspecific hybridization between probes and single-strand sequences.

FISH with rDNA and Telomeric Sequences

1. Add 200 µl of the RNase A solution to slide and incubate at 37°C for 25 min (1 µl of the RNase 10mg/ml solution in 500 µl of the 2×SSC).
2. Followed by treatment with pepsin (protocol described next).
3. Follow the steps for FISH described next.

SLIDE PRETREATMENT WITH PEPSIN

The pretreatment step of the slides is essential to achieve high-quality hybridization signals, especially in chromosomes obtained directed from bone marrow, which usually contains cytoplasm or other cellular residuals.

1. Add 100 µl of a pepsin solution 0.01% on the slides, cover with a 22×50 mm coverslip, and keep at 37°C for 5 min.
2. Remove the coverslip and incubate the slides in a coupling jar containing 2×SSC for 5 min at RT.
3. Repeat three times step 2.
4. Dehydrate the slides in ethanol series (70, 90%) for 2 min in each and 100% for 4 min (RT).
5. Air-dry and incubate the slides at 60°C for 1 h or 37°C overnight.

CHROMOSOME PAINTING

1. Incubate the slides in a formamide solution at 72°C, 1 min and 20 seconds, for DNA chromosomal denaturation.
2. Transfer the slides to a coupling jar filled with cold 70% ethanol (−20°C; 2 min) to conserve target DNA as single strands. Proceed slides into the ethanol series (90 and 100%, RT, 2 min each) and air-dry.
3. Denature the hybridization mix (containing 20 µl of the hybridization buffer and 100 ng of the labeled probe) in a thermocycler at 75°C for 10 min. (this step can be done while ethanol series is performed).
4. Add 20 µl of the hybridization mix to the slides, cover with a 22×50 mm coverslip, and incubate at 37°C for 72 h in a darkened moist chamber.
5. Remove the coverslip and wash the slides twice in 50% formamide and twice in 2×SSC at a temperature ranging from 40 to 44°C, depending on the applied probes and the phylogenetic distance between the species investigated (Note 3).
6. Add 100 µl of the appropriated detection solution (see Section 2.3), to each slide, cover with a 22×50 mm coverslip and incubate at 37°C for 1h in a darkened moist chamber. Detection steps 6 are necessary if the probes were indirectly labeled. In the case of directly labeled probes, proceed to step 7.

7. Remove the coverslip and wash 5 min for 3 times in 4×SSC/Tween = washing buffer (RT), under agitation (shaker).
8. Add 20 µl of Vectashield Mounting Medium with DAPI on each slide, cover with a 22×50 mm coverslip, and press gently. The slides are ready to be analyzed in a fluorescence microscope. For long-term storage, slides can be kept at 4°C to prolong the fluorescence intensity.

FISH WITH MICROSATELLITE SEQUENCES

Different from other tetrapods, the avian genomes have only a small amount of repetitive DNAs (4–10%). However, the importance of these repetitive sequences in sex chromosome evolution has been demonstrated in the last years.[22, 24] The following protocol has been used to map repetitive DNA in the chromosomes of birds.

1. Incubate the slides at 60°C for 1 hour.
2. Add 100 µl of RNAse Solution (1.5 µl RNAse in 100 µl 2×SSC) per slide (using a 24×60 mm coverslip). Incubate at 37°C for 1h 30 min in a moist chamber.
3. Remove the coverslip and incubate the slides in a coupling jar containing 2×SSC for 5 min at RT in a shaker.
4. Add 3 µl of pepsin (stock solution 1.0g in 50ml of water) into previously heated Water + HCl solution. Mix and homogenize the solution. Add 100µl on the slides for pepsin treatment for 10 minutes at 37°C in a moist chamber.
5. Incubate the slides in a coupling jar containing 2×SSC for 5 min at RT in a shaker.
6. Repeat step 5 twice.
7. Dehydrate the slides in ethanol series (70, 90, 100%) for 2 min in each (RT).
8. Air-dry the slides.
9. Incubate the slides in a formamide solution 70% at 72°C, 1 min and 20 sec for DNA chromosomal denaturation.
10. Repeat step 7.
11. Denature the hybridization mix (containing 11 µl of the hybridization buffer and 100 ng of the labeled probe) in a thermocycler at 80°C for 10 min. (This step can be done while the ethanol series is in progress).
12. Add the hybridization mix to the slides, cover with a 22×50 mm coverslip, and incubate at 37°C for 16 h or overnight in a darkened moist chamber.
13. Remove the coverslip and wash the slides twice in 2×SSC and once in 1×SSC at RT, five minutes each in a shaker.
14. Wash the slides once in 1×PBS for five minutes in a shaker.
15. Repeat step 7.
16. Add 20 µl of Vectashield Mounting Medium with DAPI on each slide, cover with a 22×50 mm coverslip, and press gently. The slides are ready to be analyzed in a fluorescence microscope. For long-term storage, slides can be kept at 4°C to prolong the fluorescence intensity.

Important Notes for Proceeding FISH

- Please, remember to discard the formamide solution as hazardous waste.
- If possible, try to use the formamide solution in a fume hood.
- Use low stringency temperatures (around 40°C) if you are hybridizing chicken probes in a distant species, such as Passerines. Use higher stringency temperatures (around 44°C) if you are hybridizing chicken probes in closely related species, such as Galliformes and Anseriformes.

ACKNOWLEDGMENTS

We would like to thank the Brazilian funding agencies FAPESP (Fundação de Amparo à Pesquisa do Estado de São Paulo, Process 2020/11669–2) and CNPq (Conselho Nacional de Desenvolvimento Científico e Tecnológico, Process 307382/2019–2 and 302449/2018–3) for the financial support.

REFERENCES

1. Gill, F.; Donsker, D.; Rasmussen, P. (Eds.). IOC world bird list; **2021**. www.worldbird-names.org/new/ [accessed on 08/31/2021].
2. Hackett, S.J.; Kimball, R.T.; Reddy, S.; Bowie, R.C.; Braun, E.L.; Braun, M.J.; Chojnowski, J.L.; Cox, W.A.; Han, K.L.; Harshman, J.; Huddleston, C.J.; Marks, B.D.; Miglia, K.J.; Moore, W.S.; Sheldon, F.H.; Steadman, D.W.; Witt, C.C.; Yuri, T. A phylogenomic study of birds reveals their evolutionary history. *Science* **2008**, 320, 1763–1768.
3. Isaksson, C.; Rodewald, A.D.; Gil, D. Editorial: Behavioural and ecological consequences of urban life in birds. *Front. Ecol. Evol.* **2018**, 6, 50.
4. Jarvis, E.D.; Mirarab, S.; Aberer, A.J.; Li, B.; Houde, P.; Li, C.; Ho, S.Y.; Faircloth, B.C.; Nabholz, B.; Howard, J.T.; Suh, A.; Weber, C.C.; da Fonseca, R.R.; Li, J.; Zhang, F.; Li, H.; Zhou, L.; Narula, N.; Liu, L.; Ganapathy, G.; Boussau, B.; Bayzid, M.S.; Zavidovych, V.; Subramanian, S.; Gabaldón, T.; Capella-Gutiérrez, S.; Huerta-Cepas, J.; Rekepalli, B.; Munch, K.; Schierup, M.; Lindow, B.; Warren, W.C.; Ray, D.; Green, R.E.; Bruford, M.W.; Zhan, X.; Dixon, A.; Li, S.; Li, N.; Huang, Y.; Derryberry, E.P.; Bertelsen, M.F.; Sheldon, F.H.; Brumfield, R.T.; Mello, C.V.; Lovell, P.V.; Wirthlin, M.; Schneider, M.P.; Prosdocimi, F.; Samaniego, J.A.; Vargas Velazquez, A.M.; Alfaro-Núñez, A.; Campos, P.F.; Petersen, B.; Sicheritz-Ponten, T.; Pas, A.; Bailey, T.; Scofield, P.; Bunce, M.; Lambert, D.M.; Zhou, Q.; Perelman, P.; Driskell, A.C.; Shapiro, B.; Xiong, Z.; Zeng, Y.; Liu, S.; Li, Z.; Liu, B.; Wu, K.; Xiao, J.; Yinqi, X.; Zheng, Q.; Zhang, Y.; Yang, H.; Wang, J.; Smeds, L.; Rheindt, F.E.; Braun, M.; Fjeldsa, J.; Orlando, L.; Barker, F.K.; Jønsson, K.A.; Johnson, W.; Koepfli, K.P.; O'Brien, S.; Haussler, D.; Ryder, O.A.; Rahbek, C.; Willerslev, E.; Graves, G.R.; Glenn, T.C.; McCormack, J.; Burt, D.; Ellegren, H.; Alström, P.; Edwards, S.V.; Stamatakis, A.; Mindell, D.P.; Cracraft, J.; Braun, E.L.; Warnow, T.; Jun, W.; Gilbert, M.T.; Zhang, G. Whole-genome analyses resolve early branches in the tree of life of modern birds. *Science* **2014**, 346, 1320–1331.
5. Prum, R.O.; Berv, J.S.; Dornburg, A.; Field, D.J.; Townsend, J.P.; Lemmon, E.M.; Lemmon, A.R. A comprehensive phylogeny of birds (Aves) using targeted next-generation DNA sequencing. *Nature* **2015**, 526, 569–577.
6. Kretschmer, R.; Ferguson-Smith, M.A.; de Oliveira, E.H.C. Karyotype evolution in birds: From conventional staining to chromosome painting. *Genes* **2018**, 9, 181.

7. Howell, W.M.; Black, D.A. Controlled silver-staining of nucleolus organizer regions with a protective colloidal developer: A 1-step method. *Experientia* **1980**, 36, 1014–1015.

8. Seabright, M. A rapid banding technique for human chromosomes. *Lancet.* **1971**, 2, 971–972.

9. Sumner, A.T. A simple technique for demonstrating centromeric heterochromatin. *Exp. Cell Res.* **1972**, 75, 304–306.

10. Pinkel, D.; Straume, T.; Gray, J.W. Cytogenetic analysis using quantitative, high-sensitivity, fluorescence hybridization. *Proc. Natl. Acad. Sci. U. S. A.* **1986**, 83, 2934–2938.

11. Degrandi, T.M.; Barcellos, A.S.; Costa, A.L.; Garnero, A.D.V.; Hass, I.; Gunski, R.J. Introducing the bird chromosome database: An overview of cytogenetic studies in birds. Cytogenet. *Genome Res.* **2020**, 160, 199–205.

12. Griffin, D.K.; Haberman, F.; Masabanda, J.; O'Brien, P.; Bagga, M.; Sazanov, A.; Smith, J.; Burt, D.W.; Ferguson-Smith, M.; Wienberg, J. Micro- and macrochromosome paints generated by flow cytometry and microdissection: Tools for mapping the chicken genome. *Cytogenet. Cell Genet.* **1999**, 87, 278–281.

13. Nie, W.; O'Brien, P.C.M.; Ng, B.L.; Fu, B.; Volobouev, V.; Carter, N.P.; Ferguson-Smith, M.A.; Yang, F. Avian comparative genomics: 112 reciprocal chromosome painting between domestic chicken (*Gallus gallus*) and the stone curlew (*Burhinus oedicnemus*, Charadriiformes): An atypical species with low diploid number. *Chromosome Res.* **2009**, 17, 99–113.

14. de Oliveira, E.H.C.; Tagliarini, M.M.; Rissino, J.D.; Rissino, J.D.; Pieczarka, J.C.; Nagamachi, C.Y.; O'Brien, P.C.; Ferguson-Smith, M.A. Reciprocal chromosome painting between white hawk (*Leucopternis albicollis*) and chicken reveals extensive fusions and fissions during karyotype evolution of accipitridae (Aves, Falconiformes). *Chromosome Research.* **2010**, 18, 349–355.

15. Nie, W.; O'Brien, P.C.M.; Fu, B.; Wang, J.; Su, W.; He, K.; Bed'Hom, B.; Volobouev, V.; Ferguson-Smith, M.A.; Dobigny, G.; Yang, F. Multidirectional chromosome painting substantiates the occurrence of extensive genomic reshuffling within Accipitriformes. *BMC Evol. Biol.* **2015**, 15, 205.

16. Kretschmer, R.; Furo, I.O.; Gunski, R.J.; Del Valle Garnero, A.; Pereira, J.C.; O'Brien, P.C.M.; Ferguson-Smith, M.A.; de Oliveira, E.H.C.; de Freitas, T.R.O. Comparative chromosome painting in Columbidae (Columbiformes) reinforces divergence in Passerea and Columbea. *Chromosome Res.* **2018**, 26, 211–223.

17. Furo, I.O.; Kretschmer, R.; O'Brien, P.C.; Pereira, J.C.; Garnero, A.D.V.; Gunski, R.J.; O'Connor, R.E.; Griffin, D.K.; Gomes, A.J.B.; Ferguson-Smith, M.A.; de Oliveira, E.H.C. Chromosomal evolution in the phylogenetic context: A remarkable karyotype reorganization in neotropical parrot myiopsitta monachus (Psittacidae). *Front. Genet.* **2020**, 11, 721.

18. Griffin, D.K.; Robertson, L.B.; Tempest, H.G.; Skinner, B.M. The evolution of the avian genome as revealed by comparative molecular cytogenetics. Cytogenet. *Genome Res.* **2007**, 117, 64, 77.

19. dos Santos, M.S.; Kretschmer, R.; Silva, F.A.O.; Silva, F.A.; Ledesma, M.A.; O'Brien, P.C.; Ferguson-Smith, M.A.; Del Valle Garnero, A.; de Oliveira, E.H.; Gunski, R.J. Intrachromosomal rearrangements in two representatives of the genus *Saltator* (Thraupidae, Passeriformes) and a case of polymorphism in Z chromosome. *Genetica* **2015**, 143, 535–543.

20. dos Santos, M.S.; Kretschmer, R.; Frankl-Vilches, C.; Bakker, A.; Gahr, M.; O'Brien, P.C.; Ferguson-Smith, M.A.; de Oliveira, E.H. Comparative cytogenetics between two important sonbird, models: The zebra finch and the canary. *PLoS One* **2017**, 12, e0170997.

21. dos Santos, M.S.; Furo, I.O.; Tagliarini, M.M.; Kretschmer, R.; O'Brien, P.C.M.; Ferguson-Smith, M.A.; de Oliveira, E.H.C. The karyotype of the hoatzin (*Opisthocomus hoazin*): A phylogenetic enigma of the neornithes. *Cytogenet. Genome Res.* **2018**, 156, 158–164.

22. de Oliveira, T.D.; Kretschmer, R.; Bertocchi, N.A.; Degrandi, T.M.; de Oliveira, E.H.; Cioffi, M.B.; Garnero, A.D.; Gunski RJ. Genomic organization of repetitive DNA in woodpeckers (Aves, Piciformes): Implications for karyotype and ZW sex chromosome differentiation. *PLoS One* **2017**, 12, e0169987.

23. Kretschmer, R.; de Oliveira, T.D.; Furo, I.O.; Oliveira Silva, F.A.; Gunski, R.J.; Del Valle Garnero, A.; de Bello Cioffi, M.; de Oliveira, E.H.C.; de Freitas, T.R.O. Repetitive DNAs and shrink genomes: A chromosomal analysis in nine Columbidae species (Aves, Columbiformes). *Genet. Mol. Biol.* **2018**, 41, 1, 98–106.

24. Furo, I.O.; Kretschmer, R.; dos Santos, M.S.; de Lima Carvalho, C.A.; Gunski, R.J.; O'Brien, P.C.M.; Ferguson-Smith, M.A.; Cioffi, M.B.; de Oliveira, E.H.C. Chromosomal mapping of repetitive DNAs in *Myiopsitta monachus* and *Amazona aestiva* (Psittaciformes, Psittacidae: Psittaciformes), with emphasis on the sex chromosomes. *Cytogenet. Genome Res.* **2017**, 151, 151–160.

25. Barcellos, S.A.; Kretschmer, R.; Souza, M.S.; Costa, A.L.; Degrandi, T.M.; Dos Santos, M.S.; de Oliveira, E.H.C.; Cioffi, M.B.; Gunski, R.J.; Garnero, A.D.V. Karyotype evolution and distinct evolutionary history of the W chromosomes in swallows (Aves, Passeriformes). *Cytogenet. Genome Res.* **2019**, 158 (2), 98–105.

26. Degrandi, T.M.; Gunski, R.J.; Garnero, A.D.V.; Oliveira, E.H.C.; Kretschmer, R.; Souza, M.S.; Barcellos, S.A.; Hass, I. The distribution of 45S rDNA sites in bird chromosomes suggests multiple evolutionary histories. *Genet. Mol. Biol.,* **2020**, 43, 2, e20180331.

27. Garnero, A.V.; Gunski, R.J. Comparative analysis of the karyotype of Nothura maculosa and Rynchotus rufescens (Aves: Tinamidae). A case of chromosomal polymorphism. *The Nucleus,* **2000**, 43, 64–70.

28. Barcellos, S.A.; de Souza, M.S.; Tura, V.; Pereira, L.R.; Kretschmer, R.; Gunski, R.J.; Del Valle Garnero, A. Direct chromosome preparation method in Avian embryos for cytogenetic studies: Quick, easy and cheap. *DNA* **2022**, 2, 22–29.

29. Moorhead, P.S.; Nowell, P.C.; Mellman, W.J.; Battips, D.M.; Hungerford, D.A. Chromosome preparations of leukocytes cultured from human peripheral blood. *Exp Cell Res.* **1960**, 20, 613–616.

30. López-Flores, I.; Garrido-Ramos, M.A. The repetitive DNA content of eukaryotic genomes. *Genome Dyn,* **2012**, 7, 1–28.

31. Cioffi, M.B.; Martins, C.; Centofante, L.; Jacobina, U.; Bertollo, L.A. Chromosomal variability among allopatric populations of Erythrinidae fish *Hoplias malabaricus*: Mapping of three classes of repetitive DNAs. *Cytogenet. Genome Res.* **2009**, 125, 132–141.

32. Nishida-Umehara, C.; Tsuda, Y.; Ishijima, J.; Ishijima, J.; Ando, J.; Fujiwara, A.; Matsuda, Y.; Griffin, D.K. The molecular basis of chromosome orthologies and sex chromosomal differentiation in palaeognathous birds. *Chromosome Res.* **2007**, 15, 721–734.

33. Martins, C.; Galleti, P.M. Chromosomal localization of 5S rDNA genes in *Leporinus* fish (Anostomidae, Characiformes). *Chromosome Res.* **1999**, 7, 363–367.

34. Gunski, R.J.; Kretschmer, R.; de Souza, M.S.; de Oliveira Furo, I.; Barcellos, S.A.; Costa, A.L.; Cioffi, M.B.; de Oliveira, E.H.C.; Del Valle Garnero, A. Evolution of bird sex chromosomes narrated by repetitive sequences: Unusual W chromosome enlargement in *Gallinula melanops* (Aves: Gruiformes: Rallidae). *Cytogenet. Genome Res.* **2019**, 158, 152–159.

35. Ijdo, J.W.; Wells, R.A.; Baldini, A.; Reeders, S.T. Improved telomere detection using a telomere repeat probe (TTAGGG)$_n$ generated by PCR. *Nucleic Acid Res.* **1991**, 19, 4780.

36. Yang, F.; Carter, N.P.; Shi, L.; Ferguson-Smith, M.A. A comparative study of karyotypes of muntjacs by chromosome painting. *Chromosoma* **1995**, 103, 642–652.

37. Yang, F.; O'Brien, P.C.M.; Milne, B.S.; Graphodatsky, A.S.; Solanky, N.; Trifonov, V.; Rens, W.; Sargan, D.; Ferguson-Smith, M.A. A complete comparative chromosome map for the dog, red fox and human and its integration with canine genetic maps. *Genomic.* **1999**, 62, 189–202.

38. Telenius, H.; Ponder, B.A.J.; Tunnacliffe, A.; Carter, N.P.; Behmel, A.; Ferguson-Smith, M.A.; Nordenskjöld, M.; Pfragner, R.; Ponder, B.A. Cytogenetic analysis by chromosome painting using DOP-PCR amplified flow-sorted chromosomes. *Genes Chromosomes Cancer* **1992**, 4, 257–263.

24 FISH—in Fish Chromosomes

*Francisco de M. C. Sassi, Gustavo A. Toma
and Marcelo de Bello Cioffi*

CONTENTS

DOI: 10.1201/9781003223658-24

INTRODUCTION

Representing more than half of the vertebrate species richness, fishes are a very diverse group of organisms, with approximately 36,000 valid species.[1, 2, 3] They include distinct groups such as bony (ray-finned fish, lungfish, coelacanth), cartilaginous (sharks, rays), and jawless fishes (hagfishes, lampreys).[3] Their worldwide distribution, basal phylogenetic position among vertebrates, and niche diversity make them valuable experimental models in many scientific fields, such as ecological, evolutionary, and genetics.[3] Furthermore, with the increased number of newly described sex chromosome systems (with the coexistence of both ZZ/ZW- and XX/XY-derived systems), characterization of inter- and intra-specific chromosomal rearrangements, and the implementation of novel genomic research tools, fish cytogenetics has steadily grown over the years. Many investigations on chromosomes have helped to solve taxonomic issues, elucidate meiotic behaviors, and even provided information regarding the evolution of repetitive DNA sequences, such as satellite DNAs and transposons.[4–6] In all these cases, chromosomal mapping of DNA by fluorescence *in situ* hybridization (FISH) has been an indispensable tool, allowing a more refined look into the genomic content of chromosomes, and providing a way to search for DNA sequences that may have a relevant role in chromosomal differentiation (e.g., the origin of B and sex chromosomes, identification of chromosomal rearrangements, etc.).

Despite that first applications of FISH occurred in end 1980s,[7] FISH in fishes was first applied only around 1993 using primarily rRNA genes and other tandem repeats as probes.[8–11] These protocols were mostly similar to those used in human molecular cytogenetics,[7, 12–14] with some differences in post-hybridization washes (especially in concentration of solutions and time of washes) and other minor adaptations. However, with the improvement of hybridization probes and rising numbers of FISH-derived techniques, such as genome *in situ* hybridization (GISH), comparative genomic hybridization (CGH), and whole chromosome painting (WCP), a new optimized FISH in fish protocol was designed.[15] This, instead, greatly differ from those previously used, as many changes were proposed to have a more meticulous hybridization procedure, which could be used in any fish group.

Here we present an updated fish-FISH protocol, in which we show a figure-guided protocol for the whole procedure and discuss some minor adaptations made over the years, such as adjustments for a sequential FISH and problems that have usually emerged using an overabundance of labeled probes.

MATERIAL

As usual, the standard cytogenetics equipment and reagents (centrifuge, pipettes, colchicine, methanol, etc.) are obligatory, and specialized items are listed next, including their working solutions.

FISH PROBE LABELING AND PRECIPITATION

Via Polymerase Chain Reaction (PCR)

- AmpliTaq® DNA Polymerase, Stoffel Fragment (Cat. No.: N8080038, Applied Biosystems, USA)
- Biotin 14-dATP (Cat. No.: 19524016, Invitrogen, USA)
- DNA template (~50 ng/µl)
- dNTP label mix (usually dUTP coupled with Cyanine 5, SpectrumGreen, SpectrumOrange, TexasRed plus other non-labelled dNTPs)
- Double-distilled water (ddH$_2$O) or Ultrapure Water
- MgCl$_2$
- Primer (DOP-primer or reverse/forward primers flanking the interesting section)
- SpectrumGreen dUTP (Cat. No.: 6J9410, Abbott, USA)
- SpectrumOrange dUTP (Cat. No.: 6J9415, Abbott, USA)
- TexasRed dUTP (Cat. No.: C-7631, Molecular Probes, USA)

Via Nick-Translation Kit

Commercially available label kits, usually composed of an enzyme mix (Polymerase + DNAse I), a labeling buffer, the labeling mix (usually dATP, dCTP, dGTP, and dTTP unlabelled + dUTP labeled with the selected fluorophore), PCR-grade water (for mounting the reaction), and a stop buffer (usually EDTA 0.5 M). The most common ones are:

- Biotin16 NT Labeling Kit (Cat. No.: PP-310S-BIO16, Jena Bioscience, Germany)
- Biotin Nick Translation Mix (Cat. No.: 11745824910, Roche Diagnostics, Switzerland).
- Digoxigenin NT Labeling Kit (Cat. No.: PP-310S-DIGX, Jena Bioscience, Germany)
- DIG Nick Translation Mix (Cat. No.: 11745816910, Roche Diagnostics, Switzerland).
- Nick translation mix (Cat. No.: 11 745 808 910, Roche Diagnostics, Switzerland)
- Atto550 or Atto488 NT Labeling Kit (Cat. No.: PP-305S-550, Jena Bioscience, Germany)

Precipitation

- Ethanol 100%
- Ribonucleic acid transfer—tRNA (Cat. No.: R8508, Sigma, USA)
- Sodium acetate solution (3 M, pH 5.2; e.g., Merck, Germany; store at −20°C)

SLIDE PRE-FISH TREATMENT

- 10×PBS (phosphate-buffered saline) stock for 1×PBS dilution: for 1l solution use 75.8g of sodium chloride (NaCl), 9.93g of disodium phosphate (Na$_2$HPO$_4$), 4.14g of monosodium phosphate (NaH$_2$PO$_4$). Dissolve in 1l of distilled water

- 1×PBS (phosphate-buffered saline—Cat. No.: L1825, Biochrom, Germany; store at room temperature = RT)
- Pepsin stock solution (Cat. No.: P-7012, Sigma, USA)
- Pepsin solution (0.005%): 99 µl H_2O, 10µl HCl, and 2.5µl pepsin (20 mg/ml)
- Post fixation solution (10 ml, 1% paraformaldehyde): mix 50 ml of 2% paraformaldehyde with 45 ml of 1×PBS and 5 ml 1M $MgCl_2$ (make fresh as required).
- RNAse A (Cat. No.: R4642, Sigma, USA)
- RNase solution (10µg/ml): 1.5 µl RNase A (10 mg/ml) and 1.5 ml 2×SSC.
- Ethanol 100% stock for diluted solutions (70% and 85%)

FISH WORKFLOW

- 20×SSC (Saline Sodium Citrate) stock solution, store at 4°C. For 1×SSC and 2×SSC dilutions, store at RT (Cat. No.: 15557–036; Invitrogen, USA) or prepare: 1l solution using 175.3g of sodium chloride (NaCl) and 88.23g of trisodium citrate dihydrate ($Na_3C_6H_5O_7.2H_2O$). Dissolve in 1l of distilled water.
- Deionized formamide (Cat. No.: 1 09684 2500, Merck, Germany; aliquot and store at 4°C).
- Denaturation buffer: 70% (v/v) deionized formamide, 20% (v/v) filtered double distilled water, 10% (v/v) 20×SSC; make fresh as required.
- Dextran sodium sulfate (Cat. No.: D8906–10G, Sigma, USA).
- Phosphate buffer: prepare 0.5 M Na_2HPO_4 and 0.5 M NaH_2PO_4, mix these two solutions (1:1) to get pH 7.0, and then aliquot and store at −20°C.
- Hybridization buffer: Dissolve 2 g dextran sodium sulfate in 10 ml 50% deionized formamide/2×SSC/50 mM phosphate buffer for 3 h at 70°C; pH adjusted to 7 with phosphate buffer; hydrochloric acid destabilizes buffer solution. Aliquot and store at −20°C.
- Tween 20 = polyoxyethylene-sorbitan monolaurate (Cat. No.: 10670–1000, Sigma, Germany, store at RT).
- Washing buffer (4×SCCT): 4×SSC, 0.05% Tween 20; make fresh as required.
- Vectashield Mounting Medium with DAPI FR/10ml (Cat. No.: H-1200, Vector, USA).

If an Indirectly Labeled Probe Is Used
- Avidin-FITC (Cat. No.: A2901, Sigma, USA).
- Anti-digoxigenin fluorescein (Cat. No.: 11207741910, Roche Diagnostics, Switzerland).
- Anti-digoxigenin rhodamine (Cat. No.: 11207750910, Roche Diagnostics, Switzerland).
- Detection solution 1: 994 µl of 3% NFDM/4×SSC, 2 µl of avidin-FITC working solution (1mg ml^{-1}), and 5 µl of anti-digoxigenin rhodamine (200µg ml^{-1}).

- Detection solution 2: 995 µl of 3% NFDM/4×SSC, 10 µl of streptavidin-Cy3, and 5µl of anti-digoxigenin fluorescein (200 µg ml⁻¹).
- Detection solution 3: 995 µl of 3% NFDM/4×SSC, 10 µl of streptavidin-Cy5, and 5 µl of anti-digoxigenin fluorescein (200µg ml⁻¹).
- Detection solution 4: 995 µl of 3% NFDM/4×SSC, 10 µl of streptavidin-Cy5, and 5 µl of anti-digoxigenin rhodamine (200µg ml⁻¹).
- NFDM = Nonfat dried milk powder (Cat. No.: 9999, Cell Signaling Technology, USA).
- NFDM3%/4×SSC: 40 ml 20×SSC, 160 ml ddH₂0, and 5 g of NFDM. Prepare the solution in constant agitation to dissolve the NFDM.
- Streptavidin-Cy5 (Cat. No.: 25800881, GE Healthcare Life Sciences, USA).
- Streptavidin-Cy3 (Cat. No.: S6402, Sigma, USA).

METHODS

Since fish-FISH methods were recently revisited, here we provide a review and an update on the technique. For the complete PCR cycles, conditions of amplification, and probe labeling.[15]

SAMPLES

Mitotic chromosomes can be obtained by the classical air-drying method,[16, revisited in 17] which consists of the application of colchicine to stop chromosomes in metaphase, followed by hypotonization of cells and fixation in Carnoy 2 (methanol 3:1 glacial acetic acid). The kidney is the preferred organ to obtain metaphase chromosomes, due to its hematopoietic function in adult teleosts,[18] but good preparations were already obtained from other organs such as the spleen.[19] Cell culture could also be an alternative for mitotic chromosome obtainment, using a tissue fragment that usually is muscle or fin.[20] To obtain meiotic chromosomes of fishes, the most common protocol follows the squash of the testes to liberate cells and proceed to hypotonization in KCl 0.075 M, followed by fixation in Carnoy 2.[21]

CELL SUSPENSION AND SLIDES PREPARATION

To obtain good fish results, chromosomes should be well spread (Figure 24.1). For this, the cell suspension needs to be dropped (usually 10–20 µl) into a clean, heated, and humid slide (50–60°C in a heater plate, with some humid paper). Small droplets of Carnoy 2 can also be immediately dropped over the drop with cell suspension, helping to spread the metaphase plate.

PROBES

The most important step for the success of a FISH procedure is the establishment of a probe of good quality. Such probe can be obtained by different methods, such as target DNA amplification via PCR, microdissection, and whole-genome extraction. Besides, commercially available probes (as microsatellite and telomeric sequences)

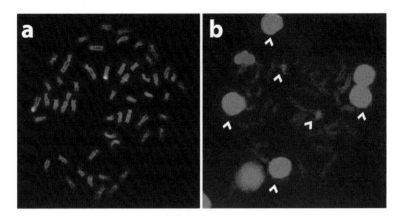

FIGURE 24.1   Examples of a good (a) and a bad (b) metaphase plate to analyze with FISH.

a) Mitotic chromosomes of *Chalceus erythrurus* (Teleostei, Characiformes) hybridized with ribosomal DNA probes (a) 5S rDNA in green, and 18S rDNA in red.

b) *Lebiasina bimaculata* (Teleostei, Characiformes) chromosomes hybridized with GC microsatellite.

Observe that in a) chromosomes are separated and well defined, while in b) they present some overlapped regions and nuclei around (arrowheads), which may interfere with the hybridization and the detection of hybridized regions.

can also be produced. We list next the methods for obtaining the most common probes used in fish cytogenetics.

Ribosomal DNA

The chromosomal localization of ribosomal genes is an important tool for the comprehension of the evolutionary process on fishes.[22] These multigene families are highly conserved and give origin to the ribosomal RNA, that actively participate in protein synthesis. They are expressed by two multigene families: the 45S rDNA and 5S rDNA. In the 45S rDNA are codified the 5.8S, the 18S, and the 28S, separated by two internal transcribed spacers (ITS1 and ITS2), two external transcribed spacers (ETS1 and ETS2), and a non-transcribed spacer (NTS). On other hand, 5S rDNA is organized into a series of repetitions, separated by an NTS, also being found as pseudogenes dispersed on chromosomes.[23] In fish cytogenetics, the most common probes of rDNA are 5S and 18S. For the obtainment of such probes, we use previously designed primers (18S from [24] and 5S from[25]) to isolate the sequences from a targeted genome, which usually is from *Hoplias malabaricus* (Characiformes, Erythrinidae), due to the previous existence of well-isolated libraries.

Genomics

More recently, genomics has become an important mechanism for probes obtainment in fish cytogenetics. Briefly, sequences from next-generation sequencing

(NGS) are obtained, and bioinformatics pipelines are applied to isolate specific libraries, where primers can be designed for the targeted sequences. Then, these primers are used in a PCR reaction to amplify the target sequence and thus be used as a FISH probe after labeling. Satellite DNAs and TE probes are the most common class of DNA used,[e.g. 26] but studies are still scarce when compared to the high diversity of fishes, and consequently of the possibilities of this technique for more DNA classes.

TRANSPOSABLE ELEMENTS (TES)

Another repetitive DNA class that represents an important target for fish cytogenetics is TEs, since they represent a highly diverse pool of sequences that can be associated with ribosomal DNAs and other DNA classes, helping them to spread into the genome.[27–30] Transposons are class II TEs that have the biggest abundance in fish genomes,[30] and are commonly mapped in fish chromosomes, such as the *Rex* family.[e.g. 31, 32] They can be PCR-isolated using already known primers, especially those isolated from *Xiphophorus* (Perciformes, Poeciliidae).[33]

SMALL NUCLEAR RNAs (SNRNAS)

In eukaryotic genomes, small RNAs have an important role in intron splicing and other RNA processing, as the maturation of primary transcripts in messenger RNA (mRNA).[34] Indeed, in fish karyotypes, they present a variable pattern of distribution, although its sequence remains conserved.[35] To obtain such probes, it can be amplified using the primers already designed for the U's snRNAs.[27, 36, 37]

MICRODISSECTION

For the whole-chromosome painting (WCP) procedure, single chromosomes (or specific regions of them) can be directly isolated by microdissection, which involves the use of a glass needle to scuff up out the chromosome from the slide, followed by its amplification via PCR and its label with a fluorescent dUTP. The complete procedure is reviewed in [38].

COMPARATIVE GENOMIC HYBRIDIZATION (CGH)

Another FISH-derived technique that is currently used in fish cytogenetics is the comparison of genomes by Comparative Genomic Hybridization (CGH). This technique is very resolutive, especially in the identification of sex chromosome systems,[39, 40] and to detect hidden variability among similar karyotypes.[41] Genomes can be obtained from the same species, varying from sexes or populations, or different species, according to the proposed objective. Total DNA is extracted and labeled via Nick Translation. Unlabeled DNA is used as a blocker (usually C_0t-1 DNA) and is hybridized together with the genomic probe to impair the hybridization in highly repetitive regions of chromosomes. Then, a specific pattern of hybridization can be obtained varying from the amount of blocker used. For intraspecific comparisons, we have great results with 10–15 μg of blocker, while for interspecific experiments, 20–25 μg.

PROBE LABELING

The probe needs to be hapten- or fluorescent-labeled for FISH. For this purpose, the isolated DNA can be directly labeled with a fluorescent molecule (e.g., Cy5, Atto 550, Atto 488) or be linked with haptens (e.g., biotin and digoxigenin), thus constituting indirect labeling. As previously described, two main techniques can be used for this: a standard PCR reaction (using designed or degenerated primers) or via Nick Translation (NT). While for the first the labeled nucleotides are incorporated during the amplicon syntheses, the second method is based on the incorporation of fluorescent/haptens-dNTPs in small aleatory gaps generated by a DNAse. When labeling by NT, it's very important to control the final size of fragments, which should vary between 200 and 600 base pairs (bp). For this purpose, several commercially NT reactions kits can be widely found.

Multicolor FISH

To hybridize different probes in a single experiment, these should be labeled with distinct fluorophores. Usually, the combination of red (Atto550, Rhodamine, Cy3, Spectrum Orange, etc.) and green (Atto488, FITC, Spectrum green, among others) probes produces good and distinguishable patterns, since DAPI (blue) is the most common counterstain used.

PRECIPITATION AND HYBRIDIZATION MIXTURE

FISH Probes

Probes need to be precipitated before the hybridization. For this, use a combination of sodium acetate 3M, tRNA, and 100% ethanol in a proportion of 5:10:100 μl for each 20 μl of the probe. Mix carefully and incubate at −20°C overnight or −80°C for 30 min. Centrifugate at 13,000 rpm for 20 min at −4°C and remove the supernatant, leaving the pellet for dry at RT. This step should proceed with caution, given that a super-dried pellet will be difficult to dissolve in the hybridization buffer. Once dry, resuspend the pellet in 20 μl of the hybridization buffer, using a vortex for better results. The hybridization buffer can be prepared before the FISH procedure and stored at −20 °C.

CGH and WCP Probes

For the FISH-derived techniques, the precipitation step needs to be differentiated. For WCP and CGH procedures, precipitate a 'blocker-DNA' (usually C_0t-1 DNA) to avoid unspecific background hybridization first, and then mount in the same tube the hybridization mixture with the hybridization buffer and the whole-chromosome probes already labeled.

FISH-FISH PROTOCOL

Day 1 (Figure 24.2):

1. Drop the chromosome preparation into clean glass slides;
2. Incubate the slides for 1h at 60°C in a glass Coplin jar;
3. Add 100 μl of RNAse solution (1.5μl RNAse in 100 μl 2×SSC) per slide (using a 24×60 mm coverslip). Incubate at 37°C for 1h 30min in a humid dark chamber;
4. In parallel put 1 ml of dd-H$_2$O + 10 μl of HCl 0.1 M to heat to 37°C and
5. put a Coplin jar with the denaturation buffer (70% (v/v) deionized formamide, 20% (v/v) filtered double distilled water, 10% (v/v) 20×SSC) to heat in a water bath to 72°C;
6. After 1h 30min in RNAse solution, remove the coverslip and wash the glass slide in 1×PBS for 5 min at RT on a shaker;
7. Add 3 μl of pepsin into previously heated 1 ml water + HCl solution (step 4) and homogenize. Add 100 μl on the still '1×PBS—humid' slide, cover with 24×60mm coverslips, and incubate for 10 min at 37 °C in a humid dark chamber;
8. Remove the coverslip and wash the glass slide in 1×PBS for 5 min on a shaker;
9. Quickly submerge the glass slide in distilled water;
10. Dehydrate the slides in ethanol 70%, 85%, and 100% for 2 min each, at RT. Completely air-dry the slides;
11. Denature the slides in the previously heated at 72°C denaturation buffer (step 5) for 3 min 15 sec;
12. Dehydrate the slides in −20°C <u>cold</u> ethanol 70% for 3 min, followed by ethanol 85% and 100% for 2 min each, at RT. Completely air-dry the slides;
13. Denature the probes in a thermal cycler at 85°C for 10 min, followed by cooling at 4°C;
14. Add 20 μl of the hybridization mixture (probes + hybridization buffer) to the slides. Cover with 24×50 mm coverslips and incubate at 37°C overnight in a dark humid chamber.

Day 2 for <u>directly</u> labeled probes (Figure 24.3a):

15. Heat a Coplin jar with 1×SSC to 65°C, remove the coverslip and wash the slides for 5 min on a shaker;
16. Transfer the slides to the washing buffer (4×SSC/Tween solution) and wash for 5 min in on shaker at RT;
17. Wash the slides in 1×PBS for 1 min at RT using a shaker;
18. Quickly submerge the glass slide in distilled water;
19. Dehydrate the slides in ethanol 70%, 85%, and 100% for 2 min each, at RT. Completely air-dry the slides;
20. Add 20 μl of DAPI + antifade onto slides, cover with a 24×50 mm coverslip and stabilize the fluorophores by leaving the slides at 4°C for 30 min before the analysis in a fluorescence microscope.

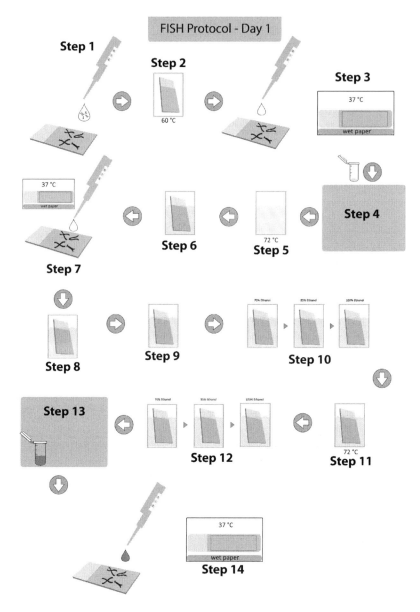

FIGURE 24.2   Illustrated fish-FISH protocol of the first day of the procedure. The steps correspond to the ones described in the fish-FISH protocol (Day 1).

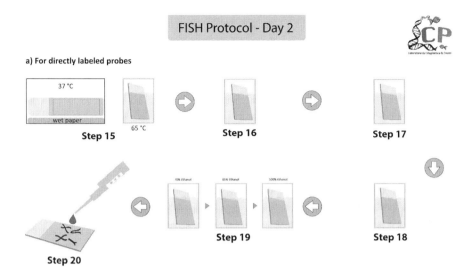

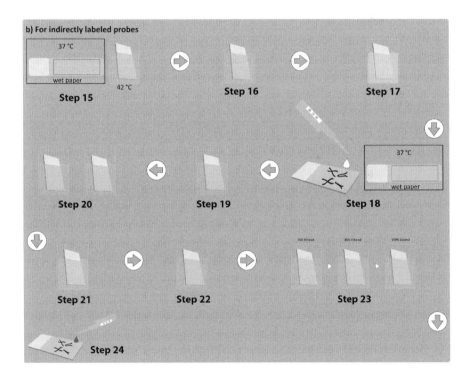

FIGURE 24.3 The second day of the illustrated fish-FISH protocol, with variations according to the label method used (a: directly labeled; b: indirectly labeled). The steps correspond to the ones described in the fish-FISH protocol (Day 2).

Day 2 for <u>indirectly</u> labeled probes (Figure 24.3b):

15. Heat a Coplin jar with 2×SSC at 42°C, remove the coverslip and wash the slides for 5 min in a shaker;
16. Wash the slides in 1×SSC at RT for 5 min on a shaker;
17. Incubate the slides in 3% NFDM/4×SSC for 5 min at RT;
18. Apply 50 µl of the detection solution (will vary according to the combination of fluorophores used), cover with 24×50mm coverslips, and incubate at 37 °C in a humid dark chamber for 1h;
19. Remove the coverslips, transfer the slides to the washing buffer (4×SSC/Tween solution), and wash for 5 min on a shaker at RT;
20. Repeat the wash as done in step 19 twice;
21. Wash the slides in 1×PBS for 1 min at RT with a shaker;
22. Quickly submerge the glass slide in distilled water;
23. Dehydrate the slides in ethanol 70%, 85%, and 100% for 2 min each, at RT. Completely air-dry the slides;
24. Add 20 µl of DAPI + antifade onto slides, cover with a 24×50mm coverslip and stabilize the fluorophores by leaving the slides at 4°C for 30 min before the analysis in a fluorescence microscope.

Variations of FISH Protocol for WCP and CGH

For the FISH-basic protocol derived techniques, some steps of the previously mentioned protocol should be modified. Given that the amount of probe is greater than in classical FISH, modifications in denaturation of probes and post-hybridization washes must occur as follows:

Day 1 for WCP and CGH:
[. . .]

13. Denature the probes in a thermal cycler at 85°C for 10 min, followed by a pre-annealing at 37°C for 30 min, followed by cooling at 4°C.
14. Add 20 µl of the hybridization mixture (probes + hybridization buffer) to the slides. Cover with a 24 × 50mm coverslip, seal the borders of the coverslip with rubber cement and incubate at 37°C overnight in a dark humid chamber.

Day 2 for WCP and CGH:

15. Heat two Coplin jars with 1×SSC at 65°C, remove the coverslip and wash the slides for 5 min on a shaker;
16. Repeat step 15 to washing the slides again;
17. Transfer the slides to the washing buffer (4×SSC/Tween solution) and wash for 5 min on a shaker at RT;
18. Wash the slides in 1×PBS for 1 min at RT with a shaker;
19. Finish as described before

Sequential FISH

Depending on the degree of chromosomal damage during the FISH technique, a second round of hybridization can be realized. For this, the previous probe-chromosome assembly must be disassembled as described next:

1. Remove the coverslip and wash the slides in 1×SSC (65°C) for 5 min on a shaker;
2. Repeat step 1;
3. Transfer the slides to the washing buffer (4×SSC/Tween solution) and wash for 5 min an a shaker at RT;
4. Repeat step 3;
5. Wash the slides in 1×PBS for 1 min at RT with a shaker;
6. Quickly submerge the glass slide in distilled water;
7. Dehydrate the slides in ethanol 70%, 85%, and 100% for 2 min each, at RT. Completely air-dry the slides.

This should be enough to disassembly the previously applied FISH probes from chromosomes and the FISH procedure should proceed from step 11 (denaturation of slides). Be careful with the time of denaturation in corresponding buffer, since chromosomes can be very fragile. This sequential FISH can be very helpful to identify shared regions of distinct probes.

IMAGE CAPTURE AND TREATMENT

The image capture should be performed in a fluorescence microscope with the respective filter combination corresponding to the light-wave excitation of probes. After being captured, images from each filter should be exported separately and overlapped in image software that allows the work in separated layers (such as GIMP or Adobe Photoshop).

ACKNOWLEDGMENTS

The authors are thankful for the funding of Fundação de Amparo à Pesquisa do Estado de São Paulo (Process numbers 2020/02681–9; 2018/14677–6 and 2020/11772–8). Additionally, we would like to thank CNPq and the Humboldt Foundation for financial support.

REFERENCES

1. Nelson, J.S.; Grande, T.C.; Wilson, M.V. *Fishes of the World*. John Wiley & Sons, New York, **2016**.
2. Salzburger, W. Understanding explosive diversification through cichlid fish genomics. *Nat. Rev. Genet.* **2018**, 19, 705–717.
3. Fricke, R.; Eschmeyer, W.N.; Van der Laan, R. (Eds.). Eschmeyer's catalog of fishes: Genera, species, references. http://researcharchive.calacademy.org/research/ichthyology/catalog/fishcatmain.asp [accessed on 01/06/2022].

4. Lehmann, R.; Kovařík, A.; Ocalewicz, K.; Kirtiklis, L.; Zuccolo, A.; Tegner, J.N.; Wanzenböck, J.; Bernatchez, L.; Lamatsch, D.K.; R. Symonová. DNA transposon expansion is associated with genome size increase in Mudminnows. *Genome Biol. Evol.* **2021**, 13, evab228.

5. Santos, R.Z.; Calegari, R.M.; Silva, D.M.Z.A.; Ruiz-Ruano, F.J.; Melo, S.; Oliveira, C.; Foresti, F.; Uliano-Silva, M.; Porto-Foresti, F.; Utsunomia, R. A long-term conserved satellite DNA that remais unexpanded in several genomes of Characiformes fish is actively transcribed. *Genome Biol. Evol.* **2021**, 13, evab002.

6. Yano, C.F.; Sember, A.; Kretschmer, R.; Bertollo, L.A.C.; Ezaz, T.; Hatanaka, T.; Liehr, T.; Ráb, P.; Al-Rikabi, A.; Viana, P.F.; Feldberg, E.; Oliveira, E.A.; Toma, G.A.; Cioffi, M.B. Against the mainstream: Exceptional evolutionary stability of ZW sex chromosomes across the fish families Triportheidae and Gasteropelecidae (Teleostei: Characiformes). *Chromosome Res.* **2021**, 29, 1–26.

7. Pinkel, D.; Straume, T.; Gray, J.W. Cytogenetic analysis using quantitative, high sensitivity, fluorescence hybridization. *Proc. Nat. Acad. Sci.* **1986**, 83, 2934–2938.

8. Pendás, A.M.; Morán, P.; Garcia-Vázquez, E. Ribosomal RNA genes are interspersed throughout a heterochromatic chromosome arm in Atlantic salmon. *Cytogenet. Genome Res.* **1993**, 63, 128–130.

9. Pendás, A.M.; Morán, P.; García-Vázquez, E. Multi-chromosomal location of ribosomal RNA genes and heterochromatin association in brown trout. *Chromosome Res.* **1993**, 1, 63–67.

10. Pendás, A.M.; Moran, P.; Freije, J.P.; Garcia-Vazquez, E. Chromosomal mapping and nucleotide sequence of two tandem repeats of Atlantic salmon 5S rDNA. *Cytogenet. Genome Res.* **1994**, 67, 31–36.

11. Phillips, R.B. Application of fluorescence in situ hybridization (FISH) to fish genetics and genome mapping. *Marine Biotechnol.* **2001**, 3, S145–S152.

12. Wiegant, J.; Ried, T.; Nederlof, P.M.; Ploeg, M.V.D.; Tanke, H.J.; Raap, A.K. In situ hybridisation with fluoresceinated DNA. *Nucl. Acids Res.* **1991**, 19, 3237–3241.

13. Wiegant, J.; Galjart, N.J.; Raap, A.K.; d'Azzo, A. The gene encoding human protective protein (PPGB) is on chromosome 20. *Genomics* **1991**, 10(2), 345–349.

14. Freije, J.P.; Pendas, A.M.; Velasco, G.; Roca, A.; Abrahamson, M.; Lopez-Otin, C. Localization of the human cystatin D gene (CST5) to chromosome 20p11. 21 by in situ hybridization. *Cytogenet. Genome Res.* **1993**, 62, 29–31.

15. Yano, C.F.; Bertollo, L.A.C.; Cioffi M.B. *Fish-FISH: Molecular Cytogenetics in Fish Species: In Fluorescence In Situ Hybridization (FISH)*. Springer, Berlin, **2017**, pp. 429–443.

16. Bertollo, L.A.C. Cytotaxonomic considerations on *Hoplias lacerdae* (Pisces, Erythrinidae). *Brazil. J. Genet.* **1978**, 1, 103–120.

17. Bertollo, L.A.C.; Cioffi, M.B.; Moreira-Filho, O. Direct chromosome preparation from freshwater teleost fishes. In: *Fish Cytogenetic Techniques (Chondrichthyans and Teleosts)*; Ozouf-Costaz, C.; Pisano, E.; Foresti, F.; Almeida Toledo, L.F., Eds., CRC Press/Taylor & Francis Group, Oxon, **2015**, pp. 21–26.

18. Zapata, A. Ultrastructural study of the teleost fish kidney. *Development. Comp. Immunol.* **1979**, 3, 55–65.

19. Oliveira, E.A.; Bertollo, L.A.C.; Rab, P.; Ezaz, T.; Yano, C.F.; Hatanaka, T.; Jegede, O.I.; Tanomtong, A.; Liehr, T.; Sember, A.; Maruyama, S.R.; Feldberg, E.; Viana, P.F.; Cioffi, M.B. Cytogenetics, genomics and biodiversity of the South American and African Arapaimidae fish family (Teleostei, Osteoglossiformes). *PloS One* **2019**, 14, e0214225.

20. Paim, F.G.; Maia, L.; Landim-Alvarenga, F.D.C.; Foresti, F.; Oliveira, C. New protocol for cell culture to obtain mitotic chromosomes in fishes. *Meth. Protoc.* **2018**, 1, 47.

21. Traut, W.; Winking, H. Meiotic chromosomes and stages of sex chromosome evolution in fish: Zebrafish, platyfish and guppy. *Chromosome Res.* **2001**, 9, 659–672.

22. Rebordinos, L.; Cross, I.; Merlo, A. High evolutionary dynamism in 5S rDNA of fish: State of the art. *Cytogenet. Genome Res.* **2013**, 141, 103–113.

23. Lafontaine, D.L.; Tollervey, D. The function and synthesis of ribosomes. *Nat. Rev. Mol. Cell Biol.* **2001**, 2, 514–520.

24. Cioffi, M.B.; Martins, C.; Centofante, L.; Jacobina, U.; Bertollo, L.A.C. Chromosomal variability among allopatric populations of Erythrinidae fish *Hoplias malabaricus*: Mapping of three classes of repetitive DNAs. *Cytogenet. Genome Res.* **2009**, 125, 132–141.

25. Martins, C.; Galetti, P.M. Chromosomal localization of 5S rDNA genes in Leporinus fish (Anostomidae, Characiformes). *Chromosome Res.* **1999**, 7, 363–367.

26. Crepaldi, C.; Martí, E.; Gonçalves, É.M.; Martí, D.A.; Parise-Maltempi, P.P. Genomic differences between the sexes in a fish species seen through satellite DNAs. *Front. Genet.* **2021**, 12, 728670.

27. Volff, J.N.; Schartl, M. Variability of genetic sex determination in poeciliid fishes. *Genetica.* **2001**, 111, 101–110.

28. Ferreira, D.C.; Porto-Foresti, F.; Oliveira, C.; Foresti, F. Transposable elements as a potential source for understanding the fish genome. *Mobile Genet. Elements.* **2011**, 1, 112–117.

29. Chalopin, D.; Naville, M.; Plard, F.; Galiana, D.; Volff, J.N. Comparative analysis of transposable elements highlights mobilome diversity and evolution in vertebrates. *Genome Biol. Evol.* **2015**, 7, 567–580.

30. Carducci, F.; Barucca, M.; Canapa, A.; Carotti, E.; Biscotti, M.A. Mobile elements in ray-finned fish genomes. *Life.* **2020**, 10, 221.

31. Ferreira, A.M.V.; Marajó, L.; Matoso, D.A.; Ribeiro, L.B.; Feldberg, E. Chromosomal mapping of Rex retrotransposons in Tambaqui (*Colossoma macropomum* Cuvier, 1818) exposed to three climate change scenarios. *Cytogenet. Genome Res.* **2019**, 159, 39–47.

32. da Silva, F.A.; Guimarães, E.M.C.; Carvalho, N.D.; Ferreira, A.M.; Schneider, C.H.; Carvalho-Zilse, G.A.; Feldberg, E.; Gross, M.C. Transposable DNA elements in Amazonian fish: From genome enlargement to genetic adaptation to stressful environments. *Cytogenet. Genome Res.* **2020**, 160, 148–155.

33. Volff, J.N.; Körting, C.; Sweeney, K.; Schartl, M. The non-LTR retrotransposon Rex3 from the fish Xiphophorus is widespread among teleosts. *Mol. Biol. Evol.* **1999**, 16, 1427–1438.

34. Marz, M.; Kirsten, T.; Stadler, P.F. Evolution of spliceosomal snRNA genes in metazoan animals. *J. Mol. Evol.* **2008**, 67, 594–607.

35. Cabral-de-Mello, D.C.; Valente, G.T.; Nakajima, R.T.; Martins, C. Genomic organization and comparative chromosome mapping of the U1 snRNA gene in cichlid fish, with an emphasis in *Oreochromis niloticus*. *Chromosome Res.* **2012**, 20, 279–292.

36. Silva, D.M.; Utsunomia, R.; Pansonato-Alves, J.C.; Oliveira, C.; Foresti, F. Chromosomal mapping of repetitive DNA sequences in five species of Astyanax (Characiformes, Characidae) reveals independent location of U1 and U2 snRNA sites and association of U1 snRNA and 5S rDNA. *Cytogenet. Genome Res.* **2015**, 146, 144–152.

37. Utsunomia, R.; Silva, D.M.Z.A.; Ruiz-Ruano, F.J.; Araya-Jaime, C.; Pansonato-Alves, J.C.; Scacchetti, P.C.; Hashimoto, D.T.; Oliveira, C.; Trifonov, V.A.; Porto-Foresti, F.; Camacho, J.P.M.; Foresti, F. Uncovering the ancestry of B chromosomes in *Moenkhausia sanctaefilomenae* (Teleostei, Characidae). *PLoS One.* **2016**, 11, e0150573.

38. Yang, F.; Trifonov, V.; Ng, B.L.; Kosyakova, N.; Carter, N.P. *Generation of Paint Probes from Flow-Sorted and Microdissected Chromosomes: In Fluorescence In Situ Hybridization (FISH).* Springer, Berlin; **2017**, pp. 63–79.

39. Sassi, F.M.C.; Deon, G.A.; Moreira-Filho, O.; Vicari, M.R.; Bertollo, L.A.C.; Liehr, T.; de Oliveira, E.A.; Cioffi, M.B. Multiple sex chromosomes and evolutionary relationships in Amazonian catfishes: The outstanding model of the genus Harttia (Siluriformes: Loricariidae). *Genes.* **2020**, 11, 1179.

40. Sassi, F.M.C.; Moreira-Filho, O.; Deon, G.A.; Sember, A.; Bertollo, L.A.C.; Liehr, T.; Oliveira, V.C.S.; Viana, P.F.; Feldberg, E.; Vicari, M.R.; Cioffi, M.B. Adding new pieces to the puzzle of karyotype evolution in Harttia (Siluriformes, Loricariidae): Investigation of Amazonian species. *Biology.* **2021**, 10, 922.

41. Sassi, F.M.C.; Perez, M.F.; Oliveira, V.C.S.; Deon, G.A.; de Souza, F.H.S.; Ferreira, P.H.N.; de Oliveira, E.A.; Hatanaka, T.; Liehr, T.; Bertollo, L.A.C.; Cioffi, M.B. High genetic diversity despite conserved karyotype organization in the giant Trahiras from genus Hoplias (Characiformes, Erythrinidae). *Genes.* **2021**, 12, 252.

25 FISH—and the Characterization of Synaptonemal Complex

Victor Spangenberg

CONTENTS

INTRODUCTION

Meiotic prophase I is the key stage of meiosis and includes several sequential sub-stages connected with synaptonemal complex (SC), which is the skeletal structure of meiotic bivalents. Therefore, studies of SC-preparations are not focused on one specific stage. On the contrary, it is a complex analysis of dynamic changes in the nuclei of germ cells of all prophase I stages of meiosis: leptotene, zygotene, pachytene, diplotene and diakinesis. In biological objects without strict synchronization in meiosis, it is usually possible to obtain all of the listed stages on one cytological preparation. The stages of assembly and subsequent disassembly of SC are associated with numerous protein markers of DNA double-strand breaks (DSB) processing (their formation and subsequent repair), meiotic recombination, chromatin remodeling and many others. Thus, the SC is a highly dynamic nucleoprotein structure. The study of protein markers of meiosis using immunocytochemical (ICC) methods makes SC-karyotyping an important extension to the analysis of metaphase mitotic chromosomes.

There is a wide variety of methods for obtaining preparations of meiotic chromosomes, i.e. total preparations of SCs. Most applicable methods for the analysis of meiotic nuclei are spread techniques.[1, 2] The main feature of spreading approaches is obtaining a flat (spread) preparation from a native three-dimensional nucleus. Notably, spread preparation of SCs should not contain too many overlapped bivalents. Although such configurations (overlapping) are unavoidable when studying

DOI: 10.1201/9781003223658-25

TABLE 25.1
Basic Morphological Criteria of Meiotic Prophase I Stages and Most-Used Protein Markers—Synaptonemal Complex Associated Proteins

Stages of Meiotic Prophase I	Synaptonemal Complex	Basic Protein Markers Used for Identification of Stages
Leptotene	formation of axial elements of chromosomes, intense DSB processing and repair.	SPO11, RAD51, γH2AFX
Early zygotene	clustering of chromosome's telomeres— formation of "chromosomal bouquet" structure. Initiation of peritelomeric/ interstitial synapsis.	SYCP3, SYCP1, RAD51, γH2AFX
Late zygotene	prolongation of chromosomal synapsis, declustering of telomere ends.	SYCP3, SYCP1, RAD51, γH2AFX
Pachytene	complete assembly of synaptonemal complexes (meiotic bivalents).	SYCP3, SYCP1, MLH1, γH2AFX(on the XY-bivalent only)
Early diplotene	initiation of homologous chromosomes desynapsis—desynaptic "forks", or interstitial regions of desynapsis.	SYCP3, SYCP1, MLH1
Late diplotene	progression of desynapsis, elongation and fragmentation of axial elements, synaptonemal complex disassembly.	SYCP3
Diakinesis	complete removal of SYCP3 from chromosome axes, but remnants near centromeres.	SYCP3

species with long chromosomes, following the spread protocols one can achieve analyzable total preparations of SCs.

Methods for obtaining total preparations of SCs were developed for mammals,[1, 2] widely used for reptiles,[3, 4] and also adapted for other groups, such as fishes,[5] insects,[6] plants,[7] and fungi,[8] with some modifications. Particular care should be taken when going for preparations from meiocytes of organisms that have strong enzymatic systems, for example fungi.[8] Ice-cold conditions or use of protease inhibitors are highly recommended, since during hypotonic shock and cell destruction, enzymes can destroy proteins of SC. Interestingly, some authors recommend to use protease inhibitors for mammalian cells as well.[9] The most useful protein markers of SC (i.e. morphologi-cal criteria for determining the stages of meiotic prophase I) are presented in Table 25.1.

IMMUNO-FISH APPROACH ON TOTAL PREPARATIONS OF SYNAPTONEMAL COMPLEXES

Immuno-FISH is a method combining immunocytochemistry (ICC) with fluorescence in situ hybridization (FISH).[5, 10] However, the combination of ICC and FISH methods applied to the same SC preparation imposes several limitations. They address key methodological differences between ICC and FISH:

I. ICC methods require keeping proteins in the most native conformation so that epitopes (both linear and 3D) are accessible by the antibodies used. Thus, buffered fixative to maintain pH is highly recommended, and all procedures of making spread SC-preparation should be carried out in ice-cold conditions.[10, 11] Using the buffered fixative allows to use ICC approaches and detect even minor foci of specific proteins involved in the processes of homologous chromosomes synapsis, DSB repair, meiotic recombination, transcription activity or silencing of chromatin. As stated before, meiocytes of organisms with strong enzymatic systems (fungi,[5] plants,[10] maybe even mammalian cells[9]) need ice-cold conditions or protease inhibitors during hypotonic shock and cell destruction.

II. FISH methods include a step of DNA denaturation in 50–70% formamide solution at a high temperature (temperature depends on the concentration of formamide used). Therefore, proteins on SC-preparations can denature during this step.

Nevertheless, in some cases, it is possible to use polyclonal and some monoclonal antibodies even after FISH as an additional round of ICC.[12, 13] Apparently, heat treatment does not destroy some of the epitopes (possibly the linear ones) and thus helps to get more data from one SC-preparation. On the other hand, protocols for unmasking protein epitopes in citrate buffer using high temperature and pressure are well-known in histological and cytological practice.[14] Many proteins could be successfully immunostained only after one round of unmasking. These facts gave rise to experiments applying certain antibodies after FISH.

Immuno-FISH studies on mammalian and reptile preparations have shown that some major proteins could be immunostained using stable fluorophores (AlexaFluor488 or AlexaFluor555) retaining partial fluorescence even after FISH. This fact must be taken into account when choosing spectral channels to avoid problems in data interpretation ICC and FISH on one spread preparation.[13, 15]

THE LOGICAL SCHEME OF IMMUNO-FISH IS AS FOLLOWS

1. ICC rounds (one or several)
2. Microscopy
3. FISH
4. Microscopy
5. Additional ICC round (if possible)
6. Fusion of ICC and FISH images.

After FISH, it is recommended to make images not only the optical channel under study, but all those available on the microscope assembly. This is important because chromatin becomes less contrasting after denaturation. The more optical channels are captured, the easier it is to colocalize the ICC and FISH images using morphological structures as reference. The most convenient for this are the SC axial structures (immunostaining of major SC protein SYCP3 is usually detectable after FISH if stable antibody conjugates are used), or secondly, chromatin contrasted with 4′,6-diamidino-2-phenylindole (= DAPI) before and after FISH.

Thus, immuno-FISH on SC-karyotypes allows to study chromatin changes (DNA and proteins) in dynamics—during successive stages of meiotic prophase I. It is important to know that meiotic bivalents are on average 10 times longer than metaphase mitotic chromosomes. Therefore, many tasks related to precise positioning of DNA-loci can be more easily solved on SC spreads. On the other hand, important discoveries on the elimination of part of the genomes also have been made using immuno-FISH on SC-preparations: germline-restricted chromosome (GRC), eliminated from somatic cells and spermatids,[16] or elimination of extra X-chromosome from 47,XXY karyotype in human.[9]

DNA FISH PROBES AND SPREAD PREPARATION OF SYNAPTONEMAL COMPLEXES

According to FISH probe's type, the methodological section can be divided into two basic parts:

1. FISH with probes for tandem repetitive DNA sequences (pericentromeric satellite DNA, telomere repeats, rDNA).
2. FISH with probes to unique DNA loci (locus-specific probes, whole chromosome probes).

The main difference of FISH on preparations of SC spreads is significantly less DNA compaction compared to metaphase mitotic chromosomes; therefore, the expected FISH-signal intensity is lower. On the other hand, in pachytene bivalents there are four copies of each DNA sequence located close to each other, and thus could provide stronger FISH-signal.[17] Figure 25.1 shows one of the hypothetical models for the interaction of four chromatin loops (DNA locus) during the assembly of the SC from leptotene to pachytene (Figure 25.1).

Short oligo-DNA probes (~18–40 bp) are mainly applicable to tandemly organized repetitive DNA-loci on meiotic chromosomes, such as telomeres, or pericentromeric satellites. To obtain a bright fluorescent signal, terminal fluorescent labeling of the 5′ end or both (5′ and 3′) ends is usually sufficient.[13] More expensive short PNA/LNA-FISH probes for tandemly repeated DNA provide reliable results and very specific signal.[13, 19] This type of probes is recommended for poorly preserved preparations or after several rounds of denaturation and washings.[15]

Locus-specific DNA FISH-probes have been successfully used to study meiosis and made it possible to visualize true chromatin loops in the structure of SCs (Figure 25.2).[12, 20–23] Probes derived from BAC clones containing dispersed repetitive DNA or so-called repeat-free (RF) probes are available for studies of human and mouse loci.[24, 25] Tyramide-FISH method has been successfully used to detect sequences shorter than 1,000 bps for the first time on mitotic metaphase

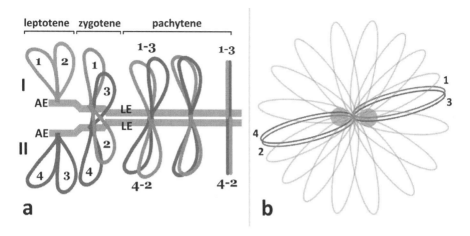

FIGURE 25.1 a) Model of interaction between chromatin loops in prophase I of meiosis. 1 and 2—sister chromatin loops of the homolog I; 3 and 4—sister chromatin loops of the homolog II. AE—axial elements of the chromosomes, LE—lateral elements of the assembled synaptonemal complex. b) Cross-section of the same model (from[18] with modifications).

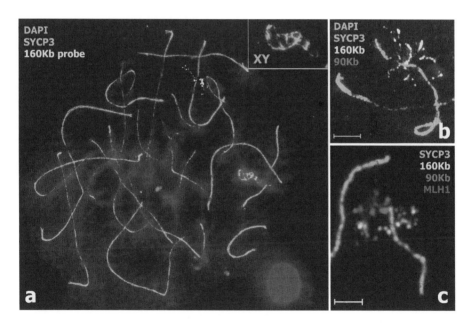

FIGURE 25.2 Immuno-FISH study of 160 Kb (yellow) and 90 Kb (violet) DNA loci on chromosome 17 in human spermatocyte I, early diplotene.

a) 160 Kb probe (yellow) on the bivalent 17.
b) 160 Kb probe (yellow) visualizes chromatin loops on the bivalent 17.
c) Immunostaining of crossing over marker, MLH1 (red) on the bivalent 17 near 160 Kb probe (yellow) Scale bar—5μm (from[18] with modifications).

chromosomes.[26, 27] This approach also showed applicability when used on preparations of SCs to detect single-copy genes.[28]

In general, FISH techniques on preparations of SCs are similar to the main protocols used on mitotic metaphase chromosomes. Particularly to those protocols where signal amplification is used (biotin-avidin, tyramide detection systems). Minor differences relate to less compaction, and hence greater accessibility of chromatin on meiotic prophase I preparation. Therefore, for example, RNAse treatment can be done in half the time compared to somatic metaphase plates.

SPREADING TECHNIQUE

MATERIALS

- Microscope slides: All preparations of primary spermatocyte's of oocytes spread nuclei (i.e. total preparations of SCs) should be done on slides with an adhesive coating, such as poly-L-lysine. This is important for attachment and preservation of chromatin during several rounds: ICC (immunostaining and washing) and FISH (denaturation, annealing, washing).
- Fixative for spreads: 1% paraformaldehyde, pH 8.4 containing 0.15% Triton X-100. Bi-distilled water with the addition of paraformaldehyde powder must be heated to 60–65°C. Add 1–2 small drops (10–15 μl) of 1N NaOH until a transparent solution is obtained (avoid boiling). Let the fixative cool, filter using filter paper, add Triton X-100, and adjust to pH8.4 with 0.1M sodium tetraborate.
- Hypotonic solution: 0.1 M sucrose (1.7g sucrose in 50ml bi-distilled water).
- Wash solution: 0.4% Photoflo (Kodak) adjusted to pH 8.2–8.4 with borate buffer. Can be stored as a 4% stock solution and diluted before use.

PROCEDURE

1. Isolate gonadal tissues and transfer to cold Eagle's medium (without L-Glutamine) or 1×PBS (pH 7.4) in a small Petri dish. Samples should be isolated shortly before preparations. Alternatively, storage is allowed in cold (+4°C) conditions in the Eagle medium (or similar) or in 1×PBS for several hours. Avoid drying of tissues. Using two forceps, disaggregate the tissue of the seminiferous tubules. In the case of ovarian tissue, cell disaggregation can be carried out mechanically between glass slide and coverslip (better to use concave microscope slide). Using a stereomicroscope, gently press the dissecting needle on the coverslip over the pressed piece of tissue and collect the suspension with a pipette. Avoid drying out and keep the temperature near 4°C using cold plate thermostat.

2. Transfer 50 μl of cell suspension from step 1 in a 1.5 ml reaction tube and continue to disrupt aggregates using pipette with a 200 μl tip with one mm shortened tip. Avoid formation of bubbles. Check a small amount of the suspension in a light microscope for large spermatocytes and in case of

overabundance of spermatids and/or fat inclusions centrifuge for 10 min at 400–500 g and remove the supernatant (for next steps use lower fraction only).

3. Add slowly an equivalent amount (50 µl) of hypotonic solution (0.1 M sucrose). Keep for 5–10 min at 4°C.

4. Repeat step 3 two times by adding 50 µl of 0.1 M sucrose every time. Keep the vial on +4°C.

5. Immerse the microscope slide with poly-L-lysine coating in cold fixative for spreading (1% paraformaldehyde, pH 8.4, containing 0.15 % Triton X-100, see Materials). The wet area can vary (24×24 or 24×50 mm) depending on the number of slides needed or suspension quality. Wipe off the fixative from the lower side of slide.

6. Using pipette, accurately touch the 20–30 µl hanging drop of the cell suspension with the surface of the cold fixative on the slide. It is better to use a not very sharp pipette tip (200 µl); this allows forming a rounded hanging drop of the cell suspension. Distribute the suspension well by turning the slide in different directions. If the suspension contains a small number of cells (oocytes), then 2–3 drops can be added per slide, but in this case, use a large (24×50 mm) area of the fixative solution on the slide.

7. Immediately after dropping, transfer the slides in a water bath at 4°C for one hour.

8. Wash three times in cold 0.04% PhotoFlo solution (pH 8.4) for 20 seconds each without agitation.

9. Place the slides to dry in a vertical position in a cold place under the air flow for 0.5–1 hour.

10. Store the slides at −20°C before use.

11. Slides should be rehydrated in PBS for 1 min before immunostaining.

CONCLUSIONS

In general, the immuno-FISH method on preparations of meiotic chromosomes is promising for working with a variety of biological objects due to the availability of antibodies with a wide range of immune homology, as well as due to the accumulation and analysis of genomic data. Thus, it is possible to study the key stage of the life cycle of an organism using both, (i) a variety of meiosis associated protein markers, and (ii) DNA markers (whole chromosome painting, unique DNA loci, satellite DNA, telomeric repeats, rDNA). In the near future, it is obvious that a large number of new protein markers associated with the SC in particular, and with meiosis prophase I in general, will become available.[29] Therefore, when using immuno-FISH-protocols, first of all highest quality preparation of spread meiotic bivalents is required, with preservation of the native form of protein markers. Second, FISH protocols should take into account the lower chromatin density in meiosis compared to the metaphase of mitosis. Overall, this means that methods of amplifying detected signals are important.

REFERENCES

1. Navarro, J.; Vidal, F.; Guitart, M.; Egozcue, J. A method for the sequential study of synaptonemal complexes by light and electron microscopy. *Hum. Genet.* **1981**, 59, 419–421.

2. Peters, A.H.F.M.; Plug, A.W.; Van Vugt, M J.; De Boer, P. A drying-down technique for the spreading of mammalian meiocytes from the male and female germline. *Chromosome Res.* **1997**, 5, 66–68.

3. Lisachov, A.; Poyarkov, N.; Pawangkhanant, P.; Borodin, P.; Srikulnath, K. New karyotype of Lygosoma bowringii (Günther, 1864) suggests cryptic diversity. *Herpet. Notes.* **2018**, 11, 1083–1088.

4. Spangenberg, V.; Arakelyan, M.; Galoyan, E.; Martirosyan, I.; Bogomazova, A.; Martynova, E.; de Bello Cioffi, M.; Liehr, T.; Al-Rikabi, A.; Osipov, F.; Petrosyan, V.; Kolomiets, O. Meiotic synapsis of homeologous chromosomes and mismatch repair protein detection in the parthenogenetic rock lizard *Darevskia unisexualis*. Molecular Reproduction and Development. **2021**, 88, 119–127.

5. Araya-Jaime, C.; Serrano, É.A.; de Andrade Silva, D.M.Z.; Yamashita, M.; Iwai, T.; Oliveira, C.; Foresti, F. Surface-spreading technique of meiotic cells and immunodetection of synaptonemal complex proteins in teleostean fishes. *Mol. Cytogenet.* **2015**, 8, 1–6.

6. Viera, A.; Santos, J.L.; Parra, M.T.; Calvente, A.; Gómez, R.; de la Fuente, R.; Suja, J.Á.; Page, J.; Rufas, J.S. Cohesin axis maturation and presence of RAD51 during first meiotic prophase in a true bug. *Chromosoma.* **2009**, 118(5), 575–589.

7. Loidl, J.; Jones, G.H. Synaptonemal complex spreading in Allium. *Chromosoma.* **1986**, 93, 420–428.

8. Mazheika, I.S.; Kolomiets, O.L. A method for preparation of synaptonemal complexes of meiotic chromosomes from basidial protoplasts of the white button mushroom *Agaricus bisporus* (Lange) Imbach. *Russ J Genet.* **2003**, 39, 283–287.

9. Sciurano, R.B.; Luna Hisano, C.V.; Rahn, M.I.; Brugo Olmedo, S.; Rey Valzacchi, G.; Coco, R.; Solari, A.J. Focal spermatogenesis originates in euploid germ cells in classical Klinefelter patients. *Hum. Reprod.* **2009**, 24, 2353–2360.

10. Sepsi, A.; Fábián, A.; Jäger, K.; Heslop-Harrison, J.S.; Schwarzacher, T. ImmunoFISH: Simultaneous visualisation of proteins and DNA sequences gives insight into meiotic processes in nuclei of grasses. *Front Plant Sci.* **2018**, 9, 1193.

11. Kolomiets, O.L.; Lelekova, M.A.; Kashintsova, A.A.; Kurilo, L.F.; Bragina, E.E.; Chernykh, V.B.; Gabliya, M.Y.; Vinogradov, I.V.; Vityazeva, I.I.; Bogolyubov, S.V.; Spangenberg, V.E. Detection of human meiotic and spermatogenetic anomalies using light, electron and fluorescence microscopy. *Androl Genit Surgery.* **2018**, 19, 24–35.

12. Pigozzi, M.I. Localization of single-copy sequences on chicken synaptonemal complex spreads using fluorescence in situ hybridization (FISH). *Cytogenet Genome Res.* **2007**, 119, 105–112.

13. Spangenberg, V.; Losev, M.; Volkhin, I.; Smirnova, S.; Nikitin, P.; Kolomiets, O. DNA Environment of centromeres and non-homologous chromosomes interactions in mouse. *Cells.* **2021**, 10, 3375.

14. Sciurano, R.B.; Solari, A.J. Ultrastructural and immunofluorescent methods for the study of the XY body as a biomarker. In: *Functional Analysis of DNA and Chromatin*; Humana Press, Totowa, NJ, **2014,** pp. 137–149.

15. Spangenberg, V.; Arakelyan, M.; Galoyan, E.; Pankin, M.; Petrosyan, R.; Stepanyan, I.; Grishaeva, T.; Danielyan, F.; Kolomiets, O. Extraordinary centromeres: Differences

in the meiotic chromosomes of two rock lizards species Darevskia portschinskii and Darevskia raddei. *Peer J.* **2019**, 7, e6360.

16. Torgasheva, A.A.; Malinovskaya, L.P.; Zadesenets, K.S.; Karamysheva, T.V.; Kizilova, E.A.; Akberdina, E.A.; Pristyazhnyuk, I.E.; Shnaider, E.P.; Volodkina, V.A.; Saifitdinova, A.F.; Galkina, S.A.; Larkin, D.M.; Rubtsov, N.B.; Borodin, P.M. Germline-restricted chromosome (GRC) is widespread among songbirds. *Proc. Natl. Acad. Sci.* **2019**, 116, 11845–11850.

17. Zadesenets, K.S.; Katokhin, A.V.; Mordvinov, V.A.; Rubtsov, N.B. Telomeric DNA in chromosomes of five opisthorchid species. *Parasitol Internat.* **2012**, 61, 81–83.

18. Spangenberg, V. Dynamics of the structural organization of the chromosomes during the meiotic prophase I in human and mouse. PhD thesis. NI Vavilov Institute of General Genetics RAS (Moscow, **2013**).

19. Lisachov, A.P.; Borodin, P.M. Microchromosome polymorphism in the sand lizard, Lacerta agilis Linnaeus, 1758 (Reptilia, Squamata). Comparat. *Cytogenet.* **2016**, 10, 387.

20. Beliveau, B.J.; Joyce, E.F.; Apostolopoulos, N.; Yilmaz, F.; Fonseka, C.Y.; McCole, R.B.; Wu, C.T. Versatile design and synthesis platform for visualizing genomes with Oligopaint FISH probes. *Proc. Natl. Acad. Sci.* **2012**, 109, 21301–21306.

21. Froenicke, L.; Anderson, L.K.; Wienberg, J.; Ashley, T. Male mouse recombination maps for each autosome identified by chromosome painting. *Am. J. Hum. Genet.* **2002**, 71, 1353–1368.

22. Bogdanov, Y.F.; Spangenberg, V.E.; Dadashev, S.Y.; Vityazeva, I.I.; Bogoliubov, S.V.; Kolomiets, O.L. Morphological manifestation of unique DNA segments in human meiotic prophase I. *Cell Tiss. Biol.* **2012**, 6, 407–411.

23. Grishaeva, T.M.; Spangenberg, V.E.; Kolomiets, O.L.; Dadashev, S.Y.; Bogdanov, Y.F. Peculiarities of chromosome organization in meiosis. *Cell Tiss. Biol.* **2013**, 7, 343–346.

24. Liehr, T. Commercial FISH probes. In: *Fluorescence In Situ Hybridization (FISH)*; Springer, Berlin, **2017**, pp. 49–61.

25. Swennenhuis, J.F.; Foulk, B.; Coumans, F.A.; Terstappen, L.W. Construction of repeat-free fluorescence in situ hybridization probes. *Nucl. Acids Res.* **2012**, 40, e20.

26. Khrustaleva, L.I.; Kik, C. Localization of single-copy T-DNA insertion in transgenic shallots (Allium cepa) by using ultra-sensitive FISH with tyramide signal amplification. *Plant J.* **2001**, 25, 699–707.

27. Kudryavtseva, N.; Ermolaev, A.; Karlov, G.; Kirov, I.; Shigyo, M.; Sato, S.; Khrustaleva, L. A dual-color tyr-FISH method for visualizing genes/markers on plant chromosomes to create integrated genetic and cytogenetic maps. *Int J Mol Sci.* **2021**, 22, 5860.

28. Kirov, I.V.; Van Laere, K.; Khrustaleva, L.I. High resolution physical mapping of single gene fragments on pachytene chromosome 4 and 7 of *Rosa*. *BMC Genet.* **2015**, 16, 1–10.

29. da Cruz, I.; Rodríguez-Casuriaga, R.; Santiñaque, F.F.; Farías, J.; Curti, G.; Capoano, C.A.; Geisinger, A. Transcriptome analysis of highly purified mouse spermatogenic cell populations: Gene expression signatures switch from meiotic-to postmeiotic-related processes at pachytene stage. *BMC Genomics.* **2016**, 17, 1–19.

RNA-FISH— on Lampbrush Chromosomes: Visualization of Individual Transcription Units

Tatiana Kulikova and Alla Krasikova

CONTENTS

DOI: 10.1201/9781003223658-26

INTRODUCTION

RNA directed fluorescence in situ hybridization (FISH) is an effective approach to localize RNA sequences in situ, visualize gene transcripts, and estimate gene expression in time and space.[1–3] In a typical interphase nucleus, transcribed genes, together with nascent transcripts at the site of their synthesis, are visible as dot-like foci.[4, 5] The only exception are several highly expressed genes with very long introns that form loop-like structures with nascent transcripts in *Drosophila* spermatocytes,[6] and in human somatic cells.[7] However, in fishes, amphibians, reptiles, birds, and certain insects, hypertranscription during oogenesis leads to decompactization of hundreds of transcribed genomic regions and transformation of the meiotic chromosomes into highly elongated lampbrush chromosomes (LBCs).[8–11] Lateral loops of LBCs representing sites of ongoing transcription serve as a model for exploration into nascent RNA synthesis and processing at the cytological level.[9, 12] Single transcription units on the lateral loops of amphibian and avian LBCs with a gradient of RNP matrix can be seen even by light microscopy.[13–15]

First in situ hybridization experiments on LBCs of newt *Notophthalmus viridescens* were performed with radioactively labeled cRNA probes detected by autoradiography.[16] In further studies, radioactive 5S rDNA probes were hybridized with non-denatured *N. viridescens* LBCs to reveal nascent RNA.[17] In these early DNA/RNA in situ hybridization experiments nascent 5S RNA transcripts were revealed on a set of lateral loops. In control preparations, RNase pretreatment eliminated 5S RNA transcripts from the ribonucleoprotein (RNP) matrix of these loops. By hybridization of the radioactively labelled DNA-probe to nascent RNA transcripts on LBCs of crested newt, *Triturus cristatus carnifex*, Varley and co-authors evidenced transcription of highly repetitive DNA during oogenesis.[18] Similarly, radioactive probes to satellite I DNA hybridized to non-denatured LBCs of *N. viridescens* without RNase digestion demonstrated intense labeling of lateral loops.[19] Next FISH method was developed and employed to detect nascent RNA on the lateral loops of amphibian and avian LBCs.[13, 20–23] Nowadays, RNA-FISH can be successfully applied to reveal newly synthesized pre-mRNA and non-coding RNA on LBC lateral loops. For example, probes made by in vitro transcription and bacterial artificial chromosome (BAC)–derived probes were effectively hybridized to transcription units of intensively expressed protein-coding genes on the lateral loops of axolotl and chicken LBCs.[15, 24]

Visualization of nascent gene transcripts on individual transcription units could be accompanied by immunofluorescent staining as described elsewhere.[25] For example, by RNA-FISH after immunofluorescent detection of the elongating form of RNA polymerase II we demonstrated that short 41 bp repeats are transcribed on the lateral loops of chicken LBCs by RNA polymerase II with hyperphosphorylated C-terminal domain.[13] For other examples of RNA-FISH after LBC immunostaining, see in references [22] and [26]. RNA-FISH can also be combined with the targeting of labeled protein components to the RNP matrix of LBC lateral loops. Besides, FISH according to nascent RNA hybridization protocol (see later) can be performed on LBC preparations after incorporation of RNA precursors such as bromouridine triphosphate (BrUTP), giving an opportunity to estimate transcription rates for individual genes.

In this chapter, we provide an effective RNA-FISH protocol that includes steps essential to preserve RNA on LBC preparations; also necessary control experiments to estimate gene expression are presented. The described methodology enables simultaneous visualization of up to 10 transcribed single-copy genomic loci on the same LBC.

DESCRIPTION OF METHODS

CHROMOSOME ISOLATION AND PRETREATMENT

Methods for obtaining preparations of LBCs and nuclear bodies from amphibian and avian diplotene oocytes are described in detail elsewhere (for example in [27] and [28], or on https://projects.exeter.ac.uk/lampbrush/protocols.htm). In these protocols, oocyte nucleus content is spread within an isolation chamber mounted on a standard microscope slide. After microdissection from the oocyte nucleus LBC spreads are fixed in paraformaldehyde (PFA) in phosphate buffered saline (PBS) followed by alcohol-based fixation. Depending on the antigen detected, other fixation protocols may be used: without ethanol or with reduced time of fixation in PFA.

Permeabilization can be omitted, since preparations of native chromosomes isolated from the oocyte nucleus by microdissection are used. The absence of permeabilization results in the preservation of native chromosome morphology.

Nascent RNA on LBC preparations exists in a complex with a number of RNA binding proteins and other RNPs; therefore, pretreatment with proteinases is not performed, as this may reduce the specific hybridization signal. For the RNA FISH protocol no ribonuclease pretreatment should be done to detect nascent RNA targets on LBC spreads. However, certain control slides imply RNase digestion step (see later in the section "Control Experiments").

RNA PRESERVATION

To achieve optimal results, RNA preservation measures must be followed, from chromosome isolation procedure to post-hybridization washings. Bench surfaces should be treated with 5% H_2O_2 or commercial RNase inhibitors. Slides and chambers for chromosome isolation, lab beakers and other glassware, as well as metal instruments should be washed in 2% 7X-detergent, thoroughly rinsed in distilled water and dried in a drying oven at 200°C or autoclaved. Deionized water used for buffers should be autoclaved. Whole chromosome isolation procedure and slide handling should be performed with sterile laboratory gloves.

CHROMOSOME FIXATION

1. Fix freshly isolated LBC preparations in 2% PFA in a PBS for 30 minutes at room temperature.
2. Dehydrate in 50% ethanol and 70% ethanol for 5 minutes each.

3. Store overnight in 70% ethanol.
4. Mark chamber positions on the opposite side of the slide by a diamond blade.
5. Remove the isolation chambers with a razor blade; avoid slide drying.
6. Dehydrate preparations in 96% ethanol.
7. Leave to dry on air.
8. Examine preparations in the microscope equipped with the phase contrast or Nomarski differential interference contrast.

PROBE LABELING

A variety of probes can be used for RNA-FISH on LBC preparations including PCR-derived DNA probes, probes obtained via nick-translation, cDNA probes biotinylated by random-primed oligonucleotide labeling, RNA probes made by in vitro transcription, and synthetic oligonucleotide or locked nucleic acid probes. Protocol for labeling DNA probes for FISH on LBC preparations by (degenerated oligonucleotide primed = DOP-) polymerase chain reaction (PCR) was given earlier in [23]. Labeling of longer probes (for example isolated BAC clone DNA) by Nick-translation can be performed as described in [29]. Oligonucleotide probes are usually labeled by terminal deoxynucleotidyl transferase.[30, 31] Probes can be labeled directly by fluorochrome-modified nucleotides or indirectly by hapten-modified nucleotides further detected by a fluorochrome. The quality of all labeled probes can be verified by DNA-FISH on metaphase chromosome preparations.

To visualize multiple gene transcripts in the same experiment, a mixture of ~10 BAC clone-based probes can be applied on a single LBC set. If gene targets localize in a relatively small genomic region (2–2.5 Mbp), BAC probes should be labeled with different haptens and fluorochromes.

RNA-FISH PROCEDURE

RNA-FISH on LBC preparations is performed according to conventional protocol[32] with modifications.[20] Stringency of the hybridization can be adjusted by varying the concentration of formamide, the incubation temperature during hybridization, concentration of Na+ in the washing buffer, and the temperature of the washing buffer. The duration of hybridization should not be greatly reduced since it may lead to underrepresentation of specific nascent RNA targets; however, in case of oligonucleotide probes, hybridization time can be reduced up to 4 hours. To detect nascent transcripts on LBC lateral loops, amplification of the FISH signal is usually not required due to the presence of multiple closely spaced RNA targets on a single transcription unit.

Re-FISH according to RNA hybridization protocol can be performed on LBC spreads after image acquisition. In this case, after coverslip removal, slides with LBCs are washed in 4×SSC (saline-sodium citrate buffer) with 0.01% Tween 20 at 37°C. Then RNA FISH procedure is applied with another probe set. However, it should be noted that RNA on the lateral loops could be partially degraded during this procedure. Thus, Re-FISH protocol can be best applied to co-localize probes which, as shown earlier, hybridize with nascent RNA.

RNA FISH Procedure for Double Stranded DNA-Probes (PCR- or Nick-Translation Products)

Hybridization Mixture and In Situ Hybridization

1. To prepare 20 µl of hybridization mixture mix 0.8 µg of labeled DNA-probe/s with 50× excess of carrier nucleic acid (salmon sperm DNA or yeast tRNA) and a competitor DNA (e.g. C_0t DNA fractions), if needed; add 5M NaCl (0.1 of the DNA mixture volume); add ice-cold 96% ethanol (2.5 of the mixture volume). Freeze for several hours or overnight.
2. Precipitate the DNA by centrifugation at 13,000 rpm at +4°C for 15 minutes. Remove supernatant.
3. Wash the pellet with 200 µl ice-cold 70% ethanol followed by centrifugation at 13,000 rpm at +4°C for 10 minutes. Remove supernatant and dry the DNA pellet on air.
4. Dissolve the DNA pellet in 10 µl 100% formamide, vortex thoroughly.
5. Add 10 µl of the mixture of 4×SSC (20×SSC: 3M NaCl, 0.3M sodium citrate, pH 7) and 20% dextran sulfate (half of the final volume of hybridization mixture), vortex thoroughly and siege.
6. Denature hybridization mixture at 95°C for 5 minutes. If a competitive DNA is added, it is necessary to adjust the pre-annealing conditions.
7. Place the tube on ice.
8. Drop 1 µl of hybridization mixture per each isolation chamber.
9. Cover all four isolation chambers with 18×18 mm coverslip. If individual hybridization mixtures are applied, each droplet can be covered with round coverslip (8 mm) or quarter sliced 18×18 mm coverslip.
10. Seal the coverslips with rubber cement.
11. Place slides into a moist chamber at 37°C, overnight.

Post-Hybridization Washings

12. Gently remove rubber cement with the forceps.
13. Place slides into a Coplin jar or 50 ml beaker with 2×SSC for 5–10 minutes to rinse coverslips.
14. Wash slides through three changes of 0.2×SSC at 60°C in a water bass with rocking.
15. Wash slides through two changes of 2×SSC at 45°C in a water bass with rocking.

Detection of Hapten Labeled Probes

16. Put the slides into a moist chamber and apply 100 µl of blocking solution (1% blocking reagent or 1% BSA in 4×SSC with 0.1% Tween 20). Immediately cover the drops with squares of parafilm. Incubate at 37°C for 40 minutes.

17. Remove parafilm and excess of the blocking solution and apply 50–100 µl of primary antibodies against a hapten or avidin/streptavidin conjugated with a fluorochrome dissolved in blocking solution. Immediately cover the drops with squares of parafilm. Incubate in a moist chamber at 37°C for 40 minutes.

18. Wash slides through three changes of 4×SSC with 0.01% Tween 20 at 42°C in a water bass with rocking.

19. Put slides into a moist chamber and apply 50–100 µl of secondary antibodies or anti-avidin/streptavidin antibodies in blocking solution. Immediately cover the drops with squares of parafilm. Incubate at 37°C for 40 minutes.

20. Wash slides through three changes of 4×SSC with 0.01% Tween 20 at 42°C in a water bass with rocking.

21. If biotin is used, repeat steps 17 and 18.

22. Rinse slides with deionized water.

23. Dehydrate in increasing concentration of ethanol: 70% and 96%, for 5 minutes each.

24. Dry on air.

25. Apply 7 µl of antifade solution (65% glycerol, 0.01M TrisHCl, 2% DABCO, pH 7.5) containing 1 µg/µl DAPI and mount 24×24 coverslips.

RNA FISH Procedure for Oligonucleotide Probes

Hybridization Mixture and In Situ Hybridization

1. Prepare the hybridization mixture containing 42% formamide, 10% dextran sulfate, 2×SSC, 10–30 ng/µl oligonucleotide probe and 50× excess of carrier nucleic acid (salmon sperm DNA or yeast tRNA). Vortex thoroughly.

2. Cover all four isolation chambers with 18×18 coverslip. If individual hybridization mixtures are applied, each droplet can be covered with round 8 mm or quarter sliced 18×18 coverslip.

3. Seal the coverslips with rubber cement.

4. Place slides into a moist chamber at room temperature, overnight.

Post-Hybridization Washings

5. Gently remove rubber cement with the forceps.

6. Place slides into a Coplin jar or 50 ml beaker with 2×SSC for 5–10 minutes to rinse coverslips.

7. Wash slides through three changes of 2×SSC at 37°C in a water bass with rocking.

Detection of Hapten Labeled Oligonucleotide Probes

Hapten labeled oligonucleotide probes are detected by the same procedure as double-stranded DNA-probes (steps 16–24). Apply 7 µl of antifade solution (65% glycerol, 0.01M TrisHCl, 2% DABCO, pH 7.5) containing 1 µg/µl DAPI and mount 24×24 coverslips.

MICROSCOPY AND IMAGE ANALYSIS

Conventional 2D fluorescence microscopy is used to examine LBC preparations after FISH. Additionally, phase contrast or Nomarski differential interference contrast (DIC) can be used to obtain images of the chromosomes themselves. Generally, no image processing is required when the transcripts are visualized on the lateral loops of LBCs. Since the images of giant LBCs consist of fragments, appropriate software is used to compile a whole chromosome image.

Because LBCs represent meiotic diplotene bivalents, four hybridization signals for each single-copy genomic loci can be seen on each LBC set. Closely spaced nascent RNA transcripts are visible as a polarized RNP matrix with a thin end corresponding to the start of transcription (Figure 26.1b–b').

CONTROL EXPERIMENTS

A combination of control experiments is necessary to accurately estimate genuine gene transcription on LBC spreads by RNA-FISH (Figure 26.1):

- DNA-RNA FISH with a positive control probe is applied to estimate RNA preservation on LBC preparations. As a positive control, a probe to gene loci previously demonstrated to be transcribed on LBC lateral loops can be used.
- DNA-RNA FISH after Ribonuclease A (RNase A) pretreatment is applied to check the specificity of probe hybridization to RNA (Figure 26.1c–c''). Other specific ribonucleases (for example RNase H, RNase R, S7 nuclease) can be applied on separate LBC preparations to establish the structure of targeted RNA.
- DNA-RNA FISH with sense single-stranded oligonucleotide probes can be used as a negative control for single-copy genomic regions transcribed from one strand.
- DNA-DNA FISH after RNase A pretreatment is performed on LBC preparations as a positive control for probe hybridization to the target sequence (Figure 26.1d–d'').
- DNA-RNA FISH procedure, but without labelled probe in the hybridization mixture, followed by post-hybridization washings and probe detection is used as a control for non-specific binding of the detection system to the RNP-matrix of LBC lateral loops.
- DNA-RNA FISH procedure without any probe followed by post-hybridization washings but without probe detection is performed as a control for autofluorescence.

Control experiments should be performed in parallel to the main DNA-RNA FISH experiment on LBC preparations obtained on the same day and under the same conditions. Images should be acquired with the same parameters as the images in the main DNA-RNA FISH experiment.

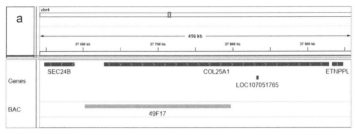

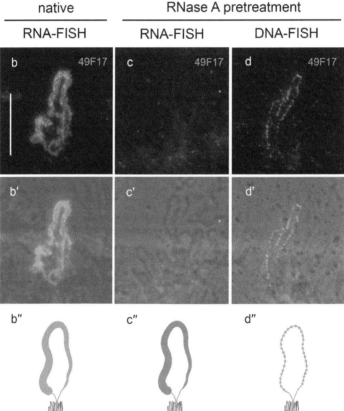

FIGURE 26.1 a–d″—Visualization of nascent RNA within individual transcription units of a single-copy gene on chicken lampbrush chromosomes.

a—A genomic region of chicken chromosome 4 occupied by BAC-clone 49F17 and overlapping with *COL25A1* gene (snapshot of the Integrative Genomics Viewer, chicken genome assembly version galGal5);

b–d′—FISH with BAC-clone 49F17 (CHORI-261 chicken BAC-library) DNA-probe (green) on lampbrush chromosome (LBC) 4 with control experiments, chromosomes are counterstained with DAPI (blue); fragments of individual homologous chromosomes (half-bivalents) with a pair of lateral loops containing transcription units of *COL25A1* gene are shown.

b–b′—RNA-FISH on native LBCs.

c–c′—RNA-FISH on LBCs pretreated with RNase A.

d–d′—DNA-FISH on denatured LBCs pretreated with RNase A.

b′–d′—fluorescent images merged with phase contrast. Scale bar—10 μm.

b″–d″—schematic drawings of the corresponding microphotographs. LBCs were imaged by a monochrome CCD camera (1.3 MegaPixel resolution) using fluorescent microscope Leica DM 4000 (Leica Microsystems).

INTERPRETATION OF RESULTS

When we interpret the results of an RNA-FISH experiment on LBC preparations, we should take into consideration the quality of spreads, RNA preservation, results of control experiments and genomic characteristics of the targeted region. Control experiments are essential to reliably establish the transcriptional activity of genes on LBCs by RNA-FISH, for example [20, 21]. It should be noted that different RNA species could be partially and nonuniformly degraded leading to variability across RNA FISH results.[33] Moreover, analysis of gene expression on LBC spreads can depend on the transcriptional activity of a certain gene, i.e. the number of nascent RNA transcripts associated with individual transcription units. In addition, nascent transcripts of tandem repeats could be still detectable on those LBC preparations on which single-gene transcripts are already undetectable due to RNA degradation.

CONCLUSIONS

Here we describe an optimized RNA-FISH protocol directed to effective visualization of transcribed genes on LBC preparations. In contrast to the basic DNA/(DNA+RNA-transcript) FISH protocol on LBC spreads,[23] critical steps necessary to detect nascent gene transcripts on lampbrush lateral loops are emphasized. The delineated protocol was successfully implemented to establish transcription of protein-coding and non-coding RNA genes on LBCs of birds,[24] and can be applied to LBC preparations of other diverse model organisms.

ACKNOWLEDGEMENTS

The research was supported by the RSF (grant #19–74–20075). Microscopy was performed in the research resource center "Molecular and cell technologies" of St. Petersburg State University.

REFERENCES

1. Chartrand, P.; Bertrand, E.; Singer, R.H.; Long, R.M. Sensitive and high-resolution detection of RNA in situ. *Methods Enzymol.* **2000**, 318, 493–506.
2. Brown, J.M.; Buckle, V.J. Detection of nascent RNA transcripts by fluorescence in situ hybridization. *Methods Mol. Biol.* **2010**, 659, 33–50.
3. Long, X.; Colonell, J.; Wong, A.M.; Singer, R.H.; Lionnet, T. Quantitative mRNA imaging throughout the entire *Drosophila* brain. *Nat. Methods.* **2017**, 14, 703–706.
4. Femino, A.M.; Fay, F.S.; Fogarty, K.; Singer, R.H. Visualization of single RNA transcripts in situ. *Science.* **1998**, 280, 585–590.
5. Ferraro, T.; Lucas, T.; Clémot, M.; De Las Heras Chanes, J.; Desponds, J.; Coppey, M.; Walczak, A.M.; Dostatni, N. New methods to image transcription in living fly embryos: The insights so far, and the prospects. *WIREs Dev. Biol.* **2016**, 5, 296–310.
6. Fingerhut, J.M.; Moran, J.V.; Yamashita, Y.M. Satellite DNA-containing gigantic introns in a unique gene expression program during *Drosophila* spermatogenesis. *PLoS Genet.* **2019**, 15, e1008028.
7. Leidescher, S.; Ribisel, J.; Ullrich, S.; Feodorova, Y.; Hildebrand, E.; Galitsyna, A.; Bultmann, S.; Link, S.; Thanisch, K.; Mulholland, C.; Dekker, J.; Leonhardt, H.; Mirny,

L.; Solovei, I. Spatial organization of transcribed eukaryotic genes. *Nat. Cell Biol.* **2022**, 24, 327–339.

8. Callan, H.G. *Lampbrush chromosomes.* Springer, Berlin; **1986**.

9. Morgan, G.T. Lampbrush chromosomes and associated bodies: New insights into principles of nuclear structure and function. *Chromosome Res.* **2002**, 10, 177–200.

10. Gaginskaya, E.; Kulikova, T.; Krasikova, A. Avian lampbrush chromosomes: A powerful tool for exploration of genome expression. *Cytogenet. Genome Res.* **2009**, 124, 251–267.

11. Krasikova, A.; Kulikova, T. *Lampbrush Chromosomes: Current Concepts and Research Perspectives.* St Petersburg University Publishing House, St Petersburg, **2019**.

12. Morgan, G.T. Imaging the dynamics of transcription loops in living chromosomes. *Chromosoma.* **2018**, 127, 361–374.

13. Deryusheva, S.; Krasikova, A.; Kulikova, T.; Gaginskaya, E. Tandem 41-bp repeats in chicken and Japanese quail genomes: FISH mapping and transcription analysis on lampbrush chromosomes. *Chromosoma.* **2007**, 116, 519–530.

14. Kaufmann, R.; Cremer, C.; Gall, J. G. Superresolution imaging of transcription units on newt lampbrush chromosomes. *Chromosome Res.* **2012**, 20, 1009–1015.

15. Keinath, M.C.; Davidian, A.; Timoshevskiy, V.; Timoshevskaya, N.; Gall, J.G. Characterization of axolotl lampbrush chromosomes by fluorescence in situ hybridization and immunostaining. *Exp. Cell Res.* **2021**, 401, 112523.

16. Barsacchi, G.; Gall, J.G. Chromosomal localization of repetitive DNA in newt, Triturus. *J Cell Biol.* **1972**, 54, 580–591.

17. Pukkila, P.J. Identification of the lampbrush chromosome loops which transcribe 5S ribosomal RNA in *Notophthalmus* (Triturus) viridescens. *Chromosoma.* **1975**, 53, 71–89.

18. Varley, J.M.; Macgregor, H.C.; Erba, H.P. Satellite DNA is transcribed on lampbrush chromosomes. *Nature.* **1980**, 283, 686–688.

19. Diaz, M.O.; Barsacchi-Pilone, G.; Mahon, K.A.; Gall, J.G. Transcripts from both strands of a satellite DNA occur on lampbrush chromosome loops of the newt *Notophthalmus. Cell.* **1981**, 24, 649–659.

20. Weber, T.; Schmidt, E.; Scheer, U. Mapping of transcription units on *Xenopus laevis* lampbrush chromosomes by in situ hybridization with biotin-labeled cDNA probes. *Eur. J. Cell Biol.* **1989**, 50, 144–153.

21. Solovei, I.; Gaginskaya, E.R.; Macgregor, H.C. The arrangement and transcription of telomere DNA sequences at the ends of lampbrush chromosomes of birds. *Chromosome Res.* **1994**, 2, 460–470.

22. Sallacz, N.B.; Jantsch, M.F. Chromosomal storage of the RNA-editing enzyme ADAR1 in *Xenopus* oocytes. *Mol. Biol. Cell.* **2005**, 16, 3377–3386.

23. Zlotina, A.; Krasikova, A. FISH in lampbrush chromosomes. In *Fluorescence In Situ Hybridization (FISH); Springer Protocols Handbooks*; Liehr, T., Ed. Springer Berlin Heidelberg, Berlin; **2017**, pp. 445–457.

24. Kulikova, T.; Maslova, A.; Starshova, P.; Rodriguez, S.J.; Krasikova, A. Comparison of the somatic TADs and lampbrush chromomere-loop complexes in transcriptionally active prophase I oocytes. *Chromosoma.* **2022**. https://doi.org/10.1007/s00412-022-00780-5.

25. Krasikova, A.V.; Kulikova, T.V. Distribution of heterochromatin markers in lampbrush chromosomes in birds. *Russ. J. Genet.* **2017**, 53, 1022–1029.

26. Kulikova, T.; Chervyakova, D.; Zlotina, A.; Krasikova, A.; Gaginskaya, E. Giant poly(A)-rich RNP aggregates form at terminal regions of avian lampbrush chromosomes. *Chromosoma.* **2016**, 125, 709–724.

27. Morgan, G.T. Working with oocyte nuclei: Cytological preparations of active chromatin and nuclear bodies from amphibian germinal vesicles. *Methods Mol. Biol.* **2008**, 463, 55–66.

28. Gall, J.G.; Nizami, Z.F. Isolation of giant lampbrush chromosomes from living oocytes of frogs and salamanders. *J. Vis. Exp.* **2016**, 118, 54103.

29. Green, M.R.; Sambrook, J. Labeling of DNA probes by nick translation. *Cold Spring Harb Protoc.* **2020**, 2020, prot100602.

30. Green, M.R.; Sambrook, J. Labeling the 3′ termini of oligonucleotides using terminal deoxynucleotidyl transferase. *Cold Spring Harb Protoc.* **2021**, 2021, prot100685.

31. Green, M.R.; Sambrook, J. Preparation of labeled DNA, RNA, and oligonucleotide probes. *Cold Spring Harb Protoc.* **2022**, 2022, pdb.top100578.

32. Young, A.P.; Jackson, D.J.; Wyeth, R.C. A technical review and guide to RNA fluorescence in situ hybridization. *Peer J.* **2020**, 8, e8806.

33. Baudrimont, A.; Jaquet, V.; Wallerich, S.; Voegeli, S.; Becskei, A. Contribution of RNA degradation to intrinsic and extrinsic noise in gene expression. *Cell Rep.* **2019**, 26, 3752–3761.e5.

27 FISH—in Insect Chromosomes

Vladimir E. Gokhman and Valentina G. Kuznetsova

CONTENTS

INTRODUCTION

Insects, i.e., the class Insecta within the subphylum Hexapoda, are an extremely diverse and species-rich group of predominantly terrestrial animals. Recent estimates suggest that there are about 5 to 15 million species of insects in the world fauna, with more than a million already described.[1, 2] In fact, these creatures play an essential role in many, if not all, terrestrial and freshwater ecosystems.[3] Given this information, one can conclude that both the morphological variation and ecological importance of insects are enormous.[1, 3] Moreover, this vast group is also substantially diverse in terms of cytology and genetics, including chromosome morphology and behavior.[4–12] For example, different types of the chromosome structure,

DOI: 10.1201/9781003223658-27

i.e., monocentric (including polytene) and holocentric (= holokinetic), are found in various members of Insecta.[4, 13, 14] This cytogenetic diversity apparently needs to be studied at the level of DNA sequences, chromosomes and entire genomes.[15, 16] To perform this research, a wide array of modern techniques is used.[13, 14] Among these methods, fluorescence in situ hybridization, or FISH, represents an essential technique on which molecular cytogenetics of insects is currently based for the most part.[17–20] With this method, fluorochrome-labeled DNA sequences (probes) are usually hybridized to metaphase chromosomes or interphase nuclei, thus providing important information necessary for physical mapping of these sequences.[18, 21, 22] The present work is aimed to review the current state and future perspectives of FISH in insect chromosomes.

SAMPLES/TISSUES

To obtain reliable results of FISH, this technique must be performed on a sufficient number of high-quality mitotic and/or meiotic divisions. In turn, appropriate developmental stages and tissues of insects must be chosen to achieve this goal. Specifically, male testes often harbor numerous spermatocytes undergoing either mitotic or meiotic divisions.[23, 24] Nevertheless, certain numbers of cell divisions can be found in the midgut as well as in the developing ovaries.[23, 25–28] Both adults and immature stages of insects can generally become a valuable source of mitoses and meioses, depending on the taxon under study.[25–29] However, developing tissues of insect larvae (and, to a lesser extent, pupae) often contain cells of varying ploidy, thus making the precise identification of the chromosome number substantially difficult, if at all possible. Exceptions from this rule include salivary glands of dipteran larvae, which contain giant polytene chromosomes, and imaginal discs of many Holometabola,[26, 30] for example, cerebral ganglia of hymenopteran and coleopteran prepupae.[7, 23] Embryos extracted from developing eggs or viviparous females also provide high-quality mitotic divisions.[25, 31] In some special cases, e.g., in many members of the suborder Symphyta (= lower Hymenoptera), unfertilized eggs can also be taken out of adult females and incubated in a moist chamber to stimulate their development.[32] It must also be noted, however, that a precise morphological identification of field-collected immature stages is often difficult, especially in holometabolous insects, but use of additional characters (for example, molecular ones) could offer a possible solution to this problem.

DESCRIPTION OF METHODS

General Notes

At present, two main groups of methods are used to produce high-quality chromosome preparations of insects, i.e., squash and air-drying techniques.[13, 31] All these methods are aimed to ensure preservation of the chromosome structure, good spreading of chromosomes as well as their reliable attachment to the slide. Nevertheless, squash technique achieves these goals by applying physical pressure to a piece of preliminarily fixed tissue, which is turned into a monolayer of cells on the slide. This preparation can be subsequently converted into a permanent one using dry ice or liquid nitrogen.[30, 33] On the other hand, air-drying technique implies initial hypotonic treatment of the tissue; after that, it is macerated and turned into cell suspension, which is then applied to a slide, fixed and air-dried.[34] Both methods have their own advantages and drawbacks. Specifically, squash technique can readily utilize material preserved with e.g. Carnoy's fixative, whereas air-drying one predominantly deals with living tissues. On the other hand, the air-drying method usually ensures better chromosome morphology, but it apparently needs additional post-fixation to keep the chromosomes attached to the slide.

In addition to the conventional FISH, a few studies of insect chromosomes were conducted using the so-called fiber-FISH.[25, 35] With this technique, FISH is performed on the stretched DNA fibers (= mechanically stretched chromosomes) obtained from interphase nuclei destroyed by either heat or chemical shock.[36] Fiber-FISH can detect DNA sequences with a substantially increased resolution of 1–400 Kb, as opposed to 1–3 Mb provided by regular FISH on metaphase chromosomes.[25, 37]

As noted earlier, probes of different types are used to perform FISH. Among these probes, both repetitive and unique DNA sequences, together with the so-called chromosome paints (see later) can be listed.[38] On the other hand, these sequences are either directly synthesized or amplified from the existing genomic material. In turn, amplification of the probes is substantially facilitated using bacterial artificial chromosomes (BACs).[38–41] Regardless of the technique used for the preparation of chromosomes and probes, the process of FISH includes certain essential stages, which can be subdivided into pre-hybridization, hybridization per se, and post-hybridization[18, 19]:

(1) preparation of chromosomes on the slide,
(2) labeling of DNA probe,
(3) denaturation of the probe,
(4) pre-treatment of the chromosome preparation,
(5) denaturation of DNA on the chromosome preparation,
(6) renaturation of DNA from the chromosomes and the probe (=DNA hybridization per se),
(7) detection of the DNA probe, and
(8) counterstaining, mounting and examination of the chromosome preparation.

FISH Protocol for Insect Chromosomes

In general, a few standard FISH protocols for various organisms were published, e.g. [21, 42]. These protocols include a number of modifications of the probe labeling (nick translation and polymerase chain reaction (PCR), direct and indirect labeling), probe denaturation, slide pretreatment and denaturation, hybridization in a strict sense, detection of hybridization sites etc. The protocol presented next is mainly based on the technique we routinely use in our lab.[24] This protocol produces strong, specific and reproducible signals for all insect orders we worked with, i.e., Odonata, Mantophasmatodea, Psocoptera, Hemiptera, Lepidoptera, Neuroptera and

Hymenoptera.[24, 28, 43–52] Nevertheless, we also provide an alternative protocol, which was used to perform FISH on chromosomes of various insects.[17]

Chemicals (in Alphabetic Order)

- 1×Taq DNA polymerase (Ref. EP0404, Thermo Fisher Scientific)
- 2-Propanol (Ref. I9516, Sigma-Aldrich)
- 10×PBS = phosphate-buffered saline: dilute 75.8 g NaCl, 9.93 g Na_2HPO_4 and 4.14 g NaH_2PO_4 in 1 l of Milli-Q water
- 20×SSC = saline sodium citrate: dilute 175.6 g NaCl and 88.2 g sodium citrate in 1 l of Milli-Q water. If necessary, use HCl to reach pH 7.0. Set up 0.4×, 1× and 2× before use
- Anti-digoxigenin-fluorescein/rhodamine, Fab fragments (Ref. 11207741910/ 11207750910, Roche)
- Antibody solution: 10 µl anti-digoxigenin-fluorescein/rhodamine (200 µl/ ml) and 990 µl Marvel solution, make fresh as required
- Avidin, fluorescein/rhodamine conjugate (Ref. A2662/A6378, Thermo Fisher Scientific)
- Biotin-11-dUTP (Ref. R0081, Thermo Fisher Scientific)
- Bovine serum albumin (Ref. 05473, Sigma-Aldrich)
- Denaturation buffer: 0.7 volume formamide, 0.2 volume Milli-Q water and 0.1 volume 20×SSC; store at 4 °C
- Dextran sulphate (Ref. 67578, Sigma-Aldrich)
- DIG-Nick Translation Mix (Ref. 11745816910, Roche)
- dNTP Mix (Ref. R0191, Thermo Fisher Scientific)
- dTTP (Ref. R0171, Thermo Fisher Scientific)
- EDTA = ethylenediaminetetraacetic acid (Ref. 03690, Sigma-Aldrich)
- Ethanol 70, 95 and 100 %
- Fluoroshield™ with DAPI (Ref. F6057, Sigma-Aldrich)
- Formamide (Ref. F7503, Sigma-Aldrich)
- Hybridization buffer: dissolve 2 g dextran sulphate in 10 ml 50 % deionized formamide/2×SSC/50 mM phosphate buffer for 3 h at 70 °C. Adjust pH to 7 with phosphate buffer. Aliquot and store at −20 °C
- Marvel solution: dilute 0.1 g milk powder Marvel in 2 ml washing buffer
- Milli-Q water (Merck)
- Pepsin solution: mix 10 µl 1 M HCl and 2.5 µl volume 20 ng/ml pepsin in 990 µl of Milli-Q water; make fresh as required
- Pepsin stock (Ref. P7012, Sigma-Aldrich)
- Phosphate buffer: dilute 20.214 g of $Na_2HPO_4×7H_2O$ and 3.394 g $NaH_2PO_4×H_2O$ in 800 ml of Milli-Q water; store at 2–8 °C
- Post-fixation solution: mix 5 ml 2 % paraformaldehyde, 4.5 ml 1×PBS and 0.5 ml 1 M $MgCl_2$; store at 4 °C
- ProLong™ Gold with DAPI (Ref. P36931, Thermo Fisher Scientific)
- RNAse A solution: add 0.05 volume RNAse A at 10 mg/ml in 0.95 volume 2×SSC and mix well; make fresh as required
- RNAse A stock (Ref. 10109142001, Roche)

- Salmon sperm DNA (Ref. D7656, Sigma-Aldrich)
- Sodium acetate solution (Ref. 71196, Sigma-Aldrich)
- Specific primers, e.g., 18S_F (5′-GATCCTGCCAGTAGTCATATG-3′) and 18S_R (5′-GAGTCAAATTAAGCCGCAGG-3′)[24]; 5S_F (5′-AACGACC ATACCACGCTGAA-3′) and 5S_R (5′-AAGCGGTCCCCCATCTAAGT)[23]; TTAGG_F (5′-AACCTAACCTAACCTAACCTAA-3′) and TTAGG_R (5′-GGTTAGGTTAGGTTAGGTTAGG-3′) (for the canonical insect telomere sequence)[24]
- Tween 20 = polyoxyethylene-sorbitan monolaurate (Ref. P9416, Sigma-Aldrich)
- Washing buffer: 0.2 volume 20×SSC, 0.8 volume Milli-Q water and 0.05 volume Tween 20

General Procedure (Based on [24])

1. Make a chromosome preparation using either air-drying or squash technique.
2. Isolate the DNA probe (either by direct synthesis, PCR or microdissection; see earlier). Prepare 50 µl of PCR mixture containing 50–100 ng DNA probe, 0.1 mM Biotin-11-dUTP, 0.1 mM dTTP, 0.2 mM other dNTPs (each), 2.5 mM MgCl$_2$, 50 pmol specific primers, 1 U 1×Taq DNA polymerase and 5 µl 10×Taq buffer. Perform PCR with the following parameters: denaturation: 1 min at 94 °C, main stage: 30–35 cycles of 30 sec at 94 °C, annealing: 30 sec at 50–56 °C (temperature depends on the length and composition of the primer), elongation: 0.5–2 min at 72 °C (temperature depends on the fragment size; for example, 1 min is enough for reliable amplification of 1 kb fragment), and final extension: 5 min at 72 °C. Precipitate labeled DNA with 5–10 volumes of 96–100 % ethanol at −20 °C and 0.1 volumes of 3 M NaCl for 30–60 min at −80 °C or 12–24 h at −20 °C. Make a DNA pellet by centrifugation at 15,000 rpm and 4 °C for 30 min. Wash labeled DNA with 500–700 µl ice-cold 70 % ethanol by centrifugation at 15,000 rpm and 4 °C for 15–20 min, then air-dry at 37 °C. Resuspend the pellet in 20–30 µl Milli-Q water. The pellet can be kept at −20 °C for a long time.
3. Put 100 µl RNAse A solution on slide and cover with coverslip, then incubate in humid chamber at 37 °C for 1 h. Remove the coverslip and wash in 2×SSC for 5 min and air-dry. Add 50 µl 0.005 % pepsin solution and cover with coverslip for 10 min at room temperature (= RT). Wash in 1×PBS for 5 min at RT. Incubate the slide in post-fixation solution in a Coplin jar for 10 min at RT. Wash for 5 min in 1×PBS. Dehydrate the slide in ethanol (70, 95 and 100 %, 3 min each) and air-dry.
4. Add 6.5–7 µl of denatured probe solution (about 100 ng of labeled probe, 50 % formamide, 2×SSC, 10 % (w/v) dextran sulfate, 1 % (w/v) Tween 20) containing 10 µg salmon-sperm DNA on the slide, cover with coverslip and seal with rubber cement. Incubate the slide on preheated hot plate at 75 °C for 5 min.
5. Incubate the slide overnight at 37 °C in a humid chamber.

6. Wash the slide in 2×SSC for 3 min at 45 °C, then in 50 % formamide/2×SSC for 10 min at 45 °C. Wash the slide twice in 2×SSC at 45 °C (10 min each), then twice in 0.2×SSC at 45 °C (10 min each). Block the reaction in 1.5 % (w/v) bovine serum albumin/4×SSC/0.1 % Tween 20 for 30 min at 37 °C in a humid chamber. Detect the biotin-labeled probe with 5 µg/ml avidin-flu-orescein/rhodamine in 1.5 % bovine serum albumin/4×SSC/0.1 % Tween 20 under coverslip. Remove the coverslip and put the slide in 4×SSC/0.02 % Tween 20 for 5 min on the shaker; repeat this step three times using new portions of the solution. Rinse in Milli-Q water briefly, pass through the ethanol series (70, 95 and 100 %) for 3 min each at RT, and air-dry. Apply 20 µl of ProLong™ Gold with DAPI and cover with coverslip. Examine under the microscope.

Alternative Procedure (Based on [17])

1. Make a chromosome preparation using either air-drying or squash technique.
2. Isolate the DNA probe (either by direct synthesis, PCR or microdissection, see earlier). Put 3.5 µl of the probe solution (suggested concentration 50 ng/µl) in 12.5 µl autoclaved Milli-Q water in a microtube, and add 4 µl of the DIG-Nick Translation Mix; mix and spin briefly. Incubate for 90 min at 16 °C. Add 1 µl 0.5M EDTA and put for 10 min at 65 °C to stop the process. Store at −20 °C.
3. Denature 2 µl of the labeled probe in 18 µl of hybridization buffer for 5 min at 85 °C, then keep the solution on ice or at 4 °C before use.
4. Put 100 µl RNAse A solution on slide and cover with coverslip, then incubate in humid chamber at 37 °C for 1 h. Remove the coverslip and wash in 2×SSC for 5 min and air-dry. Add 50 µl 0.005 % pepsin solution and cover with coverslip for 10 min at RT. Wash in 1×PBS for 5 min at RT. Incubate the slide in 100 µl post-fixation solution under the coverslip for 10 min at RT. Remove the coverslip and wash for 5 min in 1×PBS. Dehydrate the slide in ethanol (70, 95 and 100 %, 3 min each) and air-dry.
5. Put 100 µl of denaturation buffer on the slide and cover with coverslip, and then denature on preheated hot plate at 73 °C for 3 min. Remove the coverslip and place the slide in 70 % ethanol at −20 °C for 3 min. Dehydrate the slide in ethanol (70, 95 and 100 %, 3 min each) and air-dry.
6. Add 20 µl of denatured probe solution on the slide and cover with coverslip. Incubate the slide overnight at 37 °C in a humid chamber.
7. Remove the coverslip and wash the slide in 100 ml 0.4×SSC at 65–68 °C for 3–4 min in a Coplin jar in a water bath. Put the slide into 100 ml of 4×SSC/0.2 % Tween 20 for 5 min at RT in a shaker. Add 100 µl Marvel solution under coverslip at 37 °C for 10–15 min in a humid chamber. Remove the coverslip and wash the slide in 4×SSC/0.2 % Tween 20 for 2 min at RT. Detect the probe by incubation of the slide in 100 µl antibody solution for 20–35 min at 37 °C in a humid chamber.
8. Remove the coverslip and put the slide in 4×SSC/0.2 % Tween 20 for 5 min on the shaker; repeat this step three times using new portions of the solution.

Rinse in Milli-Q water briefly, pass through the ethanol series (70, 95 and 100 %) for 3 min each at RT, and air-dry. Apply 20 µl of Fluoroshield™ with DAPI and cover with coverslip. Examine under the microscope.

Use of C_0t-1 DNA (Based on the Preceding Protocol, Instead of Step 3)

Precipitate the labeled probe and C_0t-1 DNA together using a single volume of 2-propanol and 0.1 volume of sodium acetate. For better results, the use of proportions higher than 1:20 probe to C_0t-1 DNA is recommended. Centrifuge at 14,000–15,000 rpm for 20 min at 4 °C, discard the supernatant and dry the DNA pellet at RT. Dilute the pellet in 20 µl of hybridization buffer. Denature the probe solution for 5 min at 85 °C, and then perform pre-hybridization for 30 min at 37 °C. Keep the solution on ice or at 4 °C before use.

YIELDS

PHYSICAL MAPPING OF DNA SEQUENCES

Repetitive Sequences

FISH in insect chromosomes is most often used for mapping repetitive DNA sequences, in which it is especially efficient.[19] These sequences can be provisionally subdivided into the so-called satellite DNA, which usually occurs in clusters,[53] and transposable elements.[54] Among the satellite sequences, ribosomal DNAs (rDNAs) are the most frequently used as probes.[20, 43, 51, 55–57] Eukaryotic rDNA is composed of tandem repeats divided into large (45S) and small (5S) transcriptional units. In turn, these units form characteristic clusters, which usually contain hundreds or even thousands of copies.[58] The large unit contains genes encoding 18S, 5.8S and 28S rRNAs, which are separated by internal transcribed spacers.[59, 60] The small unit encodes multiple 5S rRNA genes separated by non-transcribed spacers. 5S rDNA is mostly distributed independently of 45S rDNA throughout the genome, although it often forms clusters of its own.[25, 35, 61–67] However, 45S and 5S rRNA genes are interspersed in the coleopteran *Lagria villosa* (Fabricius).[68] Nevertheless, the main rDNA clusters of insects remain the most studied, with 18S rDNA being the most frequently used probe.[24, 46, 48, 52, 61, 63, 66, 67, 69–74, etc.] The number and localization of the rDNA clusters vary among closely related forms,[33, 75–81] and, occasionally, even among particular populations and individuals of the same species. In the latter case, different sections of the 45S rDNA cistron can evolve more or less independently.[60] Moreover, certain rDNA-like sequences sometimes spread over the whole genome and can therefore be detected on every chromosome.[82]

In addition to rDNA, other repetitive sequences can also be studied using FISH. For example, histones are basic proteins, which, together with DNA, form the fundamental units of chromatin structure.[83] Histone-coding genes are usually arranged in clusters, and these clusters also can be physically mapped with this technique.[35, 61, 63, 64, 67, 74, 78, 84–86] In insects, clusters of these genes are often (but not always) co-localized with those of 5S rDNA.[61–63, 87, 88] Histone-coding genes can be arranged either in clusters of tandem repeats formed by sequences by all types (H1, H2A, H2B, H3 and H4), spaced by non-coding DNA, as individual genes or their random combinations. For example, these genes are grouped into quartets and quintets in several *Drosophila* species.[89]

Microsatellites represent short tandemly repeated DNA sequences (usually about two to six base pairs),[58, 90] which can be efficiently mapped using FISH.[35, 66, 71, 77, 91, 92] Although these sequences are predominantly associated with heterochromatic segments[53] (but see[93]), microsatellites can have diverse distribution patterns within the given chromosome set. Specifically, they can be either scattered along the whole genome or restricted to certain chromosomes and/or their segments, being either dispersed or united into clusters.[77, 93, 94]

Small nuclear DNAs (snDNAs) encode corresponding RNAs (snRNAs), which play important roles in processing pre-messenger RNA (= splicing), telomere maintenance, and a few other processes.[95] Various types of these DNAs (e.g. U1 and U2 snDNA), which are organized in clusters, can also be mapped using FISH.[64–67, 71, 96–98]

Tandemly repeated centromeric sequences of satellite DNA[53] are located in the pericentromeric regions of chromosomes.[99] FISH visualized specific sequences of that kind in certain grasshoppers, beetles and ants,[100–102] but microdissection experiments also showed that centromeric sequences can vary not only between different species, but between different chromosomes of the same species as well (see e.g.[103]).

Telomeric sequences are located in the terminal regions of chromosomes.[9] In most insects, as well as in other arthropods, TTAGG is the most widespread and apparently ancestral telomeric repeat.[48, 52, 77, 104–107] However, telomeric motifs of many insect taxa are highly variable, being often substituted by other sequences with either known or, more frequently, unknown structure.[9, 108, 109] In addition, telomeric repeats are substituted by transposons in higher Diptera, and these transposons can be detected using FISH as well (see later).[19]

Although mobile repetitive DNA sequences, i.e., transposable elements,[54] are often not arranged in clusters, they also can be revealed in insect chromosomes using FISH.[19, 25, 110–113] Sometimes, whole fractions of repetitive DNA, e.g., C_0t-1 DNA, can be physically mapped to insect chromosomes as well.[35, 66, 88, 114] This analysis can provide an insight on spreading of repetitive sequences among different chromosomes, especially within their heterochromatic regions.[65, 115]

Unique Sequences

In addition to repetitive DNA sequences, certain unique, single-copy genes also can be physically mapped on insect chromosomes using FISH. Previously, regular FISH could only reveal longer sequences.[32] Later on, sensitivity of FISH was substantially enhanced using e.g. tyramide signal amplification (TSA-FISH).[41, 74, 116]

During the last decades, an important role of chromosomal inversions in maintaining genetic diversity among and within different populations became obvious.[117] In particular, these inversions preserve certain gene complexes within the same chromosome, i.e., the so-called supergenes,[118] by preventing crossing over. A few cases of supergenes are found in insects, including ants, in which their presence is demonstrated using FISH.[119] In addition, recent research revealed that genomes of many parasitoid Hymenoptera harbor symbiotic polydnaviruses (double-stranded DNA viruses) that are integrated into insect chromosomes, and then excised from the genome and further injected into the host during oviposition.[120] Although DNA sequences of these viruses were detected in chromosomes of a particular parasitic wasp using non-fluorescence in situ hybridization,[121] similar studies could be conducted using FISH. Moreover, FISH also demonstrated that a number of physiologically important genes are sex-linked in most clades of the moth family Tortricidae.[40]

CHROMOSOME PAINTING (= MULTICOLOR FISH)

Chromosome painting mainly differs from the previously mentioned procedure in the nature and preparation of the probes. Specifically, these probes can be prepared either from entire chromosomes or from their particular sections.[16] To capture the corresponding material, the so-called microdissection is normally used. For studying insect chromosomes, glass-needle,[103, 122–126] or laser microdissection is employed.[127–129] When DNA is isolated from this material, it can be amplified using the so-called degenerate oligonucleotide primed polymerase chain reaction (DOP-PCR). This reaction performs non-specific amplification of DNA.[130] The amplified DNA then must be sheered e.g. using autoclavation, and labeled with a particular fluorochrome.[126] However, insect chromosomes usually contain large amounts of repetitive sequences, and therefore either blocking DNA (usually C_0t-1 DNA),[17] or digital techniques,[131] must be used to avoid unwanted cross-hybridization (see earlier) or to detect the specific signal using specialized software.

During the last decade, a new powerful technique of chromosome painting, i.e., the so-called Oligopaint technique, which visualizes parts of the genome using Oligopaint FISH probes,[132] was introduced into the chromosomal study of insects. This method implies use of fluorochrome-tagged custom-synthetized oligonucleotides (oligos), which specifically highlight selected chromosomes. However, this technique is based on the results of full genome sequencing, and therefore is currently available for just a few insect species (see e.g.[133]).

Study of Various Properties of Particular Chromosomes

In insects, chromosome painting is most often used to explore origin and evolution of particular chromosomes and their regions by studying their specific properties.[39, 123, 124, 127, 129, 134, 135, etc.] In particular, any chromosome and/or chromosomal segment within a given karyotype can be reliably recognized using this technique. In certain species with low chromosome numbers, as in the parasitic wasp *Nasonia vitripennis* (Walker) with $n = 5$, a study of that kind has already been conducted for every chromosome.[122] Among specific elements of the karyotype, sex chromosomes, especially W chromosomes of butterflies and moths with ZZ/ZW sex determination, as well as B chromosomes in Orthoptera are the most studied.[39, 123–125, 127, 129, 135]

In addition to FISH with known centromeric repeats (see earlier), specific regions of certain insect chromosomes are also studied using microdissection. These works include studying centromeres,[103] and heterochromatic segments.[134] This research demonstrated that centromeric regions of different chromosomes within a given karyotype could substantially differ in their DNA content. On the contrary, the latter characteristic of heterochromatic segments in stingless bees (Hymenoptera) appeared to be very similar across the whole chromosome sets.[134]

To our best knowledge, the only study of a particular fusion/fission with the so-called whole chromosome painting (WCP), was conducted a few years ago on the two cryptic species of parasitoids of the *Lariophagus distinguendus* (Förster) complex with $n = 5$ and 6.[126] To identify target chromosomes, morphometric analysis of chromosome sets of both species was performed.[134] This study showed that the only acrocentric and a smaller metacentric in the chromosome set with $n = 6$ respectively correspond to the shorter and longer arms of the largest metacentric chromosome in the karyotype with $n = 5$. This assumption was confirmed using both microdissection and WCP.[126]

In particular, the technique of Oligopaint FISH probes (see earlier) provides a novel opportunity to explore meiosis in the model organism, i.e., the domestic silk moth, *Bombyx mori* (Linnaeus), at an unprecedented scale. Specifically, a number of interesting phenomena, e.g., asynchronous pairing of homologous chromosomes, were observed for the first time using this technique.[133]

COMPARATIVE GENOMIC HYBRIDIZATION (CGH) AND GENOMIC IN SITU HYBRIDIZATION (GISH)

The CGH technique explores the number of copies of a particular DNA sequence in the test sample vs. the reference one.[22] These samples most often represent total genomic DNA, and the corresponding method is therefore called GISH. This technique can effectively highlight particular sex chromosomes in different groups of insects (see [136] and references therein). For example, GISH identified a particular segment of the neo-X element as an original X chromosome in certain true bugs (Hemiptera, Pyrrhocoridae).[128] CGH also recognized differentiated W chromosomes, which are enriched with female-specific sequences and/or common repetitive ones, in various members of the family Geometridae (Lepidoptera).[129, 137] Moreover, this method highlighted chromosomes originating from different parent taxa in a rare case of homoploid hybrid speciation in the genus *Agrodiaetus* (Lepidoptera, Lycaenidae).[138]

CONCLUSIONS

The previously mentioned results demonstrate that FISH represents a powerful tool used in the cytogenetic study of insects (Figure 27.1). In addition to physical mapping of various DNA sequences, it can highlight different genomic and chromosomal rearrangements, and therefore can outline certain pathways of the karyotype and genome evolution in this enormous group of animals.[11, 23, 25, 26, 31, 43, 44, 53, 73, 76, 82, 86, 138, 139] We expect that, either by itself or in combination with other methods, this technique will reveal many new details of the genome change at various scales of space and time.

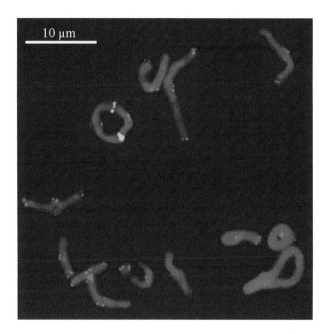

FIGURE 27.1 *Ancyra* sp. (Hemiptera, Eurybrachidae): meiotic chromosomes at diakinesis (2n = 12AA + X). Results of FISH with 18S rDNA and TTAGG telomere probes (green and red signals, respectively) are shown.

ACKNOWLEDGEMENT

The authors are grateful to Boris A. Anokhin (Zoological Institute, Russian Academy of Sciences) for his useful advice on practical aspects of performing FISH on insect chromosomes.

REFERENCES

1. Grimaldi, D.; Engel, M.S. *Evolution of the Insects*. Cambridge University Press, Cambridge; **2005**, 755 p.
2. Stork, N.E. How many species of insects and other terrestrial arthropods are there on Earth? *Ann. Rev. Entomol.* **2018**, 63, 31–45.
3. Scudder, G.G.E. The importance of insects. In: *Insect Biodiversity: Science and Society*; Foottit, R.G., Adler, P.H., Eds.; 2nd Edition, 1; Wiley Blackwell, Oxford, **2017**, pp. 9–43.
4. White, M.J.D. *Animal Cytology and Evolution*. 3rd Edition. Cambridge University Press, Cambridge; **1973**, 961 p.
5. Lukhtanov, V.A. Sex chromatin and sex chromosome systems in nonditrysian Lepidoptera "Insecta". *J. Zool. Syst. Evol. Res.* **2000**, 38, 73–79.
6. Nokkala, S.; Kuznetsova, V.G.; Maryańska-Nadachowska, A.; Nokkala, C. Holocentric chromosomes in meiosis. I. Restriction of the number of chiasmata in bivalents. *Chromosome Res.* **2004**, 12, 733–739.

7. Gokhman, V.E. *Karyotypes of Parasitic Hymenoptera*. Springer, Dordrecht; **2009**, 183 p.

8. Blackmon, H.; Ross, L.; Bachtrog, D. Sex determination, sex chromosomes, and karyotype evolution in insects. *J. Hered.* **2017**, 108, 78–93.

9. Kuznetsova, V.; Grozeva, S.; Gokhman, V. Telomere structure in insects: A review. *J. Zool. Syst. Evol. Res.* **2020**, 58(1), 127–158.

10. Kuznetsova, V.G.; Gavrilov-Zimin, I.A.; Grozeva, S.M.; Golub, N.V. Comparative analysis of chromosome numbers and sex chromosome systems in Paraneoptera (Insecta). *Comp. Cytogenet.* **2021**, 15, 279–327.

11. Kuznetsova, V.; Maryańska-Nadachowska, A.; Anokhin, B.; Shapoval, N.; Shapoval, A. Chromosomal analysis of eight species of dragonflies (Anisoptera) and damselflies (Zygoptera) using conventional cytogenetics and FISH: Insights into the karyotype evolution of the ancient insect order Odonata. *J. Zool. Syst. Evol. Res.* **2021**, 59, 387–399.

12. Sylvester, T.; Hjelmen, C.E.; Hanrahan, S.J.; Lenhart, P.A.; Johnston, J.S.; Blackmon, H. Lineage-specific patterns of chromosome evolution are the rule not the exception in *Polyneoptera* insects. *Proc. R. Soc. B.* **2020**, 287, 20201388.

13. Gokhman, V.E.; Kuznetsova, V.G. Comparative insect karyology: Current state and applications. *Entomol. Rev.* **2006**, 86, 352–368.

14. Martins, C.; Cabral-de-Mello, D.C.; Valente, G.T; Mazzuchelli, J.; de Oliveira, S.G.; Pinhal, D. *Animal Genomes under the Focus of Cytogenetics. Genetics—Research and Issues.* Nova Science Publishers, Inc., New York, NY; **2011**, 154 p.

15. Lukhtanov, V.A.; Kuznetsova, V.G. What genes and chromosomes say about the origin and evolution of insects and other arthropods. *Russ. J. Genet.* **2010**, 46, 1115–1121.

16. Liehr, T. Classification of FISH probes. In: *Fluorescence In Situ Hybridization (FISH): Application Guide; Springer Protocols Handbooks*; Liehr, T., Ed. 2nd Edition. Springer Verlag, Berlin; **2017**, pp. 43–48.

17. Alves-Silva, A.P.; Campos Barros, L.A.; Pompolo, S.G. General protocol of FISH for insects. In: *Fluorescence In Situ Hybridization (FISH): Application Guide; Springer Protocols Handbooks*; Liehr, T., Ed. 2nd Edition. Springer Verlag, Berlin; **2017**, pp. 459–466.

18. Liehr, T.; Weise, A. Background. In: *Fluorescence In Situ Hybridization (FISH): Application Guide; Springer Protocols Handbooks*; Liehr, T., Ed. 2nd Edition. Springer Verlag, Berlin; **2017**, pp. 1–14.

19. Cabral-de-Mello, D.C.; Marec, F. Universal fluorescence in situ hybridization (FISH) protocol for mapping repetitive DNAs in insects and other arthropods. Mol. Genet. Genomics. **2021**, 296, 513–526.

20. Masri, R.A.; Karagodin, D.A.; Sharma, A.; Sharakhova, M.V. A gene-based method for cytogenetic mapping of repeat-rich mosquito genomes. *Insects.* **2021**, 12, 138.

21. Schwarzacher, T.; Heslop-Harrison, J.S. *Practical In Situ Hybridization*. BIOS Scientific Publishers, Oxford; **2000**, 203 p.

22. Speicher, M.R.; Carter, N.P. The new cytogenetics: Blurring the boundaries with molecular biology. *Nat. Rev. Genet.* **2005**, 6, 782–792.

23. Cabral-de-Mello, D.C. Beetles (Coleoptera). In: *Protocols for Cytogenetic Mapping of Arthropod Genomes*; Sharakhov, I.V., Ed. CRC Press, Boca Raton, FL; **2015**, pp. 171–217.

24. Grozeva, S.; Kuznetsova, V.G.; Anokhin, B.A. Karyotypes, male meiosis and comparative FISH mapping of 18S ribosomal DNA and telomeric (TTAGG)$_n$ repeat in eight species of true bugs (Hemiptera, Heteroptera). *Comp. Cytogenet.* **2011**, 5, 355–374.

25. Camacho, J.P.M.; Cabrero, J.; López-León, M.D.; Cabral-de-Mello, D.C.; Ruiz-Ruano, F.J. Grasshoppers (Orthoptera). In: *Protocols for Cytogenetic Mapping of Arthropod Genomes*; Sharakhov, I.V., Ed. CRC Press, Boca Raton, FL; **2015**, pp. 381–438.

26. Sharakhova, M.V.; George, P.; Timoshevskiy, V.; Sharma, A.; Peery, A.; Sharakhov, I.V. Mosquitoes (Diptera). In: *Protocols for Cytogenetic Mapping of Arthropod Genomes*; Sharakhov, I.V., Ed. CRC Press, Boca Raton, FL; **2015**, pp. 93–170.

27. Angus, R.B.; Jeangirard, C.; Stoianova, D.; Grozeva, S.; Kuznetsova, V.G. A chromosomal analysis of *Nepa cinerea* Linnaeus, 1758 and *Ranatra linearis* (Linnaeus, 1758) (Heteroptera, Nepidae). *Comp. Cytogenet.* **2017**, 11, 641–657.

28. Karagyan, G.; Golub, N.; Sota, T. Cytogenetic characterization of periodical cicadas (Hemiptera: Cicadidae: *Magicicada*). *Eur. J. Entomol.* **2020**, 117, 474–480.

29. Yoshido, A.; Sahara, K.; Yasukochi, Y. Silk moths (Lepidoptera). In: *Protocols for Cytogenetic Mapping of Arthropod Genomes*; Sharakhov, I.V., Ed. CRC Press, Boca Raton, FL; **2015**, pp. 219–256.

30. Sharakhova, M.V.; Timoshevskiy, V.A.; Yang, F.; Demin, S.I.; Severson, D.W.; Sharakhov, I.V. Imaginal discs—a new source of chromosomes for genome mapping of the yellow fever mosquito *Aedes aegypti*. *PLoS Negl. Trop. Dis.* **2011**, 5, e1335.

31. Mandrioli, M.; Manicardi, G.C. Aphids (Hemiptera). In: *Protocols for Cytogenetic Mapping of Arthropod Genomes*; Sharakhov, I.V., Ed. CRC Press, Boca Raton, FL; **2015**, pp. 327–350.

32. Matsumoto, K.; Yamamoto, D.S.; Sumitani, M.; Lee, J.M.; Hatakeyama, M.; Oishi, K. Detection of a single copy gene on a mitotic metaphase chromosome by fluorescence in situ hybridization (FISH) in the sawfly, *Athalia rosae* (Hymenoptera). *Arch. Ins. Biochem. Physiol.* **2002**, 49, 34–40.

33. Grozeva, S.; Anokhin, B.A.; Kuznetsova, V.G. Bedbugs (Hemiptera). In: *Protocols for Cytogenetic Mapping of Arthropod Genomes*; Sharakhov, I.V., Ed. CRC Press, Boca Raton, FL; **2015**, pp. 285–326.

34. König, C.; Paschke, S.; Pollmann, M.; Reinisch, R.; Gantert, C.; Weber, J.; Krogmann, L.; Steidle, J.L.M.; Gokhman, V.E. Molecular and cytogenetic differentiation within the *Lariophagus distinguendus* (Förster, 1841) species complex (Hymenoptera, Pteromalidae). *Comp. Cytogenet.* **2019**, 13, 133–145.

35. Palacios-Gimenez, O.M.; Carvalho, C.R.; Ferrari Soares, F.A.; Cabral-de-Mello, D.C. Contrasting the chromosomal organization of repetitive DNAs in two Gryllidae crickets with highly divergent karyotypes. *PLoS One.* **2015**, 10, e0143540.

36. Heiskanen, M.; Kallioniemi, O.; Palotie, A. Fiber-FISH: Experiences and a refined protocol. *Genet. Anal.: Biomol. Eng.* **1996**, 12, 179–184.

37. Raap, A.K.; Florijn, R.J.; Blonden, L.A.J.; Wiegant, J.; Vaandrager, J.; Vrolijk, H.; Dunnen, J.; Tanke, H.J.; van Ommen, G.-J. Fiber FISH as a DNA mapping tool. *Methods.* **1996**, 9, 67–73.

38. Liehr, T. Homemade locus-specific FISH probes: Bacterial artificial chromosomes. In: *Fluorescence In Situ Hybridization (FISH): Application Guide; Springer Protocols Handbooks*; Liehr, T., Ed. 2nd Edition. Springer Verlag, Berlin; **2017**, 101–106.

39. Yoshido, A.; Yasukochi, Y.; Marec, F.; Abe, H.; Sahara, K. FISH analysis of the W chromosome in *Bombyx mandarina* and several other species of Lepidoptera by means of *B. mori* W-BAC probes. *J. Ins. Biotechnol. Sericol.* **2007**, 76, 1–7.

40. Nguyen, P.; Sýkorová, M.; Šíchová, J.; Kůta, V.; Dalíková, M.; Čapková Frydrychová, R.; Neven, L.G.; Sahara, K.; Marec, F. Neo-sex chromosomes and adaptive potential in tortricid pests. *Proc. Natl. Acad. Sci. USA.* **2013**, 110, 6931–6936.

41. Carabajal Paladino, L.Z.; Nguyen, P.; Šíchová, J.; Marec, F. Mapping of single-copy genes by TSA-FISH in the codling moth, *Cydia pomonella*. *BMC Genet.* **2014**, 15, S15.

42. Liehr, T.; Kreskowski, K.; Ziegler, M.; Piaszinski, K.; Rittscher, K. The standard FISH procedure. In: *Fluorescence In Situ Hybridization (FISH): Application Guide;*

Springer Protocols Handbooks; Liehr, T., Ed. 2nd Edition; Springer Verlag, Berlin; **2017**, 109–118.

43. Gokhman, V.E.; Anokhin, B.A.; Kuznetsova, V.G. Distribution of 18S rDNA sites and absence of the canonical TTAGG insect telomeric repeat in parasitoid Hymenoptera. *Genetica.* **2014**, 142, 317–322.

44. Kuznetsova, V.G.; Grozeva, S.M.; Hartung, V.; Anokhin, B.A. First evidence for (TTAGG)n telomeric sequence and sex chromosome post-reduction in Coleorrhyncha (Insecta, Hemiptera). *Comp. Cytogenet.* **2015**, 9, 523–532.

45. Kuznetsova, V.G.; Khabiev, G.N.; Anokhin, B.A. Cytogenetic study on antlions (Neuroptera, Myrmeleontidae): First data on telomere structure and rDNA location. *Comp. Cytogenet.* **2016**, 10, 647–656.

46. Kuznetsova, V.G.; Maryańska-Nadachowska, A.; Shapoval, N.A.; Anokhin, B.A.; Shapoval, A.P.; Cytogenetic characterization of eight Odonata species originating from the Curonian spit (the Baltic Sea, Russia) using C-banding and FISH with 18S rDNA and telomeric (TTAGG)$_n$ probes. *Cytogenet. Genome Res.* **2018**, 153, 147–157.

47. Lachowska-Cierlik, D.; Maryańska-Nadachowska, A.; Kuznetsova, V.; Picker, M. First chromosomal study of Mantophasmatodea: Karyotype of *Karoophasma biedouwense* (Austrophasmatidae). *Eur. J. Entomol.* **2015**, 112, 599–605.

48. Vershinina, A.O.; Anokhin, B.A.; Lukhtanov, V.A. Ribosomal DNA clusters and telomeric (TTAGG)n repeats in blue butterflies (Lepidoptera, Lycaenidae) with low and high chromosome numbers. *Comp. Cytogenet.* **2015**, 9, 161–171.

49. Gokhman, V.E.; Kuznetsova, V.G. Presence of the canonical TTAGG insect telomeric repeat in the Tenthredinidae (Symphyta) suggests its ancestral nature in the order Hymenoptera. *Genetica.* **2018**, 146, 341–344.

50. Maryańska-Nadachowska, A.; Kuznetsova, V.G.; Golub, N.V.; Anokhin, B.A. Detection of telomeric sequences and ribosomal RNA genes in holokinetic chromosomes of five jumping plant-lice species: First data on the superfamily Psylloidea (Hemiptera: Sternorrhyncha). *Eur. J. Entomol.* **2018**, 115, 632–640.

51. Golub, N.; Anokhin, B.; Kuznetsova, V. Comparative FISH mapping of ribosomal DNA clusters and TTAGG telomeric sequences to holokinetic chromosomes of eight species of the insect order Psocoptera. *Comp. Cytogenet.* **2019**, 13, 403–410.

52. Gapon, D.A.; Kuznetsova, V.G.; Maryańska-Nadachowska, A. A new species of the genus *Rhaphidosoma* Amyot et Serville, 1843 (Heteroptera, Reduviidae), with data on its chromosome complement. *Comp. Cytogenet.* **2021**, 15, 467–505.

53. Palomeque, T.; Lorite, P. Satellite DNA in insects: A review. *Heredity.* **2008**, 100, 564–573.

54. Wells, J.N.; Feschotte, C. A field guide to eukaryotic transposable elements. *Ann. Rev. Genet.* **2020**, 54, 539–561.

55. Cabrero, J.; Camacho, J.P.M. Location and expression of ribosomal RNA genes in grasshoppers: Abundance of silent and cryptic loci. *Chromosome Res.* **2008**, 16, 595–607.

56. Warchałowska-Śliwa, E.; Grzywacz, B.; Maryańska-Nadachowska, A.; Karamysheva, T.V.; Heller, K.-G.; Lehmann, A.W.; Lehmann, G.U.C.; Chobanov, D.P. Molecular and classical chromosomal techniques reveal diversity in bushcricket genera of Barbitistini (Orthoptera). *Genome.* **2013**, 56, 667–676.

57. Sochorová, J.; Garcia, S.; Gálvez, F.; Symonová, R.; Kovařík, A. Evolutionary trends in animal ribosomal DNA loci: Introduction to a new online database. *Chromosoma.* **2018**, 127, 141–150.

58. Richard, G.F.; Kerrest, A.; Dujon, B. Comparative genomics and molecular dynamics of DNA repeats in eukaryotes. *Microbiol. Mol. Biol. Rev.* **2008**, 72, 686–727.

59. Hillis, D.M.; Dixon, M.T. Ribosomal DNA: Molecular evolution and phylogenetic inference. *Quart. Rev. Biol.* **1991**, 66, 411–453.

60. Ferretti, A.B.S.M.; Ruiz-Ruano, F.J.; Milani, D.; Loreto, V.; Martí, D.A.; Ramos, E.; Martins, C.; Cabral-de-Mello, D.C. How dynamic could be the 45S rDNA cistron? An intriguing variability in a grasshopper species revealed by integration of chromosomal and genomic data. *Chromosoma.* **2019**, 128, 165–175.

61. Cabral-de-Mello, D.C.; Martins, C.; Souza, M.J.; Moura, R.C. Cytogenetic mapping of 5S and 18S rRNAs and H3 histone genes in 4 ancient Proscopiidae grasshopper species: Contribution to understanding the evolutionary dynamics of multigene families. *Cytogenet. Genome Res.* **2011**, 132, 89–93.

62. Cabral-de-Mello, D.C.; Oliveira, S.G.; Moura, R.C.; Martins, C. Chromosomal organization of the 18S and 5S rRNAs and histone H3 genes in Scarabaeinae coleopterans: Insights into the evolutionary dynamics of multigene families and heterochromatin. *BMC Genet.* **2011**, 12, 88.

63. Oliveira, N.L.; Cabral-de-Mello, D.C.; Rocha, M.F.; Loreto, V.; Martins, C.; Moura, R.C. Chromosomal mapping of rDNAs and H3 histone sequences in the grasshopper *Rhammatocerus brasiliensis* (Acrididae, Gomphocerinae): Extensive chromosomal dispersion and co-localization of 5S rDNA/H3 histone clusters in the A complement and B chromosome. *Mol. Cytogenet.* **2011**, 4, 24.

64. Bueno, D.; Palacios-Gimenez, O.M.; Cabral-de-Mello, D.C. Chromosomal mapping of repetitive DNAs in the grasshopper *Abracris flavolineata* reveal possible ancestry of the B chromosome and H3 histone spreading. *PLoS One.* **2013**, 8, e66532

65. Palacios-Gimenez, O.M.; Castillo, E.R.; Martí, D.A.; Cabral-de-Mello, D. C. Tracking the evolution of sex chromosome systems in Melanoplinae grasshoppers through chromosomal mapping of repetitive DNA sequences. *BMC Evol. Biol.* **2013**, 13, 167.

66. Palacios-Gimenez, O.M.; Cabral-de-Mello, D.C. Repetitive DNA chromosomal organization in the cricket *Cycloptiloides americanus*: A case of the unusual X_1X_2O sex chromosome system in Orthoptera. *Mol. Genet. Genomics.* **2015**, 290, 623–631.

67. Bardella, V.B.; Fernandes, J.A.M.; Cabral de Mello, D.C. Chromosomal evolutionary dynamics of four multigene families in Coreidae and Pentatomidae (Heteroptera) true bugs. *Mol. Genet. Genomics.* **2016**, 291, 1919–1925.

68. Gusso Goll, L.; Matiello, R.R.; Artoni, R.F.; Vicari, M.R.; Nogaroto, V.; de Barros, A.V.; Almeida, M.C. High-resolution physical chromosome mapping of multigene families in *Lagria villosa* (Tenebrionidae): Occurrence of interspersed ribosomal genes in Coleoptera. *Cytogenet. Genome Res.* **2015**, 146, 64–70.

69. Maryańska-Nadachowska, A.; Anokhin, B.A.; Gnezdilov, V.M.; Kuznetsova, V.G. Karyotype stability in the family Issidae (Hemiptera, Auchenorrhyncha) revealed by chromosome techniques and FISH with telomeric $(TTAGG)_n$ and 18S rDNA probes. *Comp. Cytogenet.* **2016**, 10, 347–369.

70. Maryańska-Nadachowska, A.; Kuznetsova, V.G.; Karamysheva, T.V. Chromosomal location of rDNA clusters and TTAGG telomeric repeats in eight species of the spittlebug genus *Philaenus* (Hemiptera: Auchenorrhyncha: Aphrophoridae). *Eur. J. Entomol.* **2013**, 110, 411–418.

71. Piccoli, M.C.A.; Bardella, V.B.; Cabral-de-Mello, D.C. Repetitive DNAs in *Melipona scutellaris* (Hymenoptera: Apidae: Meliponidae): Chromosomal distribution and test of multiple heterochromatin amplification in the genus. *Apidologie.* **2018**, 49, 497–504.

72. Micolino, R.; Cristiano, M.P.; Travenzoli, N.P.; Lopes, D.M.; Cardoso, D.C. Chromosomal dynamics in space and time: Evolutionary history of *Mycetophylax* ants across past climatic changes in the Brazilian Atlantic coast. *Sci. Rep.* **2019**, 9, 18800.

73. Menezes, R.S.T.; Cabral-de-Mello, D.C.; Milani, D.; Bardella, V.B.; Almeida, E.A.B. The relevance of chromosome fissions for major ribosomal DNA dispersion in hymenopteran insects. *J. Evol. Biol.* **2021**, 34, 1466–1476.

74. Provazníková, I.; Hejníčková, M.; Visser, S.; Dalíková, M.; Carabajal Paladino, L.Z.; Zrzavá, M.; Voleníková, A.; Marec, F.; Nguyen, P. Large-scale comparative analysis of cytogenetic markers across Lepidoptera. *Sci. Rep.* **2021**, 11, 12214.

75. Cabrero, J.; Bugrov, A.; Warchałowska-Śliwa, E.; López-León, M.D.; Perfectti, F.; Camacho, J.P.M. Comparative FISH analysis in five species of Eyprepocnemidine grasshoppers. *Heredity.* **2003**, 90, 377–381.

76. Nguyen, P.; Sahara, K.; Yoshido, A.; Marec, F. Evolutionary dynamics of rDNA clusters on chromosomes of moths and butterflies (Lepidoptera). *Genetica.* **2010**, 138, 343–354.

77. Travenzoli, N.M.; Lima, B.A.; Cardoso, D.C.; Dergam, J.A.; Fernandes-Salomão, T.M; Lopes, D.M. Cytogenetic analysis and chromosomal mapping of repetitive DNA in *Melipona* species (Hymenoptera, Meliponini). *Cytogenet. Genome Res.* **2019**,158, 213–224.

78. Martí, E.; Milani, D.; Bardella, V.B.; Albuquerque, L.; Song, H.; Palacios-Gimenez, O.M.; Cabral-de-Mello, D.C. Cytogenomic analysis unveils mixed molecular evolution and recurrent chromosomal rearrangements shaping the multigene families on *Schistocerca* grasshopper genomes. *Evolution.* **2021**, 75, 2027–2041.

79. Panzera, F.; Pita, S.; Lorite, P. Chromosome structure and evolution of Triatominae: A review. In: *Triatominae—The Biology of Chagas Disease Vectors*; Guarneri, A.; Lorenzo, M., Eds. Springer Nature Switzerland AG, Cham, **2021**, pp. 65–99.

80. Pereira, J.A.; Travenzoli, N.M.; de Oliveira, M.P.; Werneck, H.A.; Fernandes Salomão, T.M.; Lopes, D.M. Molecular cytogenetics in the study of repetitive sequences helping to understand the evolution of heterochromatin in *Melipona* (Hymenoptera, Meliponini). *Genetica.* **2021**, 149, 55–62.

81. Teixeira, G.A.; de Aguiar, H.J.A.C.; Petitclerc, F.; Orivel, J.; Lopes, D.M.; Barros, L.A.C. Evolutionary insights into the genomic organization of major ribosomal DNA in ant chromosomes. *Ins. Mol. Biol.* **2021**, 30, 340–354.

82. Sproul, J.S.; Barton, L.M.; Maddison, D.R. Repetitive DNA profiles reveal evidence of rapid genome evolution and reflect species boundaries in ground beetles. *Syst. Biol.* **2020**, 69, 1137–1148.

83. Maxson, R.; Cohn, R.; Kedes, L. Expression and organization of histone genes. *Annu. Rev. Genet.* **1983**, 17, 239–277.

84. Cabrero, J.; Lopez-Leon, M.D.; Teruel, M.; Camacho, J.P. Chromosome mapping of H3 and H4 histone gene clusters in 35 species of acridid grasshoppers. *Chromosome Res.* **2009**, 17, 397–404.

85. Mandrioli, M.; Manicardi, G.C. Chromosomal mapping reveals a dynamic organization of the histone genes in aphids (Hemiptera: Aphididae). *Entomologia.* **2013**, 1, e2.

86. Toscani, M.A.; Pigozzi, M.I.; Papeschi, A.G.; Bressa, M.J. Histone H3 methylation and autosomal vs. sex chromosome segregation during male meiosis in Heteroptera. *Front. Ecol. Evol.* **2022**, 10, 836786.

87. Cabral-de-Mello, D.C.; Moura, R.C.; Martins, C. Chromosomal mapping of repetitive DNAs in the beetle *Dichotomius geminatus* provides the first evidence for an association of 5S rRNA and histone H3 genes in insects, and repetitive DNA similarity between the B chromosome and A complement. *Heredity.* **2010**, 104, 393–400.

88. Cabral-de-Mello, D.C.; Moura, R.C.; Melo, A.S.; Martins, C. Evolutionary dynamics of heterochromatin in the genome of *Dichotomius* beetles based on chromosomal analysis. *Genetica.* **2011**, 139, 315–325.

89. Nagoda, N.; Fukuda, A.; Nakashima, Y.; Mats, Y. Molecular characterization and evolution of the repeating units of histone genes in *Drosophila americana*: Coexistence of quartet and quintet units in a genome. *Insect Mol. Biol.* **2005**, 14, 713–717.

90. Jonika, M.; Lo, J.; Blackmon, H. Mode and tempo of microsatellite evolution across 300 million years of insect evolution. *Genes.* **2020**, 11, 945.

91. Ruiz-Ruano, F.J.; Cuadrado, A.; Montiel, E.E.; Camacho, J.P.; López-Leon, M.D. Next generation sequencing and FISH reveal uneven and nonrandom microsatellite distribution in two grasshopper genomes. *Chromosoma.* **2015**, 124, 221–234.

92. dos Santos, J.M.; Diniz, D.; Rodrigues, T.A.S.; Cioffi, M.B.; Waldschmidt, A.M. Heterochromatin distribution and chromosomal mapping of microsatellite repeats in the genome of *Frieseomelitta* stingless bees (Hymenoptera: Apidae: Meliponini). *Florida Entomol.* **2018**, 101, 33–39.

93. Milani, D.; Cabral-de-Mello, D.C. Microsatellite organization in the grasshopper *Abracris flavolineata* (Orthoptera: Acrididae) revealed by FISH mapping: Remarkable spreading in the A and B chromosomes. *PLoS One.* **2014**, 9, e97956.

94. Elizeu, A.M.; Travenzoli, N.M.; Ferreira, R.P.; Lopes, D.M.; Tavares, M.G. Comparative study on the physical mapping of ribosomal genes and repetitive sequences in *Friesella schrottkyi* (Friese 1900) (Hymenoptera: Apidae, Meliponini). *Zool. Anz.* **2021**, 292, 225–230.

95. Matera, A.G.; Terns, R.M.; Terns, M.P. Non-coding RNAs: Lessons from the small nuclear and small nucleolar RNAs. *Nat. Rev. Mol. Cell Biol.* **2007**, 8, 209–220.

96. Anjos, A.; Ruiz-Ruano, F.J.; Camacho, J.P.; Loreto, V.; Cabrero, J.; de Souza, M.J.; Cabral-de-Mello, D.C. U1 snDNA clusters in grasshoppers: Chromosomal dynamics and genomic organization. *Heredity.* **2015**, 114, 207–219.

97. Milani, D.; Palacios-Gimenez, O.M.; Cabral-de-Mello, D.C. The U2 snDNA is a useful marker for B chromosome detection and frequency estimation in the grasshopper *Abracris flavolineata. Cytogenet. Genome Res.* **2017**, 151, 36–40.

98. Milani, D.; Ruiz-Ruano, F.J.; Camacho, J.P.M.; Cabral-de-Mello, D.C. Out of patterns, the euchromatic B chromosome of the grasshopper *Abracris flavolineata* is not enriched in high-copy repeats. *Heredity.* **2021**, 127, 475–483.

99. Plohl, M.; Luchetti, A.; Meštrović, N.; Mantovani, B. Satellite DNAs between selfishness and functionality: Structure, genomics and evolution of tandem repeats in centromeric (hetero)chromatin. *Gene.* **2008**, 409, 72–82.

100. Huang, Y.C.; Lee, C.C.; Kao, C.Y.; Chang, N.C.; Lin, C.C.; Shoemaker, D.; Wang, J. Evolution of long centromeres in fire ants. *BMC Evol. Biol.* **2016**, 16, 189.

101. Mora, P.; Vela, J.; Ruiz-Mena, A.; Palomeque, T.; Lorite, P. Isolation of a pericentromeric satellite DNA family in *Chnootriba argus* (*Henosepilachna argus*) with an unusual short repeat unit (TTAAAA) for beetles. *Insects.* **2019**, 10, 306.

102. Ruiz-Torres, L.; Mora, P.; Ruiz-Mena, A.; Vela, J.; Mancebo, F.J.; Montiel, E.E.; Palomeque, T.; Lorite, P. Cytogenetic analysis, heterochromatin characterization and location of the rDNA genes of *Hycleus scutellatus* (Coleoptera, Meloidae): A species with an unexpected high number of rDNA clusters. *Insects.* **2021**, 12, 385.

103. Fernandes, A.; Scudeler, P.A.S.; Diniz, D.; Foresti, F.; Campos, L.A.O.; Lopes, D.M. Microdissection: A tool for bee chromosome studies. *Apidologie.* **2011**, 42, 743–748.

104. Frydrychová, F.; Grossmann, P.; Trubač, P.; Vítková, M.; Marec, F. Phylogenetic distribution of TTAGG telomeric repeats in insects. *Genome.* **2004**, 47, 163–178.

105. Kuznetsova, V.; Grozeva, S.; Anokhin, B.A. The first finding of $(TTAGG)_n$ telomeric repeat in chromosomes of true bugs (Heteroptera, Belostomatidae). *Comp. Cytogenet.* **2012**, 6, 341–346.

106. Pita, S.; Panzera, F.; Mora, P.; Vela, J.; Palomeque, T.; Lorite, P. The presence of the ancestral insect telomeric motif in kissing bugs (Triatominae) rules out the hypothesis of its loss in evolutionarily advanced Heteroptera (Cimicomorpha). *Comp. Cytogenet.* **2016**, 10, 427–437.

107. Chirino, M.G.; Dalíková, M.; Marec, F.R.; Bressa, M.J. Chromosomal distribution of interstitial telomeric sequences as signs of evolution through chromosome fusion in six species of the giant water bugs (Hemiptera, *Belostoma*). Ecol. Evol. **2017**, 7, 5227–5235.

108. Mravinac, B.; Meštrović, N.; Čavrak, V.V.; Plohl, M. TCAGG, an alternative telomeric sequence in insects. *Chromosoma.* **2011**, 120, 367–376.

109. Menezes, R.S.T.; Bardella, V.B.; Cabral-de-Mello, D.C.; Lucena, D.A.A.; Almeida, E.A.B. Are the TTAGG and TTAGGG telomeric repeats phylogenetically conserved in aculeate Hymenoptera? *Sci. Nat.* **2017**, 104, 85.

110. Lorite, P.; Maside, X.; Sanllorente, O.; Torres, M.I.; Periquet, G.; Palomeque, T. The ant genomes have been invaded by several types of mariner transposable elements. *Naturwissenschaften.* **2012**, 99, 1007–1020.

111. Palacios-Gimenez, O.M.; Bueno, D.; Cabral-de-Mello, D.C. Chromosomal mapping of two *Mariner*-like elements in the grasshopper *Abracris flavolineata* {Orthoptera: Acrididae} reveals enrichment in euchromatin. *Eur. J. Entomol.* **2014**, 111, 329–334.

112. Li, Y.; Jing, X.A.; Aldrich, J.C.; Clifford, C.; Chen, J.; Akbari, O.A.; Ferree, P.M. Unique sequence organization and small RNA expression of a "selfish" B chromosome. *Chromosoma.* **2017**, 126, 753–768.

113. Amorim, I.C.; Costa, R.G.C.; Xavier, C.; Moura, R.C. Characterization and chromosomal mapping of the *DgmarMITE* transposon in populations of *Dichotomius* (*Luederwaldtinia*) *sericeus* species complex (Coleoptera: Scarabaeidae). *Genet. Mol. Biol.* **2018**, 41, 419–425.

114. Xavier, C.; Cabral-de-Mello, D.C.; Moura, R.C. Heterochromatin and molecular characterization of *DsmarMITE* transposable element in the beetle *Dichotomius schiffleri* (Coleoptera: Scarabaeidae). *Genetica.* **2014**, 142, 575–581.

115. Cunha, M.S.; Campos, L.A.O.; Lopes, D.M. Insights into the heterochromatin evolution in the genus *Melipona* (Apidae: Meliponini). *Insectes Soc.* **2020**, 67, 391–398.

116. Fominaya, A.; Loarce, Y.; González, J.M.; Ferrer, E. Tyramide signal amplification: Fluorescence in situ hybridization for identifying homoeologous chromosomes. In: *Plant Cytogenetics: Methods and Protocols*; Kianian, S.F., Kianian, P.M.A., Eds. Springer Science+Business Media, New York, NY; **2016**, pp. 35–48.

117. Kirkpatrick, M. How and why chromosome inversions evolve. *PLoS Biol.* **2010**, 8, e1000501.

118. Thompson, M.J.; Jiggins, C.D. Supergenes and their role in evolution. *Heredity.* **2014**, 113, 1–8.

119. Wang, J.; Wurm, Y.; Nipitwattanaphon, M.; Riba-Grognuz, O.; Huang, Y.-C.; Shoemaker, D.; Keller, L. A Y-like social chromosome causes alternative colony organization in fire ants. *Nature.* **2013**, 493, 664–668.

120. Fleming, J.G.W.; Summers, M.D. Polydnavirus DNA is integrated in the DNA of its parasitoid wasp host. *Proc. Natl. Acad. Sci. USA.* **1991**, 88, 9770–9774.

121. Belle, E.; Beckage, N.E.; Rousselet, J.; Poirié, M.; Lemeunier, F.; Drezen, J.-M. Visualization of polydnavirus sequences in a parasitoid wasp chromosome. *J. Virol.* **2002**, 76, 5793–5796.

122. Rütten, K.B.; Pietsch, C.; Olek, K.; Neusser, M.; Beukeboom, L.W.; Gadau, J. Chromosomal anchoring of linkage groups and identification of wing size QTL using

markers and FISH probes derived from microdissected chromosomes in *Nasonia* (Pteromalidae: Hymenoptera). *Cytogenet. Genome Res.* **2004**, 105, 126–133.

123. Bugrov, A.G.; Karamysheva, T.V.; Perepelov, E.A.; Elisaphenko, E.A.; Rubtsov, D.N.; Warchałowska-Śliwa, E.; Tatsuta, H.; Rubtsov, N.B. DNA content of the B chromosomes in grasshopper *Podisma kanoi* Storozh. (Orthoptera, Acrididae). *Chromosome Res.* **2007**, 15, 315–325.

124. Teruel, M.; Cabrero, J.; Montiel, E.E.; Acosta, M.J.; Sánchez, A.; Camacho, J.P.M. Microdissection and chromosome painting of X and B chromosomes in *Locusta migratoria*. *Chromosome Res.* **2009**, 17, 11–18.

125. Menezes-de-Carvalho, N.Z.; Palacios-Gimenez, O.M.; Milani, D.; Cabral-de-Mello, D.C. High similarity of U2 snDNA sequence between A and B chromosomes in the grasshopper *Abracris flavolineata*. *Mol. Genet. Genomics.* **2015**, 290, 1787–1792.

126. Gokhman, V.E.; Cioffi, M.B.; König, C.; Pollmann, M.; Gantert, C.; Krogmann, L.; Steidle, J.L.M.; Kosyakova, N.; Liehr, T.; Al-Rikabi, A. Microdissection and whole chromosome painting confirm karyotype transformation in cryptic species of the *Lariophagus distinguendus* (Förster, 1841) complex (Hymenoptera: Pteromalidae). *PLoS One.* **2019**, 14, e0225257.

127. Fuková, I.; Traut, W.; Vítková, M.; Nguyen, P.; Kubíčková, S.; Marec, F. Probing the W chromosome of the codling moth, *Cydia pomonella*, with sequences from microdissected sex chromatin. *Chromosoma.* **2007**, 116, 135–145.

128. Bressa, M.J.; Papeschi, A.G.; Vítková, M.; Kubíčková, S.; Fuková, I.; Pigozzi, M.I.; Marec, F. Sex chromosome evolution in cotton stainers of the genus *Dysdercus* (Heteroptera: Pyrrhocoridae). *Cytogenet. Genome Res.* **2009**, 125, 292–305.

129. Zrzavá, M.; Hladová, I.; Dalíková, M.; Šíchová, J.; Õunap, E.; Kubíčková, S.; Marec, F. Sex chromosomes of the iconic moth *Abraxas grossulariata* (Lepidoptera, Geometridae) and its congener *A. sylvata*. *Genes.* **2018**, 9, 279.

130. Telenius, H.; Carter, N.P.; Bebb, C.E.; Nordenskjöld, M.; Ponder, B.A.J.; Tunnacliffe, A. Degenerate oligonucleotide-primed PCR: General amplification of target DNA by a single degenerated primer. *Genomics.* **1992**, 8, 718–725.

131. Jetybayev, I.Y.; Bugrov, A.G.; Buleu, O.G.; Bogomolov, A.G.; Rubtsov, N.B. Origin and evolution of the neo-sex chromosomes in Pamphagidae grasshoppers through chromosome fusion and following heteromorphization. *Genes.* **2017**, 8, 323.

132. Beliveau, B.J.; Joyce, E.F.; Apostolopoulos, N.; Yilmaz, F.; Fonseka, C.Y.; McCole, R.B.; Chang, Y.; Li, J.B.; Senaratne, T.N.; Williams, B.R.; Rouillard, J.-M.; Wu, C. Versatile design and synthesis platform for visualizing genomes with Oligopaint FISH probes. *Proc. Natl. Acad. Sci. USA.* **2012**, 109, 21301–21306.

133. Rosin, L.F.; Gil, J.Jr.; Drinnenberg, I.A.; Lei, E.P. Oligopaint DNA FISH reveals telomere-based meiotic pairing dynamics in the silkworm, *Bombyx mori.* PLoS Genet. **2021**, 17, e1009700.

134. Lopes, D.M.; Fernandes, A.; Diniz, D.; Scudeler, P.E.S.; Foresti, F.; Campos, L.A.O. Similarity of heterochromatic regions in the stingless bees (Hymenoptera: Meliponini) revealed by chromosome painting. *Caryologia.* **2014**, 67, 222–226.

135. Jetybayev, I.Y.; Bugrov, A.G.; Dzuybenko, V.V.; Rubtsov, N.B. B chromosomes in grasshoppers: Different origins and pathways to the modern Bs. *Genes.* **2018**, 9, 509.

136. Ferguson, K.B.; Visser, S.; Dalíková, M.; Provazníková, I.; Urbaneja, A.; Pérez-Hedo, M.; Marec, F.; Werren, J.H.; Zwaan, B.J.; Pannebakker, B.A.; Verhulst, E.C. Jekyll or Hyde? The genome (and more) of *Nesidiocoris tenuis*, a zoophytophagous predatory bug that is both a biological control agent and a pest. *Insect Mol. Biol.* **2021**, 30, 188–209.

137. Hejníčková, M.; Dalíková, M.; Potocký, P.; Tammaru, T.; Trehubenko, M.; Kubíčková, S.; Marec, F.; Zrzavá, M. Degenerated, undifferentiated, rearranged, lost: High variability of sex chromosomes in Geometridae (Lepidoptera) identified by sex chromatin. *Cells.* **2021**, 10, 2230.

138. Lukhtanov, V.A.; Shapoval, N.A.; Anokhin, B.A.; Saifitdinova, A.F.; Kuznetsova, V.G. Homoploid hybrid speciation and genome evolution via chromosome sorting. *Proc. R. Soc. B.* **2015**, 282, 20150157.

139. Pita, S.; Panzera, F.; Sánchez, A.; Panzera, Y.; Palomeque, T.; Lorite, P. Distribution and evolution of repeated sequences in genomes of Triatominae (Hemiptera-Reduviidae) inferred from genomic in situ hybridization. *PLoS One.* **2014**, 9, e114298.

28 FISH—in Plant Chromosomes

Susan Liedtke, Sarah Breitenbach and Tony Heitkam

CONTENTS

DOI: 10.1201/9781003223658-28

INTRODUCTION

With the advent of fluorescence in situ hybridization (FISH) methods since the 1980s–1990s, we have rapidly gained insights into the structure, organization and evolution of plant genomes.[1, 2] The potential of molecular cytogenetics was also recognized in the plant breeding sector, often with a focus on hybridization, introgressions and chromosomal additions. Now, with the advances in long-read sequencing and with the new genome-editing techniques in place, the plant cytogenetics field is repositioning itself at the interface between microscopy and genomics. FISH methods are more relevant than ever, for example to better resolve the inaccessible, repetitive genome fraction,[3] the genomes' three-dimensional organization,[4, 5] but also to identify chromosomes,[6] assess haplotypes, predict recombination,[7] and determine genomic variation among individuals.[8] On top, FISH may also assist the synthetic genomics field seeking to steer chromosome mutations and to build artificial chromosomes.[9, 10] Nevertheless, the application of FISH to plant chromosomes remains challenging.

First, plant genomes can contain over 80% of repetitive DNA,[11] and up to 149 gigabasepairs.[12] Although these numbers reside at the extreme ends of the corresponding scales, they serve well to illustrate that plant genomes are often large and repetitive. Therefore, they remain difficult targets for sequence-based methods.

Second, only limited genome and haplotype information has been available for many plants, leading to less flexibility in experimental design. This gap is now closing at high speed, with many new crop and wild plant genome builds being released every year (see www.plabipd.de/timeline_view.ep). Similarly, many karyotyping tools that were originally designed for human chromosomes are now flexible in organism choice and can be used for plant chromosomes as well.

Finally, the preparation of plant chromosomes—a prerequisite for FISH—is not straight-forward. Plants contain a high amount of fibers and their cells are enclosed by a cell wall with a variable chemical composition, depending on species, tissue and developmental stage. Finding the balance between cell wall degradation and removal of cytoplasm on the one hand, and preserving chromosome integrity on the other hand, is one of the key challenges of plant cytogenetics, a process which needs to be adapted or even established for each new plant species and tissue coming to the lab.

SAMPLES/TISSUES

As most FISH experiments require spreads of mitotic chromosomes, typically, dividing plant cells are harvested. In plants, mitotic cells occur in the meristems, out of which the root apical meristems are best accessible. Hence, the wide majority of plant cytogenetics labs uses root tip tissues to prepare mitotic chromosomes, with primary roots being especially serviceable. However, for some plants this is not an option: root crops for example accumulate carbohydrates in thickened roots, and grown trees may have buried their root tips inaccessibly below the soil. For these plants, very young leaves in leaf buds may provide a suitable alternative. Examples for this are beet crops for which a barely visible young leaf can be picked from the middle of the rosette (as performed for example by[13, 14]), and trees or shrubs such as beeches or camellias, for which young shoot tips can be taken in spring season.[15, 16]

To study meiosis, tissue of the developing flower has to be harvested; i.e., the anthers for male meiosis and the ovules for female meiosis (please see for the collection of protocols for the analysis of meiotic chromosomes by[17]).

DESCRIPTION OF METHODS: FIXATION

Due to the low fraction of mitotic cells in most tissues, a number of methods are suggested to increase the amount of mitotic nuclei in the sample. Usually, prior to a fixation step, these methods involve antimitotic agents that arrest the chromosomes in the metaphase and lead to stronger chromosome condensation. Among these, some techniques use hydroxyquinoline and/or colchicine (e.g., for beets and crocus[8, 14]), hydroxyurea (e.g., for conifers[18]) or nitrous oxide under pressure (e.g., for maize[19]). In the following, hydroxyquinoline treatment prior to the fixation of leaves and roots is laid out—one of the most commonly used methodologies.

CHEMICALS

- Fixative: methanol glacial acetic acid (3:1).
- Enrichment of metaphases: 8-hydroxyquinoline.

FIXATION OF LEAF TISSUE

1. For the preparation of mitotic metaphase chromosomes, meristems of young leaves are fixed.
2. Remove approx. 1cm long leaves at the plant base with tweezers (Figure 28.1A) after 4.5–5h of light-exposure (artificial light in the climate chamber or sunlight in the greenhouse).
3. Treat with 8-hydroxyquinoline for 2–3h to block formation of the mitotic spindle and thus progression towards anaphase.
4. Fix leaf tissue in fixative at room temperature (RT).
5. Change the fixative regularly (every 30min to 2h) until green coloration of leaves is no longer visible.
6. Incubate fixed leaves at −20°C for at least one night before attempting FISH.
7. Fixed leaves can be stored at −20°C for up to six months.

FIXATION OF PRIMARY ROOT TISSUE

1. Put seeds on wet paper and incubate for some nights. Water regularly and observe germination.
2. After germination collect approx. 0.5–1.5 cm long root tips with tweezers.
3. Treat with 8-hydroxyquinoline for 2–3h to block spindle formation.
4. Fix roots in fixative for 30min at RT.
5. Replace solution with fresh fixative.
6. Incubate fixed leaves at −20°C for at least one night before attempting FISH.
7. Fixed leaves can be stored at −20°C for up to six months.

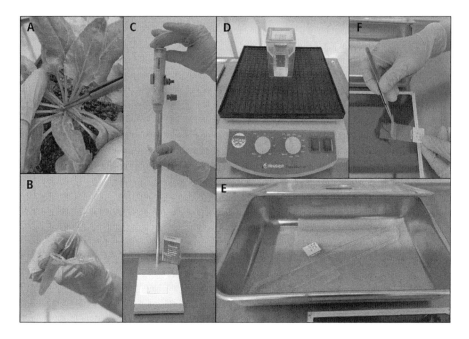

FIGURE 28.1   Photographs of selected experimental details for chromosome fixation, preparation and FISH.

A) The sugar beet plant shown harbors an approx. 1 cm long, young leaf at its base that provides suitable tissue for a chromosome preparation. After light-exposure for about 4.5–5 hours, this leaf can be taken with forceps for further experimentation.

B) With a Pasteur pipette, the nuclei pellet is suspended.

C) A mounted pipette with a height of 40–50 cm is used for the dropping method. (D) All washing steps are performed in coplin jars on a shaking plate.

D) When working with small volumes, slides are typically incubated in a humid chamber. Humidity is retained by wet tissues that are regularly changed to avoid moldiness. Slides are slightly elevated to not come into contact with the wet tissue.

E) Plastic strips are used to carefully cover the slides before any incubation in the humid chamber.

DESCRIPTION OF METHODS: PREPARATION OF MITOTIC PLANT CHROMOSOMES

As with tissue fixation, there are many ways to prepare mitotic plant chromosomes, including dropping,[20] squashing,[21] and tapping (followed by spreading using a needle).[22] For squashing and tapping, typically one slide results from a single root tip. For dropping, several slides with comparable qualities can be received from a single dropping suspension.

When establishing the chromosome preparation method for a new plant or tissue, usually, those methods are compared amongst each other. Nevertheless, depending on the chromosomal configuration of the plant, some assumptions can be already made: for nuclei with small to medium-sized chromosomes, such as beets, potato or poplar, the dropping method usually works well; for species with large chromosomes, such as conifers or crocuses, the tapping method is usually preferred.

To obtain an insight into the standard methodology of our lab, we lay out the dropping procedure next—usually the first protocol we apply to any unknown plant.

CHEMICALS

- Enzyme solution: to remove the cell wall components, the plant tissue is usually incubated in an enzyme solution. Please note that a wide range of enzyme solutions with plenty variations exist. Depending on the species, developmental stage and tissue, this solution needs to be adapted to the tissue at hands. Here, we give one example which shows typical enzyme components and concentrations, as applied to sugar beet,[13] camellia,[16] and saffron crocus[8]:

  4% (w/v) cellulase 'Onozuka' R-10 from *Trichoderma viride* (Serva 16419)

  2% (w/v) cellulase from *Aspergillus niger* (Sigma C1184)

  2% (w/v) cytohelicase from *Helix pomatia* (Sigma C8274)

  OR

  2% (w/v) hemicellulase from *Aspergillus niger* (Sigma H2125)

  0.5% (w/v) pectolyase from *Aspergillus japonicus* (Sigma P3026)

  5–20% (w/v or v/v) pectinase from *Aspergillus niger* (Sigma 17389 or Sigma P4716)

- Enzyme buffer: 4mM citric acid, 6mM sodium citrate, prepared with distilled water, pH4.5 adjusted with HCl; if autoclaved, storable for up to 12 months at 4°C.

- Fixative: methanol glacial acetic acid (3:1).

ENZYME TREATMENT

1. Wash fixed leaves with distilled water on a platform shaker at 50 rpm for 5 min.
2. Transfer the leaf onto a petri dish with enzyme buffer and continue shaking at 50 rpm for 5 min. Replace the enzyme buffer once and repeat the washing step once.
3. Incubate the leaf in a reaction tube with 100 µl enzyme solution for 3.5–4 h at 37°C. In many cases, enzyme solution, timeframe and temperature have to be modified to meet the optimal balance between digesting the cell wall and cytoplasm, but keeping the nuclei intact.
4. Macerate tissue with forceps and remove undigested components directly with the forceps or a micropistil.
5. Pre-pipette macerated tissue up and down with 200 µl pipette. Be careful to not create any air bubbles.
6. Incubate for another 15–20 min at 37°C and homogenize every 5 min with a 200 µl pipette.

WASHING AND NUCLEI PURIFICATION

1. Fill up the reaction mix with enzyme buffer (add approx. 1 ml) and centrifuge at 2,500×g at 4°C for 5 min.
2. Locate the nuclei pellet without shaking the reaction tube. The pellet can be at stuck to the side of the tube or can float on the surface.
3. Carefully remove supernatant to approx. 500 µl with a Pasteur pipette (Figure 28.1B) without disturbing the nuclei pellet, and replace it with fresh enzyme buffer.
4. Resuspend the nuclei and centrifuge again.
5. Repeat washing steps with freshly prepared fixative three times.
6. Very carefully remove the supernatant down to 200 µl (depending on size of nucleus pellet).
7. Rinse the walls of the reaction tube with 1–2 drops of fixative and resuspend the nuclei.

PREPARATION OF THE DROPPING PREPARATION

1. Use slides that have been cleaned with chromosulfuric acid (or use commercially available adhesion microscope slides). Rinse the slides in ethanol and air-dry.
2. Use a mounted pipette with a height between 40–50 cm (Figure 28.1C). Drop 13–20 µl of the cell nucleus suspension onto a slide.
3. Blow briefly and strongly over the slide, and move it with a strong, determined action downwards once.
4. Rinse the slides with fixative.
5. Air-dry slides vertically in a slide holder at RT.
6. Examine the slides under a phase contrast microscope at 100–200× magnification and evaluate the quality and density of the nuclei preparation:
    a. If the nuclei preparation is too dense, add some additional fixative to the solution. In contrast, if only very few nuclei are on the slide, let some fixative evaporate to concentrate the preparation.
    b. If the first slides are of low quality, usually the nuclei suspension will not yield any useful slides. In this case, the chromosome preparation should be attempted again. Instead, if the first slides are of high quality, with well-purified nuclei and without too much cytoplasm, it is recommended to prepare more slides from this nuclei preparation or to store the nuclei suspension for further use.
7. Choose slides with well-spread mitotic chromosomes for FISH.

DESCRIPTION OF METHODS: FLUORESCENCE IN SITU HYBRIDIZATION

EQUIPMENT

- Coplin jars: all washing steps are performed in Coplin jars (Figure 28.1D).
- Humid chamber (Figure 28.1E).

- Two shaking water baths; one heated to 37°C and the other to 42°C.
- Thermoblock/thermoshaker.
- Pasteur pipettes: disposable, plastic.
- Incubator heated to 37°C.
- Touchdown in situ system (for the controlled heating and cooling of slides).
- Phase-contrast light microscope.
- Fluorescence microscope.

CHEMICALS

- Enzyme buffer: 4mM citric acid, 6mM sodium citrate, prepared with distilled water, pH4.5 adjusted with HCl; if autoclaved, storable for up to 12 months at 4°C.
- Fixative: methanol glacial acetic acid (3:1).
- 20×SSC: 3M NaCl, 0.3M sodium citrate, fill up to 1 l with distilled water
- RNaseA: 0.1 µg/µl in 2×SSC (freshly prepared from a 10 µg/µl stock solution)
- 0.01M HCl: dilute 0.25M HCl to 0.01M HCl with distilled water
- Pepsin solution: 10 µg/ml in 0.01M HCl (freshly prepare from a 500 µg/ml stock solution).
- 4% paraformaldehyde (freshly prepared): heat 2.8g paraformaldehyde in 70 ml distilled water to 70°C in a water bath; add 50 µl 1M NaOH until the solution becomes clear; cool to RT.
- 4×SSC/0.2% Tween 20: use stock solutions of 20×SSC and Tween 20 (100%); fill up with distilled water to 1 l.
- Formamide wash solution (freshly prepared): 20% formamide in 0.2×SSC.
- Blocking solution: use 10×Block from Roche, dilute to 1×Block in 4×SSC/0.2% Tween 20
- Antibody solution (prepare fresh):
  - Dilute the anti-Dig solution 1:75.
  - Dilute the Streptavidin solution 1:200 in 1× Blocking solution.
- 4′,6-diamidino-2-phenylindole (DAPI) solution: use a DAPI stock solution (100 µg/ml in ultrapure water); dilute 1:50 in Citifluor (final concentration 2 µg/ml).

FISH with Indirect Labeling

RNase Treatment

1. Pipette 200µl of RNaseA onto each slide, cover with a plastic strip (Figure 28.1F) and place slides into a humid chamber for 30 min at 37°C (Figure 28.1E).
2. Carefully transfer the slides to a coplin jar, already filled with 2×SSC at 37°C. Only now, while the slides are submerged, remove plastic strips and wash for 3×5min at 37°C on a shaking plate.

Pepsin Treatment

3. Pour off 2×SSC and replace with pre-warmed 0.01M HCl, followed by washing for 1min at 37°C. This is to equilibrate the preparations to an acidic pH.
4. Pipette 200 µl pepsin onto each slide, cover slide with plastic strips, and place slides into a pre-warmed humid chamber for 15 min at 37°C. This will serve to further digest the cytoplasm.
5. Carefully transfer the slides to a coplin jar, already filled with 2×SSC at 37°C. Only now, while the slides are submerged, remove plastic strips and wash for 3×5min at 37°C on a shaking plate.

Fixation of Chromosomes

6. In a coplin jar, incubate the slides in 4% paraformaldehyde solution for 15 min at 37°C without shaking.
7. In a coplin jar, wash preparations in 2×SSC 3×10min at 37°C on a shaking plate.

Dehydration of the Preparations

8. In a coplin jar, dehydrate the slides in 70% ethanol for 3 min at RT on a shaking plate.
9. In a coplin jar, dehydrate the slides in 96% ethanol for 3 min at RT on a shaking plate.
10. Subsequent air-drying of vertically placed slides in a slide rack at RT.

Hybridization

The hybridization and washing stringencies are influenced by a range of parameters, e.g., the base composition of the probe, the formamide and salt concentration as well as the temperature. Stringency guide values are laid out in Table 28.1. The following protocol uses our standard procedure, which entails hybridization at approximately 76% and washing at 79% stringency (see Table 28.1).

11. Heat dextran sulfate to 70°C.
12. Hybridization solution per slide at 76% hybridization stringency:
    Formamide (final concentration 50%)   15 µl
    20×SSC (final concentration 2×SSC)   3 µl
    Salmon sperm DNA (2µg/µl)   1 µl
    10% SDS (final concentration 0.15%)   0.5 µl
    Probe (biotin-labeled) and/or   0.5–3.5 µl
    Probe (digoxigenin-labeled)   0.5–3.5 µl
    Water   *x* µl
    → *Prepare solution up to here, then mix and centrifuge down*
    <u>Dextran sulfate (50%)</u>   <u>6 µl</u>
    Total volume   30 µl
13. Prepare hybridization master mix:
    • Mix formamide, 20×SSC, salmon sperm DNA and SDS, centrifuge and distribute into the reaction tubes. The salmon sperm serves to block unspecific binding sites.

- To pipette the viscous dextran sulfate, cut about 0.5 cm of a 200 μl pipette tip. Add the dextran sulfate and mix by vortexing and spinning down.
- Denature hybridization solution for 10 min at 70°C, cool in an ice bath.
- Pipette 30 μl hybridization solution onto the marked region and cover bubble-free with plastic strip.
- Place slides in a touchdown *in situ* system and denature the chromosomes and the probe, followed by cooling down to 37°C in a controlled manner (70°C 8 min, 55°C 5 min, 50°C 2 min, 45°C 3 min, 37°C hold temperature).
- Transfer the preparations to a pre-warmed humid chamber overnight at 37°C.

Washing (All Steps in Coplin Jars)

14. Heat 2×SSC and formamide washing solution to 42°C in a shaking water bath.
15. Heat detection buffer (4×SSC with 0.2% Tween 20) to 37°C.
16. Immerse preparations in warm 2×SSC, carefully detach plastic strips, and place into 2×SSC.
17. Pour off 2×SSC and replace with pre-warmed formamide washing solution.
18. Wash in shaking water bath for 2×5 min at 42°C.
19. Wash with 2×SSC, in shaking water bath for 2×5 min at 42°C.
20. Wash with 2×SSC, in shaking water bath for 1×5 min at 37°C.

Antibody Reaction

21. Wash with 4×SSC/0.2% Tween 20 in a shaking water bath for 1×5 min at 37°C (to equilibrate the salt concentration).
22. Pipette 200 μl blocking solution onto each preparation, cover with plastic strip to block non-specific binding.
23. Place preparations in a humid chamber for 20 min at 37°C.
24. Carefully remove plastic strips, pipette 50 μl of antibody solution onto each slide.
25. Cover again and place in a humid chamber for 1 h at 37°C.
26. Remove plastic strips, wash in 4×SSC/0.2% Tween 20 in a shaking water bath for 3×10 min at 37°C. This serves to wash off the unbound antibody.

DAPI Staining

27. Pipette 15 μl DAPI solution (in Citifluor) onto the slide and carefully cover it with a large coverslip (50×24 mm). Try to avoid any air bubbles.
28. Squeeze out any excess liquid with filter paper; the cover slip should lose the ability to move freely on the slide.

Fluorescence Microscopy

29. Observe the slides under a fluorescence microscope.

TABLE 28.1
Guide Values for Hybridization and Washing Stringency in Relation to Temperature, Salt and Formamide Concentration

		Formamide Concentration [%]												
		60	**55**	**50**	**45**	**40**	**35**	**30**	**25**	**20**	**15**	**10**	**5**	**0**
Stringency T_m at 37 °C [a]	**2×SSC**	82	79	76 [b]	73	70	67	64	61	57	54	51	48	45
	1×SSC	87	84	81	78	75	72	69	66	62	59	56	53	50
	0.5×SSC	92	89	86	83	80	77	74	71	67	64	61	58	55
	0.2×SSC	98	95	92	89	86	83	80	77	74	71	68	65	62
	0.1×SSC	103	100	97	94	91	86	85	82	79	74	73	70	67
Stringency T_m at 42 °C [a]	**2×SSC**	87	84	81	78	75	72	69	66	62	59	56	53	50
	1×SSC	92	89	86	83	80	77	74	71	67	64	61	58	55
	0.5×SSC	97	94	91	88	85	82	79	76	72	69	66	63	60
	0.2×SSC	103	100	97	94	91	88	85	82	79	76	73	70	67
	0.1×SSC	108	105	102	99	96	93	90	87	84	81	78	75	72

The exemplary hybridization stringency of Tm = 76% used in this protocol is marked.

[a] based on the following equation: Tm = 81.5°C + 16.6logM + 0.41(%G+C)—0.61(% formamide)—(600/n); where M = sodium concentration [mol/l], n = number of base pairs in smallest duplex

[b] standard hybridization parameters in our laboratory

TYPICAL APPLICATIONS

FISH to plant chromosomes enables a range of different applications. Depending on the question, the respective probes, plant tissues and cell division stages need to be chosen. In the following, we showcase several examples for FISH applications in plant chromosomes (Figure 28.2):

(1) One of the earliest FISH applications has been the unequivocal assignment of chromosomes and karyotyping. For this, typically, mitotic chromosomes are prepared and hybridized with a probe cocktail producing chromosome-specific signals. These probes can be repeat-derived,[8] derived from long-insert clones,[14] or from bulked oligonucleotides.[6] As example, we follow a research question that asks for the genetic origin of a polyploid plant, namely the triploid saffron crocus. This crop's triploidization has occurred only once in history, likely in the time of the Aegean Bronze Age, and has spread since then across the globe as a clonal line.[23] Nevertheless, the parental species contribution of saffron has been debated for nearly a century. In this case, repeat-based probes have been used for comparative FISH between saffron and potential parental wild crocuses (Figure 28.2A). This FISH-based strategy clearly revealed the triploid saffron crocus emerged from cytotypes of wild *Crocus cartwrightianus*[8]; a finding that was also mirrored through genotyping-by-sequencing strategies.[24] With increasing sequence information available,

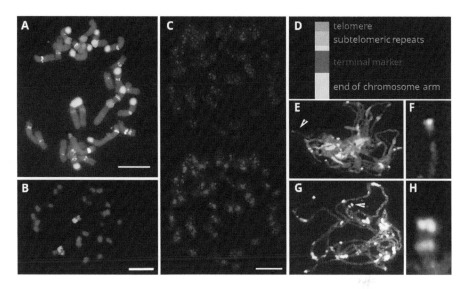

FIGURE 28.2 Application examples of FISH onto mitotic and meiotic chromosomes of plants. (A) The DAPI-stained mitotic chromosomes of triploid saffron are shown in grey. In total, 72 chromosomal landmarks are produced by hybridization of the four satellite DNAs, CroSat1 (blue), CroSat2 (green), CroSat3 (light blue), CroSat4 (red), the 18S rRNA gene (yellow) and the 5S rRNA gene (white). This six-color hybridization of repetitive DNA probes allows the unequivocal identification of saffron's chromosomes.[8] The scale bar is 10 μm. (B) The DAPI-stained mitotic chromosomes of banana (*Musa acuminata* ssp. *malaccensis*) are shown in blue. This example of chromosome painting targets the chromosomes 6 (red) and 7 (green). For this, bulked oligonucleotide probes were derived from the genome assembly and hybridized to allow the clear identification of chromosome 6– and chromosome 7–derived regions.[25] The scale bar is 5 μm. (C) The DAPI-stained mitotic chromosomes of poplar (*Populus trichocarpa*) are shown in blue. To assess the suitability of repeat-derived multilocus markers, it is important to verify their spread across all chromosomes and chromosome regions. In this example, a probe derived from the abundant transposable element SaliS-I (red) shows hybridization signals along all poplar chromosomes.[26] The scale bar is 5 μm. (D–H) In this example, the closure of the sugar beet genome assembly has been evaluated by hybridizing a terminal marker (red), the telomere (blue), and a subtelomeric satellite DNA (yellow) to meiotic chromosome spreads of sugar beet. Based on the distance between the terminal marker and the distal repeats, the closeness of the terminal marker to the telomere can be evaluated. (D) Schematic representation of the hybridized probes (red, yellow, blue) onto the ends of a pachytene chromosome (grey). (E, F) Terminal markers of the chromosome arms 1N (E, arrowhead) and 9N (G, arrowhead) have been hybridized to pachytene spreads.[14] The close-up of 1N (F) shows the overlapping of signals, indicating the closeness of the 1N terminal marker to the end of the chromosome. In contrast, the close-up of 9N shows a large distance between the marker and the telomere, indicating missing sequence information in the assembly of chromosome 9.

Source: (A) Saffron crocus FISH image reused from [8], Figure 2Q with permission from John Wiley and Sons; (B) Banana FISH image reused from[25] under CC-BY; (C) Poplar FISH image reused from[26] under CC-BY; (E–H) Figure 5 reused from[14] with permission from John Wiley and Sons.

it is also possible to generate bulked oligonucleotide probes, which are directly derived from a genome assembly. As an example, the hybridization of oligonucleotide-derived painting probes across two banana chromosomes is reproduced (Figure 28.2B).[25]

(2) Another typical FISH application in plants is the assessment of repetitive molecular markers and their distribution across all chromosomes. In the example, a specific repetitive element is hybridized onto the chromosomes of poplar (Figure 28.2C).[26] This repeat's dispersed occurrence along all chromosomes shows this repeat's suitability to act as molecular marker in a multilocus marker system.[27]

(3) Lastly, hybridization of long-insert clones onto meiotic chromosomes can help to assess the physical closure of sequence assemblies. For this, the distance between the terminal marker and the telomeric region is assessed, as was for example done for sugar beet (Figure 28.2D).[14] Whereas the assembly of chromosome arm 1N seems terminally closed (Figures 28.2E, 28.2F), the hybridization onto chromosome arm 9N suggests long genomic regions that are not covered by the assembly (Figures 28.2G, 28.2H).

CONCLUSIONS

Taken together, for the analysis of plant genome structure and evolution as well as for plant breeding, FISH is still a powerful technique.

ACKNOWLEDGEMENTS

The protocols laid out in the chapter have been established in the Dresden cytogenetics lab, initially by the late Thomas Schmidt and the late Ines Walter. We remember their efforts in setting up plant cytogenetics in Dresden and still admire their eye to detail.

REFERENCES

1. Jiang, J. Fluorescence in situ hybridization in plants: Recent developments and future applications. *Chromosome Res.* **2019**, 27, 153–165.

2. Heslop-Harrison, J.S.; Schwarzacher, T. Organization of the plant genome in chromosomes. *Plant J.* **2011**, 66, 18–33.

3. Naish, M.; Alonge, M.; Wlodzimierz, P.; Tock, A.J.; Abramson, B.W.; Schmücker, A.; Mandáková, T.; Jamge, B.; Lambing, C.; Kuo, P.; Yelina, N.; Hartwick, N.; Colt, K.; Smith, L.M.; Ton, J.; Kakutani, T.; Martienssen, R.A.; Schneeberger, K.; Lysak, M.A.; Berger, F.; Bousios, A.; Michael, T.P.; Schatz, M.C.; Henderson, I.R. The genetic and epigenetic landscape of the *Arabidopsis* centromeres. *Science.* **2021**, 374, eabi7489.

4. Perničková, K.; Koláčková, V.; Lukaszewski, A.J.; Fan, C.; Vrána, J.; Duchoslav, M.; Jenkins, G.; Phillips, D.; Šamajová, O.; Sedlářová, M.; Šamaj, J.; Doležel, J.; Kopecký, D. Instability of alien chromosome introgressions in wheat associated with improper positioning in the nucleus. *Int. J. Mol. Sci.* **2019**, 20, 1448.

5. Lysak, M.A. Celebrating Mendel, McClintock, and Darlington: On end-to-end chromosome fusions and nested chromosome fusions. *Plant Cell.* **2022**, 34, 2475–2491.

6. Braz, G.T.; He, L.; Zhao, H.; Zhang, T.; Semrau, K.; Rouillard, J.-M.; Torres, G.A.; Jiang, J. Comparative oligo-FISH mapping: An efficient and powerful methodology to reveal karyotypic and chromosomal evolution. *Genetics.* **2018**, 208, 513.

7. do Vale Martins, L.; Yu, F.; Zhao, H.; Dennison, T.; Lauter, N.; Wang, H.; Deng, Z.; Thompson, A.; Semrau, K.; Rouillard, J.M.; Birchler, J.A.; Jiang, J. Meiotic crossovers characterized by haplotype-specific chromosome painting in maize. *Nat. Commun.* **2019**, 10, 4604.

8. Schmidt, T.; Heitkam, T.; Liedtke, S.; Schubert, V.; Menzel, G. Adding color to a century-old enigma: Multi-color chromosome identification unravels the autotriploid nature of saffron (*Crocus sativus*) as a hybrid of wild Crocus cartwrightianus cytotypes. *New Phytol.* **2019**, 222, 1965–1980.

9. Beying, N.; Schmidt, C.; Pacher, M.; Houben, A.; Puchta, H. CRISPR-Cas9-mediated induction of heritable chromosomal translocations in *Arabidopsis*. *Nat. Plants.* **2020**, 6, 638–645.

10. Yu, W.; Yau, Y.-Y.; Birchler, J.A. Plant artificial chromosome technology and its potential application in genetic engineering. *Plant Biotechnol. J.* **2016**, 14, 1175–1182.

11. Schnable, P.S.; Ware, D.; Fulton, R.S.; Stein, J.C.; Wei, F.; Pasternak, S.; Liang, C.; Zhang, J.; Fulton, L.; Graves, T.A.; Minx, P.; Reily, A.D.; Courtney, L.; Kruchowski, S.S.; Tomlinson, C.; Strong, C.; Delehaunty, K.; Fronick, C.; Courtney, B.; Rock, S.M.; Belter, E.; Du, F.; Kim, K.; Abbott, R.M.; Cotton, M.; Levy, A.; Marchetto, P.; Ochoa, K.; Jackson, S.M.; Gillam, B.; Chen, W.; Yan, L.; Higginbotham, J.; Cardenas, M.; Waligorski, J.; Applebaum, E.; Phelps, L.; Falcone, J.; Kanchi, K.; Thane, T.; Scimone, A.; Thane, N.; Henke, J.; Wang, T.; Ruppert, J.; Shah, N.; Rotter, K.; Hodges, J.; Ingenthron, E.; Cordes, M.; Kohlberg, S.; Sgro, J.; Delgado, B.; Mead, K.; Chinwalla, A.; Leonard, S.; Crouse, K.; Collura, K.; Kudrna, D.; Currie, J.; He, R.; Angelova, A.; Rajasekar, S.; Mueller, T.; Lomeli, R.; Scara, G.; Ko, A.; Delaney, K.; Wissotski, M.; Lopez, G.; Campos, D.; Braidotti, M.; Ashley, E.; Golser, W.; Kim, H.; Lee, S.; Lin, J.; Dujmic, Z.; Kim, W.; Talag, J.; Zuccolo, A.; Fan, C.; Sebastian, A.; Kramer, M.; Spiegel, L.; Nascimento, L.; Zutavern, T.; Miller, B.; Ambroise, C.; Muller, S.; Spooner, W.; Narechania, A.; Ren, L.; Wei, S.; Kumari, S.; Faga, B.; Levy, M.J.; McMahan, L.; Van Buren, P.; Vaughn, M.W.; Ying, K.; Yeh, C.T.; Emrich, S.J.; Jia, Y.; Kalyanaraman, A.; Hsia, A.P.; Barbazuk, W.B.; Baucom, R.S.; Brutnell, T.P.; Carpita, N.C.; Chaparro, C.; Chia, J.M.; Deragon, J.M.; Estill, J.C.; Fu, Y.; Jeddeloh, J.A.; Han, Y.; Lee, H.; Li, P.; Lisch, D.R.; Liu, S.; Liu, Z.; Nagel, D.H.; McCann, M.C.; SanMiguel, P.; Myers, A.M.; Nettleton, D.; Nguyen, J.; Penning, B.W.; Ponnala, L.; Schneider, K.L.; Schwartz, D.C.; Sharma, A.; Soderlund, C.; Springer, N.M.; Sun, Q.; Wang, H.; Waterman, M.; Westerman, R.; Wolfgruber, T.K.; Yang, L.; Yu, Y.; Zhang, L.; Zhou, S.; Zhu, Q.; Bennetzen, J.L.; Dawe, R.K.; Jiang, J.; Jiang, N.; Presting, G.G.; Wessler, S.R.; Aluru, S.; Martienssen, R.A.; Clifton, S.W.; McCombie, W.R.; Wing, R.A.; Wilson, R.K. The B73 maize genome: Complexity, diversity, and dynamics. *Science.* **2009**, 326, 1112–1115.

12. Pellicer, J.; Fay, M.F.; Leitch, I.J. The largest eukaryotic genome of them all? *Bot. J. Linn. Soc.* **2010**, 164, 10–15.

13. Schmidt, N.; Seibt, K.M.; Weber, B.; Schwarzacher, T.; Schmidt, T.; Heitkam, T. Broken, silent, and in hiding: Tamed endogenous pararetroviruses escape elimination from the genome of sugar beet (*Beta vulgaris*). *Ann. Bot.* **2021**, 128, 281–299.

14. Paesold, S.; Borchardt, D.; Schmidt, T.; Dechyeva, D. A sugar beet (*Beta vulgaris L.*) reference FISH karyotype for chromosome and chromosome-arm identification, integration of genetic linkage groups and analysis of major repeat family distribution. *Plant J.* **2012**, 72, 600–611.

15. Anamthawat-Jónsson, K. Preparation of chromosomes from plant leaf meristems for karyotype analysis and in situ hybridization. *Methods Cell Sci.* **2004**, 25, 91–95.

16. Heitkam, T.; Petrasch, S.; Zakrzewski, F.; Kögler, A.; Wenke, T.; Wanke, S.; Schmidt, T. Next-generation sequencing reveals differentially amplified tandem repeats as a major

genome component of Northern Europe's oldest *Camellia japonica*. *Chromosome Res.* **2015**, 23, 791–806.

17. Pradillo, M.; Heckmann, S. *Plant Meiosis: Methods and Protocols*. Humana, New York, NY; **2020**.

18. Nkongolo, K.K.; Klimaszewska, K. Karyotype analysis and optimization of mitotic index in *Picea mariana* (black spruce) preparations from seedling root tips and embryogenic cultures. *Heredity.* **1994**, 73, 11–17.

19. Albert, P.S.; Zhang, T.; Semrau, K.; Rouillard, J.-M.; Kao, Y.-H.; Wang, C.-J.R.; Danilova, T.V.; Jiang, J.; Birchler, J.A. Whole-chromosome paints in maize reveal rearrangements, nuclear domains, and chromosomal relationships. *Proc. Natl. Acad. Sci. USA.* **2019**, 116, 1679–1685.

20. Schwarzacher, T.; Heslop-Harrison, P. *Practical In Situ Hybridization*. BIOS Scientific Publishers, Oxford; **2000**.

21. Schwarzacher, T.; Leitch, A.R. Enzymatic treatment of plant material to spread chromosomes for in situ hybridization. In *Protocols for Nucleic Acid Analysis By Nonradioactive Probes*; Isaac, P.G., Ed. Humana Press, Totowa, NJ, **1994**, pp. 153–160.

22. Fukui, K.; Iijima, K. Somatic chromosome map of rice by imaging methods. *Theor. Appl. Genet.* **1991**, 81, 589–596.

23. Kazemi-Shahandashti, S.-S.; Mann, L.; El-Nagish, A.; Harpke, D.; Nemati, Z.; Usadel, B.; Heitkam, T. Ancient artworks and crocus genetics both support saffron's origin in early Greece. *Front. Plant Sci.* **2022**, 13, 834416.

24. Nemati, Z.; Harpke, D.; Kerndorff, H.; Gemicioglu, A.; Blattner, F.R. Saffron (*Crocus sativus*) is an autotriploid that evolved in Attica (Greece) from wild *Crocus cartwrightianus*. *Mol. Phylogenet. Evol.* **2019**, 136, 14–20.

25. Šimoníková, D.; Němečková, A.; Karafiátová, M.; Uwimana, B.; Swennen, R.; Doležel, J.; Hřibová, E. Chromosome painting facilitates anchoring reference genome sequence to chromosomes in situ and integrated karyotyping in banana (*Musa spp.*). *Front. Plant Sci.* **2019**, 10, 1503.

26. Kögler, A.; Seibt, K.M.; Heitkam, T.; Morgenstern, K.; Reiche, B.; Brückner, M.; Wolf, H.; Krabel, D.; Schmidt, T. Divergence of 3′ ends as driver of Short Interspersed Nuclear Element (SINE) evolution in the Salicaceae. *Plant J.* **2020**, 103, 443–458.

27. Reiche, B.; Kögler, A.; Morgenstern, K.; Brückner, M.; Weber, B.; Heitkam, T.; Seibt, K.M.; Tröber, U.; Meyer, M.; Wolf, H.; Schmidt, T.; Krabel, D. Application of retrotransposon-based Inter-SINE Amplified Polymorphism (ISAP) markers for the differentiation of common poplar genotypes. *Can. J. For. Res.* **2021**, 51, 1650–1663.

29 FISH—and CRISPR/CAS9

Thomas Liehr

CONTENTS

INTRODUCTION

Fluorescence in situ hybridization (FISH) is an approach that enables the mapping of DNA-probes in nuclei and/or on chromosomes. This is an option being not available until ~50 years ago; still research and diagnostic possibilities of FISH-technique are still far from being fully utilized.[1] A major shortcut of FISH is that it normally can only be done on fixed, dead and in most cases heat denatured tissues. In other words, native chromatin structure is impaired and insights into in vivo situation of a living and metabolic active cell, is not possible.[2]

However, results similar to those obtained by normal FISH approach can be achieved on fixed and on living cells by taking advantage of the so-called "gene scissor system" based on CRISPR/Cas9, being originally a defense system of bacterial cells against foreign viral DNA. CRISPR stands for "Clustered Regularly Interspaced Short Palindromic Repeats" and Cas9 is a CRISPR-associated gene #9.[3]

The CRISPR/Cas9 system is used for targeted genomic editing, and high success rates are reported for such experiments.[3–4] However, far from all high expectations it was clear from the beginning that even rates of ~99.9% correct targeting of a gene defect in a patient is not specific enough to justify a responsible use in gene therapy.[5] Furthermore, recent research showed that CRISPR/Cas9 is able to induce chromothripsis, an effect which is definitely unwanted, deleterious and/or a starting point of malignification of cells.[6]

Still, for the topic of this book, FISH and the new possibilities in connection with the "gene scissor system" have to be mentioned. First there is (a) FISH-like labeling of living cells,[7–8] requiring a sophisticated cell-biological equipment and experience, being not related to what a "standard" molecular cytogenetic laboratory normally can do. Besides there are yet three other CRISPR/Cas9 system based approaches on

DOI: 10.1201/9781003223658-29

fixed cells. The so-called (b) "Genome Oligopaint via Local Denaturation" (GOLD) FISH assay can detect single copy sequences using Cas9-RNA and local (relatively low temperature) denaturation.[9] However, CAS-FISH-,[10] and RNA-guided endonuclease-in situ labeling (RGEN-ISL) (synonym CRISPR-FISH) protocols,[11] allow detection of repetitive DNA in fixed cells, without denaturation.

Here exclusively the protocol for CRISPR-FISH is treated.[11]

SAMPLES/TISSUES

CRISPR-FISH is done in fixed cells.[11] Here a ribonucleoprotein (RNP) consisting of a target-specific CRISPR RNA (crRNA), a fluorescent-labelled transactivating crRNA (tracrRNA), and Cas9 endonuclease are used. Several fluorochromes can be applied also in parallel in this approach.[11]

DESCRIPTION OF METHOD

The method of CRISPR-FISH is described acc. to[11].

1. To use the Alt-R CRISPR-Cas9 system, functional guide RNA (gRNA) for the target region can be constructed and commercially prepared (e.g. at Integrated DNA Technologies, https://eu.idtdna.com). Target-DNA-sequence (~20-mer) can be selected by support of web base tool Crisprdirect (https://crispr.dbcls.jp/). crRNA and fluorochrome-labelled trans-activating crRNA (tracrRNA) form the two-part guide RNA. Therefore . . .

2. dissolve lyophilized crRNA and tracrRNA-fluorochrome to 100 µM in nuclease-free Buffer 1(30 mM Hepes, pH 7.5; 100 mM CH_3COOK), each, and store separately (−20°C).

3. Mix 1 µl 100 µM crRNA and 1 µl 100 µM tracrRNA-fluorochrome plus 8 µl Buffer 1 to get 10µM gRNA.

4. Denature gRNA mix at 95°C (5 min) and hybridize for 10 min at 37°C (can be stored at −20°C).

5. Mix 1 µl gRNA from step 4 with
   – Buffer 2 =
   – 6.25 µM dCas9 protein (D10A and H840A; Novateinbio, www.nova-teinbio.com, # PR-137213)
   – 10 µl 10× buffer (200 mM Hepes (pH 7.5), 1 M KCl, 50 mM $MgCl_2$, 50% (v/v) glycerol, 10% bovine serum albumin, 1% Tween 20)
   – 10 µl 10 mM dithiothreitol
   – 80 µl double distilled water.

6. Incubate the mix from step 5 for 10 min at 26°C; store at 4°C.

7. Prepare cells acc. to standard procedures and spin them after corresponding fixation on slides.

8. Transfer slides in 1×PBS, 4°C.

9. Add to a slide each 100 µl of Buffer 2 at room temperature for 2 min.

10. Tip over the slide(s) to remove Buffer 2, add ~30 µl of fluid from step 6, and cover with coverslip.
11. Incubate at 26°C, acc. to species and DNA-target-sequence, for 2–4 hours in a humid chamber.
12. Wash slide(s) in 5 min in 1×PBS on ice.
13. Fix slide(s) 5 min in 4% formaldehyde/1×PBS on ice.
14. Wash with 1×PBS (5 min on ice)
15. Dehydrated in an ethanol series (70%, 90%, 96%; 2 min each) at room temperature.
16. Counterstain with 4′,6-diamino-2-phenylindole (DAPI) in mounting medium (Vectashield, Vector Laboratories, Burlingame, CA, USA).
17. Evaluate in a suited fluorescence microscope—laser scanning microscopy is recommended.

YIELDS

Similar results as by standard FISH using repetitive probes (as telomeric or centromeric sequences) are possible using CRISPR-FISH. The advantage of CRISPR-FISH is that tissue is not denatured. Thus, e.g. an in parallel immunostaining can be done targeting non-degraded proteins.[12]

CONCLUSIONS

Results obtainable by taking advantage of CRISPR/Cas9 "gene scissor system" system as discussed in this chapter remember pictures known from standard FISH-approaches. Still the way how they are obtained is a completely different one. Thus, the use of the designation CRISPR-FISH is traceable, but may not be the best choice in terms of potentially misleading molecular cytogenetic labs to make them think CRISPR-FISH is an easily adaptable alternative for them. The latter seems yet not to be the case, and CRISPR-FISH is only to be introduced in conditions well suited for standard FISH with a medium to high investment of time and resources.

REFERENCES

1. Liehr, T. Molecular cytogenetics in the era of chromosomics and cytogenomic approaches. *Front. Genet.* **2021**, 12, 720507.
2. Potlapalli, B.P.; Schubert, V.; Metje-Sprink, J.; Liehr, T.; Houben, A. Application of Tris-HCl allows the specific labeling of regularly prepared chromosomes by CRISPR-FISH. *Cytogenet. Genome Res.* **2020**, 160, 156–165.
3. Adli, M. The CRISPR tool kit for genome editing and beyond. *Nat. Commun.* **2018**, 9, 911.
4. Li, H.; Yang, Y.; Hong, W.; Huang, M.; Wu, M.; Zhao, X. Applications of genome editing technology in the targeted therapy of human diseases: Mechanisms, advances and prospects. *Signal Transduct. Target. Ther.* **2020**, 5, 1.
5. Uddin, F.; Rudin, C.M.; Sen, T. CRISPR gene therapy: Applications, limitations, and implications for the future. *Front. Oncol.* **2020**, 10, 1387.

6. Leibowitz, M.L.; Papathanasiou, S.; Doerfler, P.A.; Blaine, L.J.; Sun, L.; Yao. Y.; Zhang, C.Z.; Weiss, M.J.; Pellman, D. Chromothripsis as an on-target consequence of CRISPR-Cas9 genome editing. *Nat. Genet.* **2021**, 53, 895–905.

7. Wang, H.; Xu, X.; Nguyen, C.M.; Liu, Y.; Gao, Y.; Lin, X.; Daley, T.; Kipniss, N.H.; La Russa, M.; Qi, L.S. CRISPR-mediated programmable 3D genome positioning and nuclear organization. *Cell.* **2018**, 175, 1405–1417.

8. Wang, H.; Nakamura, M.; Abbott, T.R.; Zhao, D.; Luo, K.; Yu, C.; Nguyen, C.M.; Lo, A.; Daley, T.P.; La Russa, M.; Liu, Y.; Qi, L.S. CRISPR-mediated live imaging of genome editing and transcription. *Science.* **2019**, 365, 1301–1305.

9. Wang, Y.; Cottle, W.T.; Wang, H.; Feng, X.A.; Mallon, J.; Gavrilov, M.; Bailey, S.; Ha, T. Genome oligopaint via local denaturation fluorescence in situ hybridization. *Mol. Cell.* **2021**, 81, 1566–1577.

10. Deng, W.; Shi, X.; Tjian, R.; Lionnet, T.; Singer, R.H. CASFISH: CRISPR/Cas9-mediated in situ labeling of genomic loci in fixed cells. *Proc. Natl. Acad. Sci. U. S. A.* **2015**, 112, 11870–11875.

11. Ishii, T.; Schubert, V.; Khosravi, S.; Dreissig, S.; Metje-Sprink, J.; Sprink, T.; Fuchs, J.; Meister, A.; Houben, A. RNA-guided endonuclease—in situ labelling (RGEN-ISL): A fast CRISPR/Cas9-based method to label genomic sequences in various species. *New Phytol.* **2019**, 222, 1652–1661.

12. Ishii, T.; Kiyotaka, N.; Houben, A. Application of CRISPR/Cas9 to visualize defined genomic sequences in fixed chromosomes and nuclei. In: *Cytogenomics*; Liehr, T., Eds. Academic Press, New York, **2021**, pp. 147–153.

Index

Page locators in **bold** indicate a table. Page locators in *italics* indicate a figure.